Descartes Was Right!

Souls Do Exist

and

Reincarnation Proves It

Descartes Was Right !

Souls Do Exist

and

Reincarnation Proves It

A CHALLENGE TO RETHINK DUALISM

Casimir J. Bonk

Order this book online at www.trafford.com
or email orders@trafford.com

Most Trafford titles are also available at major online book retailers.

Printed in Victoria, BC, Canada.

ISBN: 978-1-4269-2497-2 (soft)
ISBN: 978-1-4269-2498-9 (hard)

Library of Congress Control Number: 2009913598

Our mission is to efficiently provide the world's finest, most comprehensive book publishing service, enabling every author to experience success. To find out how to publish your book, your way, and have it available worldwide, visit us online at www.trafford.com

Trafford rev. 1/28/2010

www.trafford.com

North America & international
toll-free: 1 888 232 4444 (USA & Canada)
phone: 250 383 6864 • fax: 812 355 4082

Apology to a Soul

Why have I denied you
If you are me?

Why have I misunderstood you
If from earth you'll make me free?

Forgive me, now that I know
How it was meant to be.

Contents

Illustrations

Tables

ACKNOWLEDGEMENTS

Were it not for the work of Dr. Ian Stevenson, this book would never have been started. The crucial evidence of the soul's reality is confirmed by the scientific investigations of reincarnation, which by the very definition of the term implies the migration of a personality from one body at death to a new body. Little recognition has been given to Dr. Stevenson by his peers for this historic revelation of the true nature of the human being. Although he did not expand this research beyond his medical goals into exploration of the effects of his findings into the metaphysical domain of dualism, the scientific nature of his work is evidence for support of the Cartesian argument for the soul's existence. I am deeply thankful to Dr. Stevenson for unknowingly providing the solid foundation for my ability to develop this scientifically proven phenomenon into the argument for dualism's replacement of monism.

I also acknowledge that the work of Dr. Jim Tucker adds credibility to the scientific nature of reincarnation research. By working with Dr. Stevenson and reporting the results of the investigations he has alerted the scientific community to accept reincarnation as real rather than assuming it to be a paranormal illusion.

Developing the arguments against the materialistic version of the human nature was made easy by the elaborate and consistent descriptions of the neuroscientific and cognitive research with conceptual explanations. I am indebted to those neuroscientists and neurophilosophers who wrote the books which have provided the pro and con arguments which showed that monism was not a proven version of how the human being behaves. I am thankful to these professionals for the information I needed to prepare this challenge to their materialistic thinking.

For assistance in showing me some of the errors in my thinking through critique and discussion, I acknowledge the help of Richard and Nedra Nesbit. As a retired engineer and technical reviewer, Richard's comments about this engineering venture into philosophy were pertinent and useful.

Thanks to all the friends and acquaintances who, through their comments during our discussions unknowingly contributed to my long deliberations on what all the evidence means. Their contributions, though individually meaningful at the time, have lost identity in time.

Last, but far from least, are my thanks to my dear wife Emily who gracefully gave up togetherness time as I sat at the computer into the late evening hours. Her constant reminders to quit for the night and go to bed were great reminders but not always heeded as they should have been. Her greatest pertinent contribution was in giving the final printout one more review and to my great relief finding those errors that crept in as I neglected her nocturnal advice.

PREFACE

When I wrote my first book *What If? In a Rental Car* about the meaning of life, I used an analogy that the human body is like a rental car with the soul as the driver. When the earthly trip is over, the rental car is returned to the "rental agency" while the driver continues the journey into eternity. I did not attempt to prove the soul's existence. The soul's importance was its effect on the meaning of life if the answer to the question, "what if the soul is real" was either positive or negative. Little did I realize that this second book would rationalize the analogy into reality. By discovering that scientific evidence of the soul's existence had been produced, I felt that the driver-car portion of the analogy was confirmed. However, since the evidence of the soul's existence was proven through the verification of reincarnation, not only was the auto portion made credible but even the rental car aspect became more meaningful since reincarnation is the soul's use of different rental cars for the different earthly trips. I conceived the analogy at my mother's funeral, many years before discovering the reincarnation evidence. I did not know at the time that the evidence was already in published form. At that time I was an aerospace engineer using rental cars rather than planning on writing books. But the analogy was part of my philosophy of life, the belief that I had a soul that was a separate entity unlike my body.

Why is this book necessary? With retirement from the engineering profession, my philosophic leaning motivated me to write the first book. During the writing of that book, I wanted to understand why the meaning of life appeared to be different for the secular community than it was for the spiritually inclined communities. It became obvious that the secular world, and even the religious to a certain extent, have succumbed to science's claim of having the responsibility for explaining natural laws, including those affecting life and death. Consequently, I had to learn what these scientific explanations were regarding life, death and human behavior. Through extensive reading I realized that there was a consistent scientific mindset that denied any immaterial influences on human behavior. But the denial, lacking any supporting evidence, seemed to be more of a convenience for avoiding the need to analyze the physical/non-physical interactions. Since science is able to explain the physical functioning of the body, physical explanations for all of human activities seemed adequate. According to the neurophilosophers, human behavior is controlled strictly by the physical brain. I was amazed, however, at the lack of scientific confirmation of the concepts that attempted to explain how this occurs. The inability to justify the concepts was, and continues to be, vindicated by the optimistic reliance on science's eventual solution of the relevant problems. I decided that these misleading and unsupported concepts needed exposure.

What makes this exposure possible is the *scientific* evidence of the soul's reality through confirmed investigations of reincarnation. Reincarnation defines the soul as the migrant between bodies and the soul allows reincarnation to occur. The initial disclosure of this evidence was published in 1966 but it has been ignored because reincarnation and the scientific concept of physicalism are directly opposed. The conflict forces the argument for dualism into a confrontation with the scientific community that is currently (2009) completely enamored of a totally physical explanation for human behavior. This continuing dedication to the physical approach is a confirmation that the scientists are unaware of the soul evidence or are committed to disregarding it as an undesirable and damaging obstacle to the pursuit of their materialistic objectives. How could the research effort continue with such a deconfirming obstacle blocking further efforts since scientists can only analyze physical matter and are incapable of dealing with immaterial entities and their properties? Such a dilemma demands that the soul issue remain dead for them. This book is a plea for a return to reality.

Aware as I was of this impending confrontation, why did I continue writing this book? To a retired engineer who gets much exercise by jumping to conclusions and possible connections, I realized that here was the evidence for the existence of the soul, physical evidence of a non-physical being that was deemed an impossibility. However, the scientific reporting of the reality of reincarnation only provided evidence based on medical research investigating birthmarks and defects. There were no claims by the researchers associating this evidence with dualism. This may have been the reason why the neuroscientific community failed to note, or ignored, the evidence. The more I read, the more I realized that since no one had even attempted connecting reincarnation to dualism, I had an opportunity to resurrect the issue of Cartesian dualism as a conflict between physical causality and *immaterial reality.* Dualism has been denigrated by scientists as a mythical unsubstantiated belief. By raising the issue, I could call for rethinking what the philosopher René Descartes (1650) had claimed about dualism. The required soul evidence Descartes needed was now available and I could initiate the eventual resolution of the dualism debate.

There comes a time in any controversial issue when a conceptual perception is upended by newly discovered evidence. (The scientific landscape is strewn with corpses of false theories.) Since science continues to deny the soul's existence, the evidence produced by a renowned medical researcher is a firm basis for an opposing argument. Surely, I reasoned, scientists will accept such scientific evidence even though it is generated in another discipline.

The argument, however, has to be made pertinent to dualism. This is where my experience as an engineering systems analyst qualifies me to write this book. It is the use of engineering logic, rather than philosophic logic or scientific rationality, that unearths the false assumptions and conclusions imbedded in the physical approach to immaterial reality. To an engineer a problem is a reason to search for a solution.

My contributions in organizing the argument are threefold. First I reveal the scientific evidence of the reality of reincarnation. Although this evidence was disclosed over forty years ago, no scientist has applied it to the mind-brain and soul-body problems. Is this because the definition of reincarnation conflicts with materialism? Since scientists ridicule paranormal phenomena as illusions of the brain, the physical evidence of reincarnation should have alerted them to a possible tilt in their basic reasoning. This did not happen as they continue their denials of any immaterial entities.

My second contribution is in relating reincarnation to the mind-brain problem by showing how the properties of the soul, as demonstrated in the investigated cases, conflict with and deconfirm the neuroscientific concept of brain-generated human behavior.

The third contribution is an unprecedented explanation of dualism given that now there is certainty of the soul's participation. I take advantage of the opportunity to present original ideas of how this dual nature can overcome some of the unsolved scientific gaps. Obviously, I cannot supply the answers to this never-before attempted explanation of dualism as a functional system because the *non-physical laws* of integrating the two entities are not known. But their effects are known to the scientists through their cognitive research of human behavior. Since the possibility of a body-soul union has not been addressed, the rethinking of dualism by philosophers and scientists must address the integration of both physical and non-physical properties in the human being.

There is no other book that fills the physical/non-physical gap between the positions of the two sides of the debate because no other book relies on the soul's reality. By introducing the evidence of the soul, I have closed this gap. The recognition and acceptance of this evidence can change the mind-brain research from a purely physical study to an integrated one in which the soul's contributions to human behavior will be analyzed. The neuroscientists admit (as I quote them in the book) that there is no theoretical framework for understanding how the brain works. That is a major admission of the error that allowed the fractionized approach in the ongoing research. A framework of some sort of system analysis should precede a major area of inquiry. If a soul had been considered as a possible integral part of the human being, a complete analysis could have considered how such an entity could utilize the physical neural system and brain in directing the activities of the human assemblage. But this was not done. Therefore, the assumed theories of physical causality need to be reconsidered. The sensory environmental stimuli to the physical brain must now be perceived by a non-physical mind and processed differently to obtain physical body activity. The effects of pure thought and the ensuing commands for activation to the brain must be explained. It will be a new branch of scientific analysis that must include the knowledge of physiological and psychological research with traditional philosophic viewpoints. The opening quotation of this book's Introduction by a mind-brain researcher anticipates such a need for a uniquely different type of future cognitive research.

I have found no book that ventures into the problematic world of the soul's control of human behavior through interaction with the brain. Such an approach is only found in philosophical literature and that without any analysis of how that comes about. But now, some neuroscientists are claiming that there is a non-physical element in the human make-up that must be accounted for in explaining the mind-brain problem. These claims are based on experimental testing which should be scientifically recognized. The reports of these experiments and arguments against physical causality bolster my confidence that my observations of the lack of credibility of the concepts of physicalism are confirmed.

For the secular community, which is mostly apathetic to the idea of a human dominated by a soul, the psychological effects of such a positive identification cannot be envisioned. From the impact on social problems to the search for world peace, the idea of an immortal soul being responsible for its actions will affect all human rationality (it already does in the Eastern cultures).

My objective in calling for a review of dualism under these newly exposed conditions is not destructive but constructive in attempting to point the way to a realistic explanation of human behavior.. There must be a redirection in the search for understanding the how and why of human activity because the current research only leads to inconclusive and misleading solutions. Behavior is not physically caused but by a combination of physical and non-physical control. By describing the evidence that is pertinent and why it should be considered, I give the reasons for its applicability. By presenting some of the arguments that make dualism a complex issue within the historical context, I paint the overall picture of how we arrived at the present impasse. By analyzing some of the conflicting conceptual approaches, I demonstrate why the assumed theoretical bases are invalid. By showing how the new evidence can fill some of the conceptual gaps, I propose that an integrated analysis of the mind-brain (soul-body) problem can be initiated with new thinking. Finally, through a suggested system approach, I propose a possible starting point of an integrated analysis with flow charts of the system elements.

If this hasn't been done by competent scientists and scientific philosophers, rational metaphysical philosophers, and the ever-alert skeptics, what are my qualifications in making this controversial challenge? I am not a physicist, physiologist, psychiatrist, cognitive scientist, or theologian; I am merely a retired aerospace engineer and student philosopher. This, I'm sure, will be a major criticism against the validity of the challenge, a good example of attacking the messenger while disregarding the message. I am not a philosopher, at least not one with published works that would lend quotable credibility to my philosophic views. While pursuing an engineering degree at Catholic University in Washington, D.C., I began studying philosophy in a required course, the philosophy of science. Since obtaining my degree and while engaged in engineering work, I continued my study of philosophy and viewed my life from a philosophic perspective. I am a philosopher to the extent that I seek the truth about the universe and my existence. Philosophy may reveal and explain problems but it is engineering that solves them. Like Descartes, I think, and my thinking is based on engineering logic and a continuing desire for knowing who I am and how that affects the meaning of life for me. I use the mind that neuroscientists claim I do not have, a mind that is separate from the brain. With it I find inconsistencies in their concepts, assertions, and assumptions and that qualifies me to write this book. My goal is to add to the credibility of Cartesian arguments with positive evidence, in other words, to supply the evidence that was not available to Descartes.

I have read many books that describe the positions of scientists, philosophers, and professors and have selected quotations of their opinions or conceptual solutions to define their positions correctly. My interest in any subject has always been to understand objectively. My background in aerospace engineering allows me to analyze technical work and to try to elaborate on it through flow diagrams and system analyses. (To me, dualism is an integrated system of physical and non-physical components.)

Another qualification is that I am an observer of the debate, not a participant. I do not have to abide by the traditional and persisting views of either side. Throughout my extensive literary research and application of logic, I have been struck, and at times amused, by the implied authenticity of conclusions based on overly optimistic assumptions. I have become a skeptical reader. My books are bloodied with red penciled notes disagreeing with, not only the statements, but also the illogical reasoning behind those statements.

So, can this book do anything to bring the two sides together? At least it can initiate an awareness that there are missing links in the analyses of the body-soul and mind-brain problems and that some new thinking will lead to better than assumption-based conclusions. Overturning the convenient scientific denial of the soul's reality will undoubtedly raise major resistance to any analytical redirection. For those scientists who already have doubts about the current approaches, the revelation of the soul's reality may be an opportunity to begin a new integrated approach about how the analyses of human activity and behavior should be structured.

For whom is this book intended? It is information for a broad general audience but it is specifically directed at the scientific and philosophic communities because of its effect on their ongoing materialistic research of human nature and behavior. It is not a rigorous confrontation with the existing analytic and philosophic concepts proposed by the proponents of physicalism. The problem's complexity demands simplification, not a belabored array of physiological and technical detail. My purpose is to try to connect the existing dots and supply the missing ones to produce a total picture. Understanding the conflicts and interactions is far more important than a display of detail that can be found in books written for detailed explanations.

The book is also intended for the average person who is interested in the reality of reincarnation, the true nature of behavior, body-soul interactions, and the meaning of life. It is important for the average person to know how the soul's existence is verified scientifically and to know how science has misled everyone to "think" that the physical brain is the means of thinking. Every one should appreciate the relevance of the soul with its mind to human behavior. It should be a revelation to that portion of the secular society that is not prepared psychologically to having a soul imposed on it. The effects of relativism clash with such acceptance, which imposes moral responsibility on the personalized self.

With the average reader in mind, I decided to leave out most of the complex details of the human brain and the nervous system. The exclusion of such detailed material helps in maintaining the flow of the thematic material. A suggested reading list contains adequate material for the motivated pursuer of such specific knowledge. For the readers who are already familiar with the involved disciplines, the inclusion of excessive detail would be annoying.

Although a spectator of, but not a participant in the debatable conflicts, I will admit that I am convinced that Cartesian dualism is reality but that does not prevent me from trying to show the true nature of the debate. I make my position clear throughout, even in the book's title. Mine is a skeptic's approach to understanding the meanings of the various claims as presented in the published literature. The conflicting issues are pitted one against the other using the precise published quotations for supportive accuracy.

I am aware of the religious opposition to a belief in reincarnation but I do not indulge in an argument on this point, although I am forced to acknowledge its existence. My use of reincarnation as scientifically derived evidence of the soul's reality is keyed to the need for the review of dualism as a scientific undertaking, not as a revision to religious dogma.

It should be clear from this preface, that the intent of the book is directed at presenting a comprehensive picture of the contrast between monism and dualism. The introduction of the conclusive evidence of the soul's reality demands its proper recognition by the scientific community with the resultant admission that a redirection of the thinking about dualism should occur. Unpredicted findings have a special role in science. When they upset ongoing scientifically acceptable theories, they create an immediate negative attitude toward acceptance. But if sufficiently validated, the findings contribute to the store of knowledge that will eventually lead to a better understanding of the universe, which should benefit humanity.

To maximize the benefits of reading this book, keep a pen or pencil (preferably red) ready at all times to underline and make marginal notes about important statements and conflicting criticisms and to jot down ideas and comments, complementary or contradictory. By doing so, you become a co-author of the book because you are, in effect, changing the text to your thinking. The more you do this, the more you shape your concept of how you will eventually understand your true nature. My only regret is that I will not be able to discuss those comments with you.

Introduction

There is an Elephant in the scientific laboratory but the neuroscientists can't see it because they are facing in the opposite direction and refuse to turn around for fear of seeing it.

"If a non-physical mind exists, the research project for the next century should be to explore the impact of such non-physical influences – where in the brain does such influence occur and what laws are broken."[1]

In the above statement, David L. Wilson, a researcher in mind-brain interactions, is aware of the impact a separate non-physical mind can have on neuroscientific research. This scientific conclusion supports and emphasizes the need for rethinking dualism because there is scientific evidence that such a mind does exist.

The evidence of a soul's reality, with a non-physical mind, presented in this book affirms that such a century of exploration can and should begin now.

Thousands of books written by neuroscientists and cognitive scientists try to explain human mental states as properties of the physical brain on the assumption that the soul, therefore the mind, does not exist. Science has based its neurophysical and neurobiological research on this baseless, but convenient, assumption. It has wasted intellectual and financial resources on a naïve assumption without even attempting an analysis of the possible effects of a non-physical entity on the functioning of the physical brain and nervous system. This omission is unscientific, an analytic departure from the normal procedure of assessing other possibilities since a soul's reality has always been a philosophic position. Now with the evidence of the soul, there is a major gap to be filled with no past studies to rely on as the jump start for that research project that David Wilson claims is conditionally imminent.

The evidence of the soul's reality is documented in scientific reports of investigations and verifications. This evidence proves several issues that affect not only the scientific and philosophic communities but every human being. These issues are:

1. Reincarnation, demonstrating spiritual continuity, is shown to be real through *physical first person spontaneous* accounts by the reincarnated soul of past lives, accounts that are subsequently verified.
2. Therefore, the soul, as the link between the old and new physical bodies with its memory and other non-physical functional properties, *is a reality. Every human being has a soul.*
3. The existence of the soul confirms a spiritual life after earthly death. Therefore, every human being is committed to an eternal existence regardless of contradictory beliefs.
4. Scientists try to explain mental activity and behavior as properties of the physical brain initiated solely by body sensor physical stimuli. The basic assumption for this approach is a denial of the soul's existence. The soul's reality negates this conceptual approach.
5. The mind is part of the soul, not of the physical brain. Each is a distinct entity.
6. The mind controls the brain for physical activity. Some experimental evidence of this has been documented.
7. Thinking, free will, emotion, and memory are properties of the soul, not the brain. Scientists try to explain memory as a physical function with a physical storage location. But reincarnation evidence demonstrates that memory of past life experiences resides in the soul, not in the body, because the soul retains it whereas it would be non-existent if it were part of the dead body.

8. Skills learned in a previous life, physical and linguistic, are retained in the reincarnated soul-mind because they are demonstrated in some reincarnated bodies without any instructions.
9. It is possible the soul is consciousness.
10. Dualism replaces monism, confirming the philosophy of René Descartes.
11. There is a logical approach to analyzing the duality realism in a transition from a soulless concept to the realism of a body controlled by a soul.

These are the main issues addressed in this book. What makes the reincarnation evidence so compelling is that it is **physical evidence of a non-physical entity**. It is what science has asserted to be impossible, an assumption that has been exploited by science as a convenient base for substituting materialist concepts for immaterial properties. However, the scientific reports of the soul's reality, implemented by deconfirming exposure of naive conceptual claims of physical causality, challenge this exploitation.

The challenge is a logical and factual, not idealistic, confrontation with the scientific attempt to eliminate the soul, and consequently the mind, as real non-physical entities of the human being. A likely reaction to this dismantling of the materialistic viewpoint and its claimed scientific support will be a violent attack on the credibility of the evidence. But the evidence is physical evidence, which the scientific community is obliged to review objectively as being in its domain even though the evidence negates part of the ongoing research. Science's dedication to materialism may make acknowledgement of its errors in denying the non-physical part of the human being extremely difficult. However, there are neuroscientists who are espousing negative views about the ongoing neuroscientific and cognitive research. What is lacking in their viewpoints that would support and make their cases complete is the physical evidence and contradicting analyses that are revealed in this book.

What effects will the disclosure of the soul's reality produce? They will be widespread, not restricted to scientific research but encompassing all of humanity. The following list is representative of these effects:

1. The worldview of man's destiny must change with the certainty of an afterlife.
2. The definition of human nature and behavior must incorporate the new reality.
3. The concept of human behavior through physical causality will no longer be credible.
4. Monism is replaced by dualism.
5. Personhood must be redefined for social and legal reasons.
6. Philosophers can better define metaphysical effects with factual certainty.
7. The theological negation of reincarnation must be reversed.
8. A resurgence of religious fervor is possible due to the certainty of an afterlife.
9. The mindsets of neural researchers must accept non-physical reality.
10. Secular acceptance of scientific positivism will be curtailed due to science's deliberate misleading explanations of the nature of Homo sapiens.
11. New research about the effects of dualism on the mind-brain problem must begin.
12. A holistic approach for the new research must be defined.
13. A leader, either individual, academic, or corporate, must emerge to organize the new neurorealistic research effort.

The bottom line of my challenge to rethink dualism is that the confirmation of the soul as part of the human being validates dualism and negates monism. To continue the research of human nature and human behavior demands a new discipline in which the guidelines for the study must be defined by the unique problems of dualistic integration, not by unsupported idealistic concepts. Such unsupported idealism has caused confusion about human behavior and the meaning of life because it is viewed in terms of invalid conceptual explanations and optimistic scientific promises of eventual vindication. The introduction of the soul into cognitive scientific research should unveil a broader, more positive, and exciting approach for *final* resolution of the true nature of Homo sapiens.

What will you find in this book to support the above statements? After a historical and defining background of the nature of dualism (Chapter 1) and some precautionary advice on objective appraisal of philosophical and scientific claims (Chapter 2), the evidence is presented in Chapter 3. Both physical and non-physical evidences are included. With this information you will be the judge of the validity of the material in the subsequent chapters which include philosophy, argument, contradictions, explanations, analogies, and finally a suggested framework for rethinking dualism. It is a progressive venture into the reasons why the challenge to rethink dualism is necessary.

With this preliminary overview of what to expect, let us proceed to a more detailed explanation of the book's contents.

DESCARTES WAS RIGHT! You do have a soul! It doesn't matter whether you know it or not; whether you believe it or not; whether you care or not; or whether you argue against it. The evidence in this book will convince you or, at least, make you doubt about your disbelief. Since the seventeenth century when René Descartes, a French philosopher (1596-1650), introduced the philosophy that a spiritual soul is part of the human being, an ongoing debate has been unable to resolve the allegation. Although there are also other related issues, the principal contention between the advocates of materialistic monism and metaphysical dualism is the existence of a soul. This is the basic issue regardless of various concepts of either ism. If Descartes had the evidence that now exists, the debate would never have started.

There are various interpretations of the controversy, like the materialist inclusion of dual states in the monist approach, which tend to confuse the basic issue. A simple diagram, Figure I.1, illustrating the fundamental differences between the two isms will, hopefully, bring everyone to the same starting line in this controversial book. (The book is controversial because it attacks ideals and the extensive efforts that have been expended to justify those ideals.)

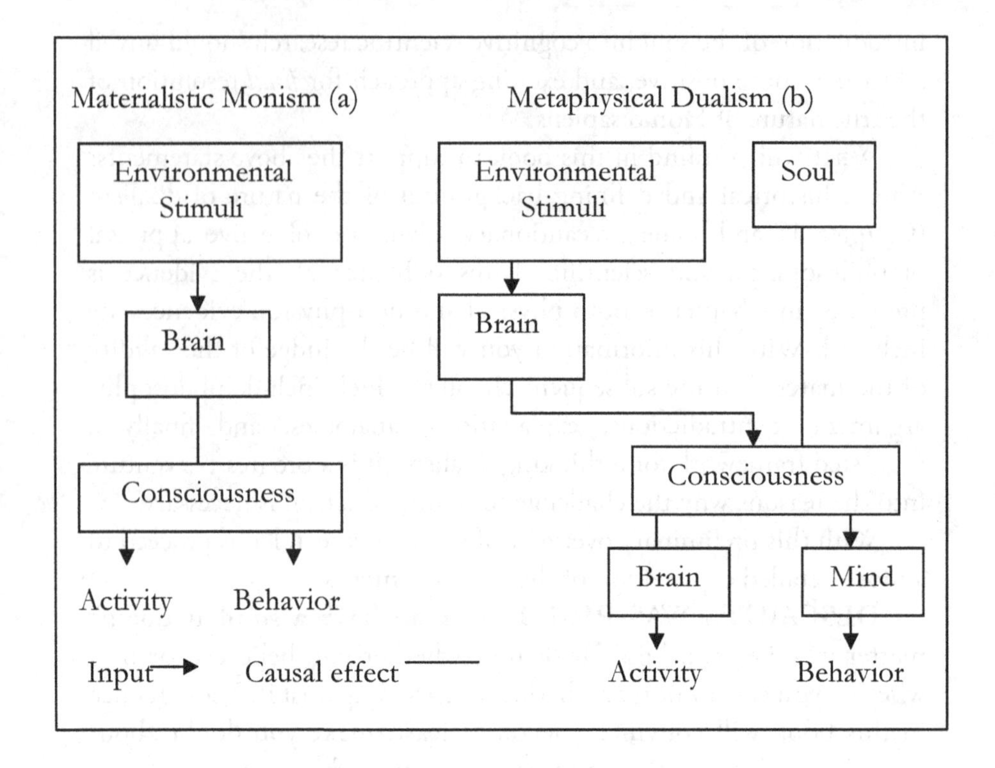

Figure I.1 Basic Ismic Differences

Figure I.1 is my understanding of the basic issues from the varied published material I have read. The simplicity of the diagram masks the complexity of the sublevels of the diagram.

In the monist approach, (a), the brain is the unopposed initiator and activator of all human activity, mental and physical. It creates the mental states, in other words, it generates consciousness since there is no other entity involved, using as inputs the external stimuli sensed by the five senses. All feelings, desires, decisions, etc. are properties of consciousness, which are then the inputs for human behavior and activity. But what initiates the entire process? It is the stimulation of the brain by the outer environment which the brain accepts passively, discriminates and translates into command signals. The definition of a mind in monism is inconclusive or non-existent.

In contrast, the metaphysical approach, (b), depends on the spiritual entity, the soul, for initiation of consciousness and its resulting properties. The brain still receives the environmental stimuli as inputs and processes them, but the soul uses that processed input in its consciousness to evaluate it, to decide on an action, to override any unnecessary automatic response, and to generate the commands for the brain to initiate activity. Note that the brain is still processing the physical information, the inputs from the senses. A pure thinking process does not depend on the brain because it is completely non-physical. When a mental decision is made, the command then goes to the brain to generate the motor signals that will produce physical activity. A mind to brain input line is not shown due to the unknown but experienced process of neurophysical interaction. It is possible that the soul does not use consciousness because it *is* consciousness, which creates the mind to perform the mental activities. The mind is a vital element of the soul. Actually, I feel strongly that the mind, the soul, and the "person" (I) are one entity, like in Descartes' famous statement, "I think, therefore I am." The body is for my use to accomplish my earthly mission, whatever I choose it to be.

The problem between the ismic approaches is the lack of understanding about how each is implemented. The concepts of monism are based on assumptions that knowledge of the physical nervous system including the details of the brain elements defines how human cognition functions. As this book shows, however, it is the dualist side with a soul that can assign where the properties of cognition belong.

Prior to the seventeenth century, the generally accepted traditional belief was that a soul was part of the human makeup. The requirement for scientific verification had not been imposed on philosophic rationalizing. When Descartes wrote the *Meditations*, he provided the base for the debates about the soul and the existence of God. The other contributing factor to the evolving debate was the developing ability of the scientists to explain the natural physical laws. With this development came the need for verification of the new theories and concepts about the Universe and the human being. Although dualism dealt with non-physical entities, the same strict demands for physical proof, as in physical situations, were imposed on any claims of non-physical reality. These demands remain today, allowing science to proceed with a materialistic approach that all human activity is caused only by the physical brain and the physical environment, uncluttered by any need for considerations of a soul.

The attitude that science knows for a fact that the soul does not exist is evident in the books dealing with neuroscience, cognitive science, and neurophilosophy. An example of the latter, which continues the legacy of this scientific belief, is the recently published (2002) book, *Brain-Wise.* The author, Patricia Churchland, a neurophilosopher wastes no time in highlighting this point by making the denial in the first words of the opening paragraph of the Introduction!

> *Core Question*
> *Bit by experimental bit, neuroscience is morphing our conception of what we are. The weight of evidence now implies that it is the brain, rather than some non-physical stuff, that feels, thinks, and decides. That means there is no soul to fall in love. We do still fall in love, certainly, and passion is as real as it ever was. The difference is that now we understand those important feelings to be events happening in the physical brain. It means that there is no soul to spend its post mortem eternity blissful in Heaven or miserable in Hell.*[2]

What can you expect of a book that begins with the elimination of what should be a major contributor to a *complete* analysis of the human mind, even before introducing what the book is intended to disclose? One can be sure that the objective throughout the book will be not only to discredit any metaphysical implications but mainly to promote and justify physical causality as the only explanation for human feeling, thinking, and deciding. The major reason for being so open about this approach is the nearly universal acceptance of the neuroscientific community's stated claim that *absolutely no physical evidence* exists of the soul's reality. The claim is made without any supporting evidence that it is a known fact.

Author Churchland then begins introducing the book's material. By page 42 of Chapter 1 on Metaphysics, subconscious reality assumes control and she writes:

> "........*For example, feelings of unshakeable convictions or absolute certainty may accompany consideration of the hypothesis that the mind is a nonphysical substance. Descartes, as we saw in chapter 1, seems to have enjoyed such a certainty and to have believed it warranted a specific conclusion.*
> '*Feelings of certainty, however, are no guarantee of truth. They can, of course, motivate* testing *a hypothesis for truth. They can motivate continuing a research project even in the face of scoffers. But feeling certain that a hypothesis is true is, sadly, all too consistent with falsity of the hypothesis.*" (The word testing in regular font indicates italics in the original.)

I could not believe what I was reading. If the same logic in these statements is applied to her earlier denial of the soul's existence, she was just admitting that the certainty of her denial also was, "sadly, all too consistent with falsity". Substitution of soul terminology in the last four sentences shows the consistency in the two opposite views.

The use of such biased arguments, without the realization that the same rationale applies to one's own uncertainty, illustrates the arrogant scientific use of scientism in the assumed hypotheses of the brain's physical dominance of human mentality. This is, sadly, the prevailing attitude I have detected in the materialistic research of physical causation.

Typical denials by scientists and scientific philosophers of metaphysical intervention have an aura of authenticity because they have been made by scientifically oriented professionals. The Western culture has accepted these denials since it has succumbed to the belief that science can provide all the answers to natural universal phenomena, not realizing that science can provide *absolutely no* answers to the laws or realities of the metaphysical universe.

It is obvious from these denials that the neuroscientific community is not interested in finding any evidence that refutes their claims. Any evidence that is brought to light about the nonphysical is promptly discounted by attributing it to psychic charlatans, brain damage, fantasy, extrasensory perception (ESP), outright fraud, or simply efforts at notoriety. Such responses, when delivered in derisive terms, quickly conclude any persistence in the issue. But there is always the offer for reconsideration of the soul issue if some evidence happens to surface.

Such an offer, expressed in clear terms by another representative of the scientific community, was made by Dr. Paul Churchland in his book, *Matter and Consciousness*, published in 1999, (ninth printing):

> *If someone really does discover a repeatable parapsychological effect, then we shall have to reevaluate the situation, but as things stand, there is nothing here to support a dualist theory of mind.*[3]

The original edition of the book was published in 1984. My reason for mentioning the dates is that in 1966, a book was published as part of a series of the American Society for Psychiatric Research, which reported the first in a long series of investigations of reincarnation that proved the existence of the soul. (This evidence is presented in the Chapter on Evidence.) The significance of the dates is that in spite of the intention to reevaluate the situation, no effort was or has been made (2009) to find, evaluate, comment, or in any way to even speculate about the appearance of this evidence. So much for the credibility of scientific self-serving verisimilitude. It is clear, and understandable, that the scientists cannot have their years of inadequate and misleading analyses be shown false. But that is not what the above quote or other promises implied.

What is this evidence that reveals an entity that appears to be a soul? It is a total of about 2500 investigated cases of reincarnation in various countries. Only twenty of the initial cases are reported in this first (1966) book, *Twenty Cases Suggestive of Reincarnation*, by Dr. Ian Stevenson. Definite proof of only one case would be sufficient to debunk (author pun intended) the scientific claims that all human behavior is controlled by a physical brain with no interface with a non-physical entity. But you may ask, and rightly so, how does reincarnation relate to dualism and the cognitive effort of science? The concise answer, to be elaborated in much detail later, is that in reincarnation a personality leaves a body at death and is born in another body. The link between the two bodies is the same personality with a memory of what happened in and to the previous body. Such a transfer of the same personality refutes two major neuroscientific assumptions and analyses. The first assumption is that memory is a physical element of the brain. If it were a physical property, the memory of the previous body would remain in that body, whereas it comes across with the carrier of that memory, the entity that must be the soul. The second analytical point, also an assumption by the materialists, is that a nonphysical mental action cannot initiate and control a physical bodily activity: it must be initiated by physical stimuli. But the transferred soul in the reincarnated body *controls* the behavior of an undeveloped child by demanding recognition as the previous personality and by wanting to return to that family (case presented in Chapter 3). But these memories of the past life and motivating desires to return are not physical stimuli of the present environment; they are residuals of experiences in an environment that is not available to the subject's brain in the new environment. In some cases, the reincarnated child exhibits mature skills from the past life without any training. The knowledge of the skills resides in the memory but to activate the physical actions to perform the skills, the nonphysical mind-soul must command the brain to generate the motor commands for the physical organs to perform the activity. According to the claims of cognitive scientists, such control is impossible. To have it happen in a slowly maturing child with no previous training in that skill, as has happened, is a complete negation of the scientific claim. This is the evidence the neuroscientists, and neurophilosophers in particular, do not want to encounter. But they must because it is ***physical first person spontaneous evidence*** of a non-physical entity.

Physical evidence is not the only factor in the determination of what constitutes human nature. Experiential subjective evidence, which cannot be analyzed objectively in a third person mode, also has a bearing on reality. A mental (paranormal) event can only be described and evaluated as a first person experience. When validated by competent medical and psychiatric personnel and witness's confirmation, such experiential evidence, being unique in that it cannot be repeated by the same person, must be allowed some credibility. Instead it prompts immediate discrediting criticisms.

These are the evidential factors that exhort me to challenge the scientific community to rethink dualism. Now comes the dilemma. Since the evidence is now available in scientific form, the scientists are obliged to recognize it, according to their promises, and rethink their analyses *about the effect of a soul-mind on human behavior* with respect to intelligence, rationality, consciousness, free will, desire, etc. Dr. Jeffrey M. Schwartz emphasizes this point with respect to free will in his book, *The Mind and the Brain* (2002), in which he describes his research in the treatment of obsessive-compulsive disorder (OCD):

> *There, I propose that the time has come for science to confront the serious implications of the fact that directed, willful mental activity can clearly and systematically alter brain functions: that the execution of willful effort generates a* physical force *that has the power to change how the brain works and even its physical structure. The result is directed neuroplasticity.*[4]

Note that Dr. Schwartz uses the term *mental activity*, which is non-physical, as the generator of the commands, and that it is the *willful effort* that generates a physical force. These are two physical-non-physical interactions that cognitive scientists claim are impossible. This type of contradiction casts serious doubt about the credibility of the evidence that science is using in shaping the analyses of human behavior.

Any new analytical effort should be addressed as a unified problem for resolution of the nature of Homo sapiens. Now, it is treated as a physical-only issue because the mental state is considered to be physically oriented. Two reasons why a reevaluation will be difficult are obvious. The first is that science does not know how to analyze the nonphysical: its feet are set in the concrete of physical analysis. The second is the uncompromising, persistent denial of the soul's reality, as merely a belief, a myth to be discounted as "folk psychology". This facile elimination of the soul through a simple denial overlooks a major deficiency on the part of the disclaimers. They too are not able to supply any evidence that the soul does not exist, thereby showing that they also fail in doing what they demand the dualists to do.

In addition to making the challenge, a secondary purpose of the book is to provide other aspects of the debate: the history of the evolution of the debate, a look at the skeptics' array of question begging techniques to evaluate conceptual and actual disclosures, arguments with philosophical differences, technical contradictions, applicable analogies, and an original plan for the necessary basic elements of a redirected system analysis. This material is for the reader's appreciation of the conceptual conflicts with reality, what must be changed, and why.

An abbreviated historical summary of dualism in Chapter 1 will have to suffice since the history is not that crucial to the purpose of this book. Chapter 1 also addresses monism and dualism with their different versions adding complexity to the understanding of precisely what is meant in the continuing debate. What is substance dualism? What is meant by dualistic monism? Without understanding these differences, the arguments and concepts about mental states versus brain states and properties become a confusing mix of semantic doublespeak.

Just as philosophers use esoteric language to control the reception of their points of view, scientists also have their tactics to promote their individual theories. In the dualism debate, the incompatibility between the scientific ability to analyze only physical properties of the brain-body and the need to include the non-physical mental states, calls for the definition of some rules that will lead to rational and justifiable evaluations of the various theories. There are basic rules proposed by Descartes for logically searching for meaningful and credible answers. They are stated in Chapter 2 along with other methods of "thought control" I have detected in my attempts to read objectively, that is, without prejudice or bias. The subtleties of these tactics can and often do sneak by unnoticed. How often have you counted the number of "ifs" and "assume that" in the disclosure of a concept or an analogy that evolves into an impression that the concept is valid?

Having prepared you for reading carefully to detect thought control, I move to Chapter 3 which will give you experience in trying to find a use for those tactics in my presentation of the crucial subject, the evidence for and against both monism and dualism. Chapter 3 is of the greatest importance in the challenge to the materialistic monists because it provides the evidence they claim cannot be produced. What physical evidence can a non-physical entity provide to satisfy the scientific requirements? Some ground rules for acceptance are examined. The evidence is of several types: documented investigative reports, scientific test reports, confirming observational information including participating comments, and medical observations. The evidence has the potential of slowing the ongoing neuroscientific analyses of the mind-brain relationships and constitutes the basis for the review of Cartesian dualism. If the scientists accept the validity of the evidence, can or will the entire scientific community abide by the claims to reevaluate the scientific base?

In Chapter 3, you may experience what could be a surprising revelation about who you are and what effect that will have on your life. This should apply not only to the average reader, but more specifically to the deniers of the soul's reality.

In Chapter 4, I will ask you to try to forget the details of the evidence presented in Chapter 3, retaining only the claim that the soul is real. I address some of the conflicts and arguments of the current debate with philosophical comments about possible resolutions. It would appear that the chapter order is reversed: 4 should come before 3. The intent of the reversed order is to give the reader an idea how the arguments are proceeding with the denial of a soul and how they would or could be different if the results of Chapter 3 had been included in the arguments. This would not have been possible if the disclosure of the evidence was made after this chapter on arguments.

The first four chapters constitute the background survey of where the debate remains as of the writing of this book.

In Chapter 5, there are many conflicts and contradictions, not only in the philosophy of the debate but also in the technical concepts which are substituted for the inability to explain monism. Some of these conflicts are by direct refutation of conceptual approaches, technically, scientifically or philosophically. Included are some specific issues of physical causality, neural dominance, concepts of memory, free will and consciousness, and attempted explanations of the mind/brain problem. As in the previous chapter, the knowledge that the evidence for the reality of the soul exists, will allow these conflicts to be viewed as having completely different resolutions. Thus using the hindsight of Chapter 3 to understand the impact on the material of Chapter 5 becomes evident.

Chapter 6 on memory, Chapter 7 on free will, and Chapter 8 on consciousness have the same revealing potential; of upsetting not only the technical conceptual work of the cognitive scientist but also the neurophilosophical base of the philosophers. Memory plays an important role in thinking, learning and knowledge acquisition.

Without substantiation, it is claimed to be physical rather than non-physical. Free will is a controversial scientific and philosophical issue. There is no cohesive definition of consciousness. If the soul's reality is finally recognized and accepted with respect to these three issues, I believe there will be an explosion in the technical analysis of the human neurological system, cognitive reevaluation, and philosophical rethinking. It should also have a major impact on the worldview of the meaning of life, with the dual nature, for the human being. But that is a subject for another book.

Taking a brief relaxing deviation from the factual to the rational, Chapter 9 delves into the possible contributions of analogy to the understanding of the dualist concept. An analogy does not prove anything. Through similarities it points out the compatibility of apparent ideological differences. The first analogy is that of an auto and its driver. The interplay of the mental driver with the physical car mimics so well the mental-physical interactions of daily life, including any deficiencies caused by brain damage. The other analogy is that of the comparison of the hardware and software of a computer to the relationship of the mind with the physical body, especially the brain. But this analogy with the evidence of the real soul moves the comparisons to a different level, one that clarifies the skeptical criticism of the existing computer-brain analogies.

It is easy to criticize the efforts of anyone trying to arrive at a solution to a problem. The underlying difficulties may or may not be comprehended with the result that criticism is easy. The proposal for the solution to the problem is often the part of the criticism that is conveniently left unaddressed. This is unacceptable in a serious challenge as this book is. It is, therefore, incumbent on me to provide at least a "such-as" solution for the direction that future analyses of the human nature should take. This is the subject of Chapter 10 with recommendations for and flow diagrams of a basic system, which can be expanded in various conceptual implementations. As a retired systems analyst, I am able to predict the path to be followed in the logical search for the answers to the integration of body and soul as a unitary system. In such a complex system, the only initial approach can be the definition of the areas of interaction and the intended effects. But even with this unembellished information, were it to trigger the creative imagination of some neuroscientist or philosopher, it might be the starting point for the major task. This attempt at defining the dualism system is a unique "first ever" since the soul's reality has never been scientifically recognized. The inevitable first step must be acceptance of the soul.

The Conclusion summarizes why the evidence of reincarnation leading to the reality of the soul is credible, how the evidence is applied to the mind brain problem, the effects of the application, and the possible repercussions of this application.

The format of the book is intended to produce a cohesive structure through a likeness to a judicial trial. The first three chapters (Part I) comprise the background and evidence phase that provides the necessary information to inform the jury (you). Chapters 4 through 7 comprise Part II, the argument and philosophy phase, which attempts to show why current solutions are inadequate and argues for recognition of dualism as the true nature of the human being. The third part with an explanation of the nature of the verdict explains how the sentence of rethinking dualism can be carried out.

An epigraph adds thematic progress to each chapter as it builds the case for the challenge. Due to the importance of memory in the deconfirming arguments, an Elephant is the logical representation of the soul with its memory.

The primary and basic message of this book is: given the unrecognized, but authentic and valid, evidence of the reality of the soul as part of the human being, it remains for the scientific community to accept this scientific evidence and redirect its totally physical approach to one which includes the *effects* of the soul on human actions and behavior. Members of the scientific community have promised to do so if and when the evidence of the soul's reality becomes available, as it has since 1966.

Part I

The Case – Background and Evidence

The Monism-Dualism debate is a complex array of philosophies, concepts, beliefs, and convictions. To understand these complexities for an appreciation of the problem that eventually must be resolved, it is essential to have an overview of the controversy. A quick trip through the last three centuries is sufficient to get the flavor of the verbal conflicts. It is not only the content of the developing rhetoric but also the biases associated with conceptual solutions that assure the probability that the controversy will continue. The evidence and the logical proposals in this book, whose purpose is to light the path to a verifiable solution, may be sufficient to overcome the adamant defense of the current position. Part I, therefore, is the statement of the Case in this imaginary trial of the Cartesian philosophy.

In part I, in addition to the definition and background of the debate, which is Chapter 1, other material is presented on how to judge the presentations of the prosecution and the defense. Chapter 2 is a description of the rhetorical and misleading flavors that are injected into the arguments in an attempt to control the minds of you, the jury. The main content of Chapter 3 is the evidence which will either convince the jury or result in lingering doubt about its ability to change the direction of scientific research. The prosecution's confidence level is high that the critical evidence is sufficient to convince the Court in Parts II and III that a rethinking of dualism is mandatory to resolve the hopeless debate.

1

The Monism-Dualism Debate

Descartes described the Elephant but
he couldn't find one to prove it existed.

I think, therefore, I am. With these words, René Descartes (1596-1650), a French philosopher and mathematician, opened the gates of *modern* philosophic thought. His addition to the existing concept of philosophy was the individual's ability to reason rather than interpret what was caused by the laws of Nature. In so doing, he developed the theory that since matter cannot think, there must be a mind in man that does the thinking. He did this by disclosing in minute detail how he started his rationalizing; by doubting everything he had previously believed in order to find a starting point for his philosophic credibility. The result was that he could be sure of only one thing — he could think, which led to that basic statement. Although the famous statement is constantly quoted, and argued about in spite of its clarity, it is nevertheless incomplete without a precise definition of thinking. In his rudimentary approach for identity, Descartes also questioned what thinking meant. In his words, taken from *Meditation II*:

> But what then am I? A thing which thinks. What is a thing which thinks? It is a thing which doubts, understands (conceives), affirms, denies, wills, refuses, which also imagines and feels.[5]

In the dualism debate with man's physical attributes pitted against his mental abilities for the governance of behavior, the conflicting interpretations of thinking are crucial. Thinking, as an active mental function in dualism, is initiated by the mind. But the materialist approach rejects the presence of a non-physical mind, substituting a created mental state by the brain, which uses sensory stimuli to initiate the creative act. The conflicting effects of the meaning of thinking are obvious.

Prior to the seventeenth century, there was disagreement, but no major controversy, about the nature of the human being. Philosophers proposed their views, theologians insisted on their teaching, but regardless of the culture of any civilization, the idea of some sort of spirit residing in the human body was an accepted belief. The Greek philosophers had different versions of such a spirit. Plato and Socrates identified the soul as the entity that thinks, decides, acts, and is independent of the flesh and blood in which it is a prisoner. As a separate substance it was immortal. Aristotle, on the other hand, identified the soul as the form of the organization of the living being and that it did not survive its death.

The ancient Egyptians believed in the immortality of the human spirit to the extent of not only providing provisions for the journey into the hereafter but also in the case of the pharaohs and nobility, killing slaves to accompany and serve these special spirits. The American Indians honored the spirits of their dead.

In the Eastern cultures of ancient India and China, the belief in a human soul was so strong that they believed in the reincarnation of the soul if the life that was just terminated was not worthy of retention in the spirit world. People strove to live lives that would preclude their reincarnation. That belief in reincarnation is still prevalent today (as will be shown in a later chapter).

As a unique happening in the history of humanity, one that has had an effect on all human thinking since then, the awaited Messiah of Biblical prophesies became a real human being. This messenger of God came from the spiritual universe with information about that universe. Whether He was, or is, accepted as that messenger, His unexplained power over natural law supported His claim to that messianic mission. What is pertinent to the search for the reality of the soul is the evidence

He supplied through His teaching and His deeds about the reality of that foreign universe. Just as we accept the statements by any foreigner about the conditions of the land he comes from, the information that Jesus Christ left with us about the non-physical domain, must be given equal credence in our search for the reality of the non-physical spirit, the soul. Jesus Christ in his three years of teaching often referred to the spiritual existence of the past Israelite prophets. He promised to prepare places in the heavenly Kingdom for His disciples, implying an afterlife, and brought several dead bodies back to life, indicating that there were souls that could return to give life to the dead. In the religion that was based on Christ's teaching, several philosophers rationalized the concept of the soul through logic, or faith, based on His revelations. Thomas Aquinas, whose philosophies made a lasting impression on Western culture, rationalized the dual nature of man.

When Descartes wrote his *Meditations on First Philosophy*, in which he logically and objectively deduced that a separate entity, a soul, is part of the human being and that God exists, he submitted this work to the learned community for comment or dissent to which he could respond. The comments he received were in the form of Objections to which he responded. However, the eventual reaction to Descartes' work initiated what is known as modern philosophy. Dualism, the dual nature of the human being, became a major subject for debate, which continues to this day. The existence of God is also a major issue, but except for its relevance to the existence of the soul, is not usually associated with the dualism controversy and is not part of this book. The existence of God has its own set of arguments different from those of the dual human nature.

When man began to understand the physical natural laws, he also began to depart from the adherence to religious beliefs and to develop a dependence on his own relativistic beliefs of individualism. Galilei Galileo (1564-1642) and Isaac Newton (1643-1727) postulated the basic laws of physics, which were to become the cornerstones for building scientific rigor and positivism (I use the term positivism in the sense that it is a warning against non-acceptance by theology and metaphysics). As more of the laws of nature were explained, science became a domineering factor in the development of Western culture.

In the process, the reasoning of Descartes about the united body-soul and the existence of God was minimized by this scientific dominance.

The theory of Darwinian evolution, with its insistence on natural selection as the cause of man's development, including his brain, contributed to the departure of the belief in a spiritual soul (and mind), as well as the departure from the traditional belief of man's creation by a Supreme Being. With the release from the idea of a soul that was meant for a life after death, the materialistic philosophy touted by some of the early Greeks reappeared and found a following, which has become the philosophy of the modern culture. It is within this materialistic conditioning that the neuroscientists have found a responsive society to their theory of physical predominance over any belief in spiritual reality. Critics, however, argue that the scientific approach cannot explain the human ability to have non-physical thoughts, desires, and emotions, and the ability to override the normal physical reactions to environmental sensory inputs. In other words, consciousness cannot be explained as a purely physical operation. Philosophers, the religious community, and even some scientists, still adhere to the dualistic concept. Thus the debate remains active and unresolved.

Until the time of Isaac Newton, the requirement for scientific verification had not been imposed on philosophic rationalizing. With the development of the ability to explain the natural laws came the need for verification of new theories and concepts about the universe and the human being. Although dualism dealt with non-physical entities, the same rigorous demands for proof, as in the physical situations, were imposed on any claims of non-physical existence. This general requirement was never made specific as to what type of evidence would be acceptable. How could physical evidence be generated of something that has no physical substance, occupies no physical space, and cannot be detected by any test equipment? The scientific community assumed it was safe in its denial of the spiritual existence. This belief in scientific certainty continues to skew the debate in its direction. Thus the debate continues with both sides refusing to compromise. This refusal continues on the assumption that there is no scientific evidence to support the existence of the soul. If scientific evidence were to be found, would the scientific community even admit the error of disclaiming the reality of the soul, much less incorporate that information in the analyses of human nature?

Some philosophers are content to abide by the fundamental description of dualism but the majority prefer to promote their own diverse concepts. Differences in interpretation of terms such as *substance* versus *essence* seem more important than defining the effects of a soul on scientific analyses, should any evidence confirm the existence of the soul. Even if the extreme of the dualistic approach, a separate soul as an integral part of the human being, were to be studied, the study must consider how the soul is able to utilize the physical neural network in directing the activities of the human assemblage. If the opposite approach is analyzed, the study must explain how a physical system can produce non-physical emotions, desires, rational choices of responding to or overriding the effects of physical stimuli.

Theologians, in general, affirm the belief in a soul as a separate entity with an afterlife following the earthly death. Their belief relies on philosophical and ontological Biblical arguments, neither of which is accepted by the scientists.

The average human does not carry on this debate, and more likely is unaware of it or unconcerned about it. Such unconcern, however, may mask his interest in the basic issue since it affects the meaning of life for all human beings. The outcome of the debate, therefore, if conclusive, should have an impact on all social relationships. This should be sufficient reason to become interested in the body-soul effects.

What is the basic difference between monism and dualism? The materialists assert that human reaction to the environment and the consequent behavior are controlled by the physical brain without immaterial participation. The dualists insist that a separate entity, the soul, is the reason for man's ability to think, decide, and have feelings in reacting to the world environment.

Monism is the theory that the human being is composed of one, and only one, substance, the brain-body configuration. There is a variation of this theory in which the substance is a neutral material that makes up both a mind and a brain. The difference is in how this neutral material is arranged in each. But this common material is not one entity as it is in the basic monistic theory. Instead it is made up of many entities of the same kind from which different properties can be identified.

Dualism's clarity is complicated by monism's references to two natures but not two entities in the human being. Some of these variations are described below. In this book, the basic definition of dualism is the Cartesian version of body and soul as separate entities.

There are also semantic differences in the general usage of the term dualism. It may be merely the characteristics of a physical item (milk as food and a liquid), non-physical properties (short and long term memory), or of concepts (holistic and dualistic interactionism). Much of the misunderstanding in any examination of a controversy lies in the lack of clear definition of terms, which should narrow the boundaries of the specific debate. For a clear understanding of the terms and the various meanings, let's look at some examples from the dictionaries and encyclopedias, which define typical dual relationships.

In the Oxford Universal Dictionary, Third Edition,

Dualism,

a. The state of being dual; Twofold division, duality.
b. Gram. The fact of expressing two in number.
c. A system of thought which recognizes two independent principles.

   Spec. Philos. The doctrine that mind and matter exist as independent entities opposite to idealism and materialism.

   The doctrine that there are two independent principles, one good and the other evil.

   Theol. The (Nestorian) doctrine that Christ consisted of two personalities.

   Chem. The theory, now abandoned, that every compound is constituted of two parts having opposite electricities.

In the Dictionary of Philosophy (revised Second Edition),

> Dualism. A theory concerning the fundamental types into which individual substances are to be divided. It asserts that substances are either material or mental, neither type being reducible to the other. Dualism is to be distinguished first from monism and then from pluralism. The latter is a theory about the number of substances rather than their type, and states that there exists more than one substance. Some pluralistic philosophies, such as Cartesianism, are dualistic, but others are not. For example, Berkeley asserts the existence of a plurality of substances, but he is a monist in saying that these are all of the same kind, in that all are mental substances.

In the above definitions, dualism is either state, fact, system, doctrine, condition, view, concept or theory. An argument could develop on this issue alone: Which version should be used in a proposed debate. In the body-soul analyses, the different versions concern conceptual interpretations of how the human being can operate physically and mentally with only physically identified properties. The following definitions are intended to clarify the differences between the various interpretations.

Substance Dualism

Substance Dualism is essentially the Cartesian version in which there are two substances, one physical and the other non-physical. The definition of substance here is that which has an independent existence. The physical part, the body, occupies space and can be measured. The non-physical part is that which does not occupy space but produces effects that can be experienced, such as thinking, feeling, deciding and willing.

Property Dualism

Property Dualism is that version in which there is no other substance than the physical brain. But this strictly physical brain has a set of unique properties, which cannot be found in any other physical object. These properties are nonphysical but their explanation is couched in terms that a physical system can accommodate. They are associated with consciousness and intelligence, that is, they constitute the ability to think, to learn, to distinguish between physical qualities of objects, and so on. A puzzling feature of this concept is that although these mental phenomena are caused by the brain, they are supposed to have no causal effect; they do not constitute volitional motivation leading to activity. If that is so, then the physical brain is the motivator, but it must rely on the conscious thinking and desires of these mental properties to generate the motor commands for the bodily physical activities. The property dualists have difficulty explaining this puzzle. The need to physicalize the human system reveals the resultant inconsistencies and contradictions.

Epiphenomalism

Epiphenomalism is a version of property dualism with some differences from that stated above. Although property dualism only includes the nonphysical properties without explaining their generation, epiphenomalism states that all mental events are the effects of physical events but they do not cause other physical or mental events. Conscious experiences do exist because mental states are necessary for certain events but they are not like material processes and play no role in human behavior.

Interactionist Property Dualism

Interactionist Property Dualism differs from property dualism in that the properties do have a causal effect which means that they are responsible for behavior. It is a two way correlation; the mental state causes physiological events and vice versa. Desire and volition, therefore, do cause behavioral activity: an answer to the question posed in property dualism.

Integrative Dualism

Integrative Dualism proposes that the person and the body are not metaphysically identical and are separable. However, they are integrated to operate as a unit through interactions of their individual functions. They are a single unit mentally and physically. Note that in this concept it is the person that is the non-physical entity, not a mental property, or mind-stuff, or epiphenomenon.

Holistic Dualism

Holistic Dualism emphasizes that during earthly life, body, mind, and spirit are functionally integrated and constitute an existential entity. This entity should be thought of as an inseparable operating unit without considering how each separate part affects the other parts of the unit. (This holistic approach is addressed in Chapter 10 in the system description of dualism.)

Emergent Dualism

Emergent Dualism proposed by William Hasker, a Professor of Philosophy, argues not only against the physicalism of materialistic concepts but also against the Cartesian and Thomistic versions of substance dualism. His version uses the confirmed results of neurophysiological research for the critical role of the brain and nervous system in mental processing. However, he theorizes that these mental properties "emerge" only when certain necessary constituents are configured in a certain arrangement through evolutionary development. This is brought about by the organization and functioning of the nervous system which creates a "field of consciousness".[6] This field in turn modifies and directs the functions of the physical brain. These emergent properties involve causal powers that are not evident in the absence of consciousness. Through this emergence of a mind under certain conditions he avoids the automatic physical causality of behavior by having the emergent properties control the free will of a conscious being.

These are a few of the many versions of dualism that contribute to the complexity of the dualism debate. Isn't it obvious from the above examples that unless the exact definition of dualism is specified for any debate the ensuing arguments could go on (and do) without any conclusive answers? There is the other side of the debate that isn't too simple either. There are versions of monism, but to limit the confusion, three definitions will suffice for the purposes of this book.

Single Substance

Philosopher Spinoza (1632-77) proposed that there is only one infinite substance. Mind and matter were merely attributes of this infinite substance. George Hegel's (1770-1831) philosophy was a form of monism in that he believed that there could only be one thinking substance, an agent that thinks of objects that lead to truth.

One Intrinsic Nature
One intrinsic nature has to do with the mind-body relationship. Minds and bodies posses the same intrinsic nature but they differ in the way that the common material of their composition is arranged. This common material is not considered to be one entity as it would be in the first definition, but composed of many entities of the same kind.

Pluralism
Pluralism is the theory which maintains that there are many properties of reality (being). The uniqueness of these properties is that they cannot be reduced to monism or dualism because they are all mental.

There are many reasons for the dualism controversy. Among these are: history, philosophy, science, theology, definitions, semantics, reality, analogies, beliefs, arrogance, stubbornness, Darwinism, Creation, paranormal experience, common sense, etc.

With so many different approaches and conflicts concerning the mind-brain and body-soul problems, what does the debate boil down to? It began with Descartes and has arrived at the historic point when Descartes' philosophy can be verified but at a significant scientific and philosophic price. That situation was created by the application of scientism to a problem that is not strictly a scientific one. The dualism-monism conflicting situation is summarized in the following points:

1. Descartes said that there is a soul in each human body that can think and it is a separate entity from the body.
2. Descartes could not verify his claim because physical existence of a non-physical entity could not be demonstrated in the laboratory or at any other location. Since then the dualists have not been able to find any conclusive physical evidence, any that could comply with the critical demands for visible and body related disclosure. Paranormal experiences have been the only valid evidence since they are the only means of contact with any non-physical entity, and

such evidence is obviously unacceptable to the physically oriented materialists.

3. The concept of a soul was, and is assumed currently, to be only a myth; a religious illusion to satisfy the human desire for immortality. Taking advantage of this situation the materialist scientists have indulged in extensive analyses to support their basic approach of physical causality. Ironically, without any supporting evidence for their denial of a soul's reality, this approach too is only a mythical convenience.
4. With this assumption of the soul's non-existence, the neuroscientists and cognitive researchers have analyzed and explained the human brain and nervous system functions as completely physical and hope to prove that this physical combination is responsible for human rationality and behavior. They have an advantage in this argument: By using physiological and neural research results that can be demonstrated, their concepts seem credible. For the unexplained gaps, the implied physical evidence also seems adequate. These scientific data of the human physical system are unquestionably correct but the extrapolation of their effects on the non-physical functions is invalid.
5. Although this approach has many unsolved problems, like the creation of mental states, the definition of memory, free will, and consciousness, and their effects on human behavior, the scientists optimistically claim that the problems will *eventually* be solved through the intensive neuroscientific and cognitive studies.
6. These studies are conceptual explorations of physical data that are possible demonstrations of physical causality but they are based on assumptions of extrapolated reality. Furthermore, there is no functional description of how the separate conceptual elements can be integrated into a viable system that can produce mental states by physical causality.
7. Due to the lack of supporting evidence of the scientific approach, even dissenting neuroscientists doubt that physicalism will ever explain the mind-brain problem.

There are many gaps in the attempts to provide conceptual solutions in monistic problems. In conceptualizing the solutions to the elements that contribute to the body-brain interactions, the theorists complicate (perhaps intentionally) the understanding of their proposals by the semantic methods they use. An example of such a crutch is the neuroscientific reliance on representations. A representation is a brain's created pattern of the outer world in a form that can be used by the neural network for translation into activity and behavior. It is a semantic substitution for the word *idea* or *though*t. So why make the change? Because it can be better adapted to a physical description of a brain activity rather than struggling to explain how an immaterial mind operates. It is an invention that displaces a common-usage term with a concept that can be manipulated around any unsolvable problem. The explanation then takes on the aura of the solution and from then on, the concept is used as the solution. The debate is further complicated by arguments and proposals that logically make no sense, like the denial of the mind's reality, while logic demands the use of a mind to *create* the argument for the absence.

How will all this change since there is scientific evidence that the soul is a reality, the admission of which should redirect the nature of cognitive research? The following points are only conjectures of possible activity.

1. Cognitive scientists and neuroscientists will refuse to accept the findings of the reincarnation investigators because for their research efforts to stay alive, the issue of the soul must remain dead. Criticism of such unwanted information will label it fantasy, hallucination, extra sensory perception (ESP), psychic fraud, or a desire to generate public attention.
2. A review of the evidence, if initiated and conducted by a competent authority, acceptable to all parties, will find the evidence scientifically verifiable. Additional verifying studies will be proposed.
3. The scientific community will be forced to accept the evidence because it is valid scientific reporting according to scientific standards. Since physical evidence is more reliable than promissory conceptual verification, a review of the

ongoing analytical efforts should be initiated to determine the effects (not the process, as a start) of the soul on the brain and the neural system. Accepting the soul's evidence will negate most of the conclusions and concepts that have been based on the false assumption of man without a soul. The ongoing research could come to a dead halt since scientists do not know how to analyze an immaterial effect on a physical entity. A new philosophically enhanced science will evolve.

4. Social philosophy will be forced to take into account the dual nature of the human being and its effects on social behavior. The importance of religion in human behavior will increase.
5. The Eastern and Western cultures may find a common ground for resolution of philosophical and religious differences. The probability of social globalization may improve since the definition of personhood as a socially leveling criterion will be evident.

When (and if) the above steps occur, the future of Western civilization is unpredictable. The Eastern philosophies maintain a belief in the soul, which is reflected in the behavior patterns of the Eastern cultures. Whether Western culture will change is problematic.

Until any of the above actions occur, the problem of how the mental and neurophysical processes interrelate, which currently do not include the metaphysical aspects of the union, will not begin to be solved.

The need for dualism in religious and moral fundamentals is critical. Without a soul that is destined for immortality, the moral constraint on man's behavior becomes irrelevant. Some philosophers maintain that these religious restrictions on man's rational motivation in achieving life's goals are part of his subconscious desire for a union with a Supreme Being. In direct opposition, philosopher Nick Bamforth believes that such restrictions must be overcome to allow man to act as he feels is right.[7] Man must be able to live as he desires without the imposed unnatural restraints to fully express himself in the "safety of a harmonious group", he says, in order to connect with the Unity, the Oneness of the Universe. If this is the "liberation" that the moral revolution initiated in the 1960's (same principle), then it is obvious from its product, the decadent modern culture, what this freedom from restraint produced. It did not develop into a safe harmonious group; instead it became an unsafe destructive group. The tolerance of *any* activity and the loss of the respect for human life are prime examples of this moral relativism. The needed reform in the secular community revolves around the dual nature of man; not in the conflict between rationality and intuition, but in the belief that there is a responsible eternal entity that controls the physical activity of the human being.

The inclusion of the following chapter may seem superfluous because it does not contribute any pertinent material to the debate. However, its importance cannot be measured by this criterion. Its main thrust is in developing the insight in how to evaluate the material that does have an effect on the debate. With scientific positivism pervasive throughout the neuroscientific and neurophilosophic material about denying dualism, much of it based on assumptions about unexplained problems, it is important to understand the techniques used to gain acceptance of the theoretical work. This foray into the concepts of mind control can be of benefit, not only in any related reading, but also in evaluating what I present in the remainder of the book. Actually, not only am I trying to affect your thinking as you proceed but I am also stating the reasons for the challenge I'm making to the scientific community. It is mandatory to rethink the effects of the soul on human behavior in a concise and objective manner. It is easy to criticize, but difficult to submit a solution to the criticism. I hope that I can pass the test for I believe that a major rethinking effort is required and in the last chapter, I propose a path leading to that effort.

2

What Do You Mean When You Say…?

Would you understand what an Elephant looks like if you only read about it?

A genius is one who can simplify a complex problem and supply a reasonable solution. That's the positive angle: the ability to explain a thought, idea, or concept so clearly that the average mentality can understand it. There is a corollary to that: a genius is one who can convince anyone about anything whether it is true or not. It is the latter corollary that bothers us average readers when we read material prepared by experts recognized in their field of expertise. Are we being given factual information or some other predetermined cause that may not be fully understood but must be explained convincingly? That's when we are at their mercy. How can we distinguish the valid from the implied facts without taking a course in *Meaning and Logic*?

Are there criteria for objective disclosures of concepts, viewpoints, experiences, proposals, or happenings? Yes, and they are necessary because controversy over proposed theories and claims is the inevitable reaction by the human mind to anything that is not conceived by that mind. Such controversy can lead to better understanding or to perpetual debate. The debate about dualism is the result of the latter.

At times, claimed objectivity is subjectivity in disguise, as in presenting or analyzing only one side of an issue or criticism. Various techniques are used to divert attention from the weak points of an argument like introducing distracting (often meaningless) analogies;

inventing new terms for assumed facts that later become verifying points; using computer models based on assumed data; and exploiting similar meanings of words and terms. Seeing through these deceptive tactics comes only after an intentional effort at recognizing them and reading material that may contain them. My intent, beginning with this chapter, is to help you do both in this book due to its controversial nature.

Philosophers, scientists, test managers, commercial promoters, and especially politicians use language that will most benefit their causes. How can a listener or reader search through the verbiage for the real messages that can affect one's planned use of the information? How does one wade through the esoteric terminology, invented expressions, subtle inconsistencies, and scientific unsubstantiated assertions to determine the meaning and sincerity of the message? In a determined effort to find the real or intended message one must resort to several techniques and play one message against others in a matching game to find analytical consistency.

In more mundane cases like the daily news in newspapers, on the radio or on TV, the material is often deliberately slanted editorially to sway the judgment of the recipient. It is very difficult to find the logical certainty of the message unless one trains the intellect to be aware of the lingual and verbal tactics. The uses of specific words and phrases like "as you well know" and "first of all..." are key indicators of evading the issue or of what is appropriately called the "spin".

How often have you been misunderstood in what you said, which led to a heated discussion, or an argument, only to learn later that it could have been avoided if the meaning had been clarified in the beginning? How often have you heard arguments that were illogical but accepted by a passive audience? How often have you been confused by the semantic ploy of doublespeak? Are you able to fathom the meaning of any statement to your satisfaction when it is presented in confusing language or conceptual extrapolations? Can you detect an attempt to mislead (control) your thinking for acceptance, rather than merely to communicate with you? The politicians and the media are experts at creating impressions through the choice of specific terms and double meanings. All the above are examples of the possible pitfalls that you may encounter in your efforts to isolate the true message in any presentation, literary or vocal. This chapter is based on my experiences in stumbling through those pitfalls and in being rescued by my experiences in engineering logic.

The philosophic debate about dualism is replete with intentionally misleading tactics that are used in argument and accusation. Are there logical rules that should be applied equivocally to all sides in the debating game that supposedly searches for the truth? Yes there are, and one set of such rules was proposed by René Descartes, the alleged founder of modern philosophy, in his *Rules for the Direction of the Mind.* It is an explanation of his method of reasoning, which led to the philosophy of dualism. That same method can be an aid in our search for the difference between monism and dualism.

Descartes postulated 21 rules including what the end of study should be, methods of performing studies, and examples for clarifying ideas. Although all the rules apply to logical reasoning through any problem, three rules relate more closely to what this book is attempting to present. These three rules follow with brief reasons for their choices.

Rule III. In the subjects we propose to investigate, our inquiries should be directed, not to what others have thought, nor to what we ourselves conjecture, but to what we can clearly and perspicuously behold and with certainty deduce; for knowledge is not won in any other way.

When we engage in original thinking, what others have thought should be rejected. In studies that have produced verified physical data, the information cannot be ignored. Instead it should be the base upon which new thinking evolves. Our own conjectures can be so implanted that we ignore criticisms even when they are supported by facts. Deduction is the progressive development of known data to a logical conclusion.

Physical causality, which is the scientific answer for the denial of dualism, is based on incomplete understanding of human consciousness, as is shown in a later chapter. Conceptual and philosophic thinking to explain memory and free will are based on assumptions that conveniently explain these conscious functions. The certainty, according to this rule, is not present in these approaches and the warning of the rule applies.

RULE V. Method consists entirely in the order and disposition of the objects toward which our mental vision must be directed if we would find out any truth. We shall comply with it exactly if we reduce involved and obscure propositions step by step to those that are simpler, and then starting with the intuitive apprehensions of all those that are absolutely simple, attempt to ascend to the knowledge of all others by precisely similar steps.

This rule must be observed if investigations are to lead to truth. However, the rule is ignored by many in undertaking an investigation of a complex issue by attempting to solve it without first breaking it down into its simpler components. Comprehension of the total complexity may come from identification of the possible mutual effects of the simpler parts. Intuition implied here is not the unsupported imaginative kind but that which a clear and conscious mind, free from doubt, employs. First insights are derived through intuition; conclusions are firmed by deduction.

In the study of dualism, the elimination of the existence of a separate non-physical entity simplifies the explanation of physical causality. But the mental vision should have been directed to the *possibility* of a dualistic human makeup. The oversimplification was wrong according to Rule V because there are two intuitive elements. The result is an incomplete analysis. In searching for the truth about the complete issue, all factors must be considered in their simplest forms regardless of any convenient escape of non-existence. As the assumed non-existent entity has become a reality, the damage to the incomplete analysis is major.

RULE VII. If we wish our science to be complete, those matters which promote the end we have in view must one and all be scrutinized by a movement of thought which is continuous and nowhere interrupted, they must also be included in an enumeration which is both adequate and methodical.

In an analytic or synthetic process that involves several steps, it is necessary to maintain the effect of each step on previous and succeeding steps, not merely on adjacent steps. There should be no gaps in the rational chain of intermediate conclusions. Any break in this chain, even in minor considerations, will reduce the credibility of the final outcome. Enumeration of all steps is necessary to assure that the path of reasoning can be traced at any time. Even if the analysis leads to an unsolvable end, that is a valid conclusion, because barring a breakthrough, the statement can be made that human knowledge is incapable of solving such a problem. Such a situation may develop upon inclusion of the non-physical soul as an essential contributor to human behavior because the mind as a separate entity and interface with the brain involves non-physical laws which are not understood. But the fact remains that the mind and the interface exist as part of the complex human system. The movement of thought can come to a standstill and must be reinvigorated with new thinking.

The continuous movement of thought, according to Rule VII, calls for special care to avoid confusion caused by mixing physically related issues with non-physical explanations or concepts. Nor can interruptions in rational thought be tolerated if they are caused by jumps to conclusions sans supporting evidence. These are major gaps in the flow of reasoning because if any one of the assumptions is not true, the entire analysis is flawed. Just as scientists demand proof of the soul's existence, they also ignore any evidence that it does exist. Basing the entire materialistic approach on such a biased presumption makes it vulnerable to a disastrous collapse as the reality of the soul is established.

The above rules also apply to our attempts at answering the continually perplexing questions of who we are. When candidate answers are derived, their credibility should be evaluated by how they are described, not only technically and logically but also by linguistic usage. The latter may contain misleading conclusions while maintaining the credibility of the logical and technical analyses. The use of many technical or philosophic terms can divert or lose the main rational line for the average reader.

What other technique do we have for probing the intent and integrity of a concept, scientific or otherwise? Above all else, we must become skeptics, using those tactics that force any analytic or conceptual approach to be justified by rational rigor. We can uncover how well the concept's originators are prepared to satisfy our demands for justification. We do this by subjecting their conceptualizing to our standards for credibility. On the other hand, if we are defending an idea or a concept and are subjected to skeptical demands for justification of those ideas and the supporting data, we can learn much about how secure we are in our thinking.

How do we develop the standards and the ability to detect attempts at skillful misrepresentation? There are many techniques to do this but they all have one thing in common: they are language-oriented. Symbols and diagrams are forms of language as they clarify what cannot be adequately described with words. Among the verbal techniques are: precise definition of words and intent, verbal specificity in conceptualization, awareness of the use of "if", definition of representative terms, semantic clarity, interpretation, interpretive justification, questioning conceptual relativity, use of common sense perceptions, analogical validity, detection of scientism, detection of circularity, technical clarity, questioning the logic of philosophical arguments, and exposition of double standards. These skeptical processes should not denigrate or ridicule conceptual thinking; instead they should strive to strip the deceptive and irrational logic from any unsupported or supposedly justifying claims.

The above list of skeptics' tactics, though rather large in number, when integrated mentally, creates a state of mind that questions each assertive statement. In philosophic terms, it alerts an inquisitive mind to the need for questions (question-begging) or the satisfaction that there is an adequate explanation, a non-question-begging conclusion. Within this list are a number of tactics that can continue a question-begging situation ad infinitum. But within this list is also the solution to this endless questioning tactic that can terminate it: a demand for the skeptic's definition of his acceptance standards to which the responses can be addressed. This forces the skeptic to limit the reasonable, not impossible, demands and to submit to acceptance if they are met. An example is the impossibility of physical verification of non-physical phenomena. But in the dualism debate, that is a major demand.

The cognitive science's position in the dualism controversy is couched in esoteric terms, neurological and physiological, that may be difficult for the average person to follow. Not only do the standard scientific and philosophic terms tax the average ability to comprehend the arguments, but the use of unproven theories as fact, subject to future confirmation, adds to the confusion. Frequently, this use of unsubstantiated facts as supportive evidence circumvents the mandatory scientific ground rule that all analyses must be backed by firm evidence. Laxity in this regard in claiming probable verification at some future time is tolerated by the scientific community because it stops the opposing spiritual argument by a simple, unilateral, and unsupported denial of that spiritual entity. As unbalanced as this is, the theorists and analysts further complicate the ability to understand their concepts by inventing terms and expressions whose meanings are strictly limited to their explanations. These added complications break the reader's train of thought and force the retention of the meanings of these terms in subsequent usage.

To prepare your mind for the task of critically reading everything that follows in this book about the dualism debate, the following descriptions of the items listed above should help you to mentally integrate them into a questioning attitude about the described material as well as my comments and original contributions. This attitude will assist you in deciding how this game of dualism ends on your scoreboard.

Precise Definition of Words and Intent

Use of words with multiple meanings is a reliable tactic for creating a deceitful impression. Such a word is *mind*, which is causing much confusion in neuroscientific work and, most decidedly, in the dualism debate. The cause is an imprecise distinction between mind and brain. The following list is an example of how we use the many meanings of a word and find no difficulty in understanding the various meanings, mainly because of the contexts in which they are used.

Mind/brain	often used synonymously
Mind my business	tend to my business
Mind your own business!	leave me alone
Do you mind if I…?	may I?
Never mind	forget it
Mind your manners	behave
Mind what I say	listen and obey
Piece of one's mind	bluntly given opinion

Mind stuff	mental phenomena
Mindless	lacking intelligence or good sense
The greatest mind of the century	recognized intelligence
Mind the kids	baby-sit
That blows my mind	overwhelms my rationality
Follow your mind not your heart	rationality over emotion
Make up one's mind	decide
Mindful	attentive

Why can one word be used in so many different ways and still not be misunderstood? It is the common function of mental activity to connect the word with the contextual application. The brain is often confused with the mind even in the same contextual debate.

Consciousness is another word that has several interpretations, which are contributing to the mind/brain/separate entity controversy. It is discussed in Chapter 8.

Use of words with similar, but precisely different, meanings is also common when the intent to create an impression is deliberate. An example of such deceptive usage is in the use of the word occupation instead of liberation in the criticism of the United States post-war activities of the second Iraq war. The media kept referring to the presence of the United States Army and the Coalition Forces as an *occupation*, instead of *liberation*, because both terms refer to a controlling presence of troops in a foreign country. Occupation, however, means full control of the defeated country. But in Iraq, the war was one of liberation, after which the restoration of order was a cooperative, not a controlled, operation. Never in an occupied situation was the defeated country allowed to rearm soon after the cessation of hostilities and to reinstate the members of the defeated army to use weapons to assist the liberating forces. Nor was any defeated country ordered to set up its own independent government as soon as possible. The intended use of the word *occupation* by the media and the critics was obvious but never contested. The lack of criticism is an indication of how little attention is given to the precise meaning of words in the reception of information.

Verbal Specificity in Conceptualization

What does this term mean? Because concepts are original ideas or views about existing theories, the need for new terms or new meanings of words becomes necessary. For invented words, the exact meaning should, and most of the time is, given. However, the specific limitations on the words or terms are not always given. This gives rise to later usage that can slant the meaning in a desired direction.

During my engineering career, I conceived the idea of light intensification for a reflector through a parabolically configured optical unit. I called it a light pipe. It would function as a pipe in conducting light energy from one end to another. However, it did not have the shape of a conventional pipe or the emptiness of its interior. Without completely describing the differences in the shape, material, and the lack of an empty interior, a general description would be meaningless, simply because I used a common term for a specific and unusual application. But the functional applicability of the transmission characteristics of a pipe added significantly to the understanding of the conceptually functional requirement.

In the dualism debate, considerable use of conceptualization demands much verbal specificity to reduce the effects of psychobabble (I presume the meaning of psychobabble is clear without explanation). After the words and terms are adequately defined, a skeptical awareness is necessary to watch for deviations from those definitions in the continuing conceptual revelation.

Detection of the Use of *"if"*

What a difference this word makes in an argument. It derails the line of thought because it transforms the credibility of the subject under discussion from factual to speculative. No longer is anything that follows real; the implication of reality has been introduced although it is subject to conjecture and interpretation. The discussion imperceptibly changes course and becomes irrelevant if it continues with additional ifs. In reading a scientific disclosure, note when there are if's that tend to mislead through substitution of a speculation for a portion of the main line of reason

The use of "if" is very common in philosophical arguments. At times the speculative diversions are comical and nonsensical when they disregard the limitations imposed by natural laws. Consider the argument about the existence of more than one universe. As a philosophical subject it has a wide range of sub-arguments. But in the end, it is an irrelevant point since it has no practical application in our trying to understand the reality of our own universe. Besides, the existence of another universe cannot be proved.

But the word "if" also has a practical use when it is used in a conditional mode. In this mode it presents options leading to a choice or decision, "If we meet at five o'clock, we can have dinner together." This sets up a yes or no condition. Its use in such instances is unavoidable and leads to a positive conclusion.

Definition of Representative Terms

In addition to inventing new words, there is a practice of avoiding the use of generally accepted terms by substituting representative terms, which have associated meanings. Through constant use of these representative terms the reader is led to become accustomed to their usage and to think of them as having the same meaning as the general terms.

In dualism arguments, the term mental state is often substituted for mind. Through continued use, we are then led to the compatibility of mental states with brain states. The next step is to argue that they are so similar, or so closely related, that they really are physical activities leading to the materialistic conclusion that the entire thinking process is physical. Somewhere in this game, the concept of a distinctly individual mind has been left on the sidelines.

A contrasting tactic is to refer to a word used by someone in the past and then build a case against that usage by using one's own definition without explaining the context of the previous application. This is especially common in politics.

Conscious experience is an activity that is difficult to explain because it is not understood. Therefore, terms such as, conscience, self-consciousness, cognition, or self-representational capacity are used, thereby filling the gap between incomprehension and awareness of its vital involvement in human behavior. This use allows the discussion of related subjects without having to define the basic term, which within present knowledge cannot be defined or explained, especially as a physical property.

Semantic Clarity

In any of the previous four methods for controlling the thought process of a reader or debater, it is obvious that the semantic clarity is obscured.

Semantics is the use of language to convey meaning. It can be objective or purposive. By objective I mean the presentation of as clear a path as possible to the understanding of the word or thought. Purposive, however, is selecting that meaning of a word or term to intentionally slant the meaning for the desired effect in furthering the intended line of thinking. Syntactically it may be correct but the impression that purposive meaning gives may be subtle enough to be overlooked. It is especially effective with such leading phrases as, "It is obvious that..." or "As you well know...." For example: "In your own experience, haven't you felt that conscience is merely the ability to choose what you want to do, rather than an understanding of what is right or wrong?" The relativistic slant is obvious in this example, but the intent is to divert the thinking from the moral meaning of conscience. With respect to the previous section on the use of interpretive terms, this example illustrates how the two techniques can be used effectively toward defining a concept of conscience: the substitution of a term and the slanting of the resultant meaning. The next step would be to use this interpretation in combination with other plausible semantic excursions to an intended result.

A two or three year old child knows no semantic rules but it can make its meaning clear.

There are conditions under which the meaning of a word or term generally has an acceptable base. In scientific work the use of language is bounded by technical requirements. In philosophy, semantics is especially troublesome because the freedom of thought needs a freedom of expression in trying to explain approaches to the imponderable problems of reality. The good and the bad of those two conditions can be found in scientific philosophy. The clarity in these conditional areas can only be ferreted out through skeptical question-begging.

In the secular community, social meanings are often difficult to pin down because of the traditional acceptance of ideas and motivations. Take for example the concept of the soul with which we must deal in dualism. Due to the increasing secularity of society, the metaphysical effects of an afterlife on earthly life are being replaced by the materialistic demands of here and now. Consequently, the language used to give meaning to current thought is dependent on this secularism. Meanings other than secular are looked at with ridicule and inconsequential understanding. Metaphysical and religious references are shunned. How can true meaning through semantic rigor be found if the search area has been cordoned off?

There is an outgrowth of this intentional semantic reluctance to refer to things or activities objectively. It is called political correctness and it has brothers and sisters in the areas of religion, ethnicity, the feminist movement, racial equality, immigration, etc. The fear of treading on human sensitivities has blurred the rightful literary demand for clear expression by substitution of semantically obscure terminology.

There is semantic danger in using terms of one discipline in applications to another. Trying to explain the properties of the non-physical mind in physical terms, like "mind-stuff", is impossible principally because the laws of mind functioning are not understood. Another example is comparing the mind-brain relationship to a computer's hardware-software operation. The dynamic ever-changing mind state is completely different from the permanent physical nature of the designed computer arrangement which requires an outside operator to even make it active.

Verbosity is another culprit in semantic obscurity. Expressing a thought concisely allows subsequent elaboration but avoids misrepresentation. Once there is a meeting of the minds on the intended direction of subsequent elaboration, the introduction of digressing commentary is counterproductive input. General and President Dwight D. Eisenhower often quoted the definition of an *intellectua*l as someone who used more words to say less than he knew. Are philosophers examples of the Eisenhower definition? Their use of ambiguous comparisons, continuous use of if's, use of irrelevant analogies, and references to past intellectual or historic errors, which may not be applicable, places them in a precariously qualifying position.

The search for semantic correctness in the dualism debate requires dedicated awareness of the meaning of the language being used. Is its purpose to justify a preconceived approach on an assumed basis or to describe it on a logical basis? How does this differ from interpretation or interpretive justification?

Interpretation

Interpretation does not determine fact. Interpretation is merely opinion, which can have little justifying value, but is often used in that manner. How can a researcher, even a scholar knowledgeable in the subject, infer the meaning of someone else's thoughts? How often have you heard the opening statement, "What the author really meant by that was…"? How can you know what I mean except by what I say? If I don't express myself clearly, then how can you possibly know what I should have said to satisfy your interpretation? This is why I wish I could have discourses with the readers of my books.

Interpretation is valid when intended to clarify an ambiguous definition of a word and when its meaning is not clear from the context. However, it must be made clear that the problem is with the specific word, not the entire idea with which the word is used.

Take the word "fire". In discussing the heating quality of fire, there is no doubt about the interpretation of the thought. When the command fire is given on the rifle range, the meaning is also clear. When the exclamation "Fire!" is heard in the kitchen, everyone in earshot takes immediate action. But to say that a person should be fired from his professional position requires some interpretation: for what reason, based on whose recommendation, whether it is a specific or general case of firing, etc.

Resort to interpretation, therefore, calls for questioning the reason for its use.

Interpretive Justification

What do I mean by interpretive justification and how valid is it? It is the use of accepted facts to imply their applicability to theories or concepts through interpretive liberty. The validity of the method can only be measured by the degree of success in achieving the purposive intent of the user. When a city council is petitioned for the installation of a stop light at an intersection where several accidents have occurred, the justification is based on the interpretation of the facts that predict an increase in the accident rate. Will the building of a nearby mega-mall increase the flow of traffic at that intersection? But in a case of demanding immunization for an entire community due to the infection of a single individual demands interpretative justification of the risk of a major increase in the infection rate. Unless the conditions and reasons are specified, the justification must be critically questioned.

In the dualism debate, the extrapolation of known physical neural interactions to explain concepts of non-physical properties like memory, learning and volition is an example of invalid interpretive justification.

Questioning Conceptual Relativity

When several concepts are proposed for a specific problem, it is evident that not all are credible. To distinguish the credibility of any one concept requires a set of standards against which to judge the concepts. Which set of standards? Those that you must set: those that will convince you to believe the applicability and credibility of the concept. In Chapter 6 there are several concepts about memory. Since memory has not been adequately identified either in function or brain location, these concepts appear to be valid. Are extrapolations of physical data and unwarranted conclusions used to make the concepts seem viable? Or are there admissions of only conceptual philosophy? Either way conceptual approaches are inconclusive.

Questioning conceptual relativity is a main issue in this book. With the introduction of vital new evidence, the present concepts become vulnerable to major change. They are presented in their current form to emphasize their shortcomings as concepts based on assumptions, when factual evidence is available to invalidate the concepts.

Use of Common Sense Perceptions

Common sense, which is not given credit for being so common in the scientific world, is a relative gauge of the scientific overuse of seeming overpowering arguments. Denying the validity of common sense is a rejection of the main difference between man and the lower animals—the use of rational powers. These powers, though possibly lacking in adequate knowledge of the technical reasoning involved, can nevertheless detect an imbalance in the technical rationality based on own experience. In other words, common sense can often demand a more convincing justification. This is the skeptic's main offensive weapon. I am willing to consider any concept or theory for its worth until it becomes illogical. To me, the use of common sense is then justification for rejection of that theory or concept.

The scientific community derides the idea of common sense as "folk psychology". Yet it is common sense that directs the scientific effort in the directions that will probably yield the most results. How can a philosophy of intuitive intent be ridiculed while intuition is being used to generate the very essence of the analytical effort?

Analogical Validity

An analogy never proves a point. Its purpose is to increase understanding. However, it can also be used to mislead through inappropriate comparisons. In such use it is intentionally deceptive and catching the deception is a means of uncovering the intent.

In my previous book, I compared the similarity of the body-soul union with that of an auto-driver unit. Although the interface between soul and body cannot be explained, the controlling function is similar to that of the driver controlling the car by his hands and feet. There are many other similarities, physical and operational, but the main purpose of this analogy is to show that there is a similarity even though the body-soul interface cannot be explained.

However, comparing the brain-body relationship to a computer hardware-software combination, as mentioned above, is misleading because there is a missing element in the analogy. The software (the brain) is a program in which the decision paths are set. The computer (the body) executes the commands given by the software. This analogy can explain human behavior as a response to the programmed responses to physical stimuli. This is intended to help understand physical causality. However, what is missing in the analogy is the ability to override the responses given any number of reasons to do so. Those reasons can only be initiated by an operator who is not part of the analogy but who makes the computer behave through the keyboard. He is outside of the comparing scenario of the analogy. Any changes to the software have to be made by a programmer, another operator removed from the analogy. The analogy with the missing operators, therefore, is invalid, but for the passive recipient of the analogy, it plants the seed of credibility to the concept of physical causality.

Another technique of deceptive argument similar to the use of an analogy is the use of a straw man. By setting up a straw man that is then easily demolished, the false impression is given that the point regarding the main issue has been made. An example that is often used in problems associated with consciousness and the mind is the *zombie* argument. Assume (notice the "if" effect at the beginning) that a zombie exists that is a perfect clone of you except for an inability to experience certain feelings like pain, hot, cold , etc. If it is possible to conceive of such a being, then it must be logically possible. But if it is logically possible, then the absence of feeling in the zombie shows that consciousness, which includes feeling, is not part of the brain, since the zombie has a brain but not the feelings. What an utterly ridiculous argument that begins with absurd inconsistencies and uses them to attempt a logical conclusion.

Detection of Scientism

What is scientism? It is the theory that investigative methods used in the natural sciences should be applied in all fields of inquiry.

In the natural sciences all investigation is physical. But even in the physical issues, like the evolution of the universe where scientific theory and philosophy mix, scientific methods alone cannot dictate how answers are to be theorized. An example is the use of the indeterminate nature of quantum theory, as observed in experiments, in trying to extend the similarity of randomness of neutron behavior to the indeterminate nature of free will in human behavior. To extrapolate the uncertainties of quantum theory to the supposedly obvious uncertainties of physical causality magnifies the error by adding scientific uncertainty to metaphysical uncertainty.

Scientific methodology for investigating an impersonal physical world through repeatable experimentation in a test lab is not applicable to investigating unrepeatable personal activities like near-death experiences (NDE). Since the phenomena are experienced by different people, the criterion of repeatability by the same test specimen should not be applied for denial of the occurrence. Although natural occurrences involve physical aspects, the involvement of the non-physical mental activity removes it from the methodology for physical investigation. The inability to investigate by physical means should add emphasis to the personal testimony of each occurrence. The similarities between the various incidents constitute the response to the repeatability requirement

The evolutionist, Charles Darwin, adhered to the principles of physical investigation by merely accepting what his investigations revealed and applying those facts to theorize about the physical effects on the natural selection of species. But the modern Darwinists have applied, and expanded, Darwin's principles to deny the metaphysical aspects of creation. This incompatibility of applying the investigative techniques of natural development to metaphysical concepts of original creation is a prime example of scientism. Even Darwin believed and said in the closing statement of the *Origin of Species* that he believed the Creator created the *first living cell from which all living matter evolved.* This final statement indicated that all his physical investigation only allowed him to sum it up in an inescapable belief, which was beyond the bounds of his physical research.

Physical experiments with non-living material, such as metal alloys or compounds, can be contrived to produce exact repeatability. A planned set of experiments with human beings searching for statistical results of mental responses, using physical test methods, cannot assure repeatability due to the subject's mental decision processes that are unpredictable. Therefore, even planned physical tests that rely on rational responses cannot comply with the demands of scientism to use physical test methods in all physical type tests. Thus tests using physical stimuli cannot predict human behavior which demands decisional processing.

I can accept the results of scientific tests for what they show up to the point when they are interpreted abstractly to include immaterial aspects. I then become a skeptic.

Detection of Circularity

Circularity is the stating of a presumed fact and then using that fact as supporting evidence for the argument. It is often undetected when a lengthy text intervenes between the original statement and the subsequent use in the argument.

Here is an example of circularity: Does God exist? Yes. Why can you say so? Because it is in the Bible. But what makes the Bible credible?

Because it is divine revelation. Note the wording and the development. The initial argument is about God, a Being, and existence. What follows is a discussion of the Bible, which could be of some length, setting up, for instance, the historic accuracy of the Bible, then leading to the additional biblical importance due to the divine revelation and the connection of this revelation to other revelations. It can be played many ways but the circularity remains.

Another example with a dualistic flavor, concerns the soul. When I say I believe I have a soul, I am stating a belief. But because I say I believe I have a soul, there must be such a thing as a soul that I can conceive of. Because I can conceive of such a thing, a thing as a soul must exist. I cannot believe something that I cannot conceive could exist. This is circular reasoning. There has to be better, and more convincing, evidence of the existence of a soul.

Technical Clarity

Pursuing a methodical analysis to reach a credible solution is a demanding task requiring dedicated persistence. It should, according to the Descartes Rule V, consist of progressive steps that are a discernable path from premise to conclusion. Because it is a task that doesn't always have a direct path to the conclusion, some steps may be omitted or slanted conveniently to assure that the conclusion justifies the intended result. The invalid conclusion will be obvious if these omissions or invalid assumptions are detected.

Technical evaluation of a proposed solution as a method of searching for the truth is an attempt at finding the discrepancies and inconsistencies in the analytical work. But how can any detected flaws be isolated without performing a rigorous step by step review of the total analysis? One method is to describe the detected flaw and use a contradictory solution to reveal the discrepancy, with a comparison of the validity of both versions. In the dualism debate, the controversial issues are conceptual since researchers are principally involved in trying to prove the theory of physical causality but are far from having the answers. The technical evaluations, therefore, can only be of the conceptual type, not of factual physical descriptions, which by researcher admissions cannot be verified. The problem is the interaction of the immaterial mind with the physical brain and the interface through which the interaction occurs. This is the area that defies explanation or evaluation within present knowledge.

For the average reader, technical evaluation with its details can be laborious reading. In this book it will only be used in Chapter 5 and the extent of the content will be to pick specific concepts or claims and by comparison show conflicts or deficiencies in logic. Direct quotations will be the justifications for the claims and for dissenting opinions. A commentary on the salient points will accompany each comparison. It will then be your decision about the validity of the claim or concept. For anyone wishing to perform additional evaluation, references will be given for all the quotations.

There is a need for technical evaluation of the world's growing environmental problems, such as: global warming, deforesting, extinction of animal life, water pollution, radiation hazards, etc. The issues should not be biased with politically correct agendas but be verifiable analyses according to Descartes' Rule V.

Questioning the Logic of Philosophic Arguments

Philosophic arguments are rational searches for the meanings of reality and being. These main subjects allow endless pursuit of ideas and interpretations. How can you engage any philosopher's mind to argue his philosophy? To agree is passively easy but to dissent requires an ability to first understand and then to analyze. The main tool should be formal logic but that is not in everyone's toolbox of analytical techniques. Without logic the recipient is at the philosopher's mercy. So how can the average person engage the philosopher's mind? It can be with the commonsense that accrues with life's experiences. Although science derides the use of common sense, it is a necessary and important asset of every human being. With this tool anyone can pry apart any concept when commonsense indicates a tilt in the rationale according to one's own perception of the correct rationale. Common sense includes all the methods described above as an integrated skeptical attitude that not only rips apart erroneous thinking but also establishes one's own philosophy, which is then the main combatant against all other philosophies. That skeptical attitude, however, must be based on a "no matter where it leads" resolve. With such a personal philosophy, no one is unarmed in the battle of seeking the truths about life's meaning and one's own being. This is the purpose in delving into the truth about dualism; it is the revelation and recognition of one's place in the universe.

As an example of the philosophical attempt to understand the reality and purpose of the universe (whether for man or only with man), the introduction of a scenario including another universe, or many universes, is a meaningless diversion. It adds nothing to the basic issue of man's nature. Its intriguing possibilities could constitute another debate. However, to consider whether a spiritual universe exists as a temporary or permanent residence for souls would be pertinent to the debate about human nature. A redefinition of the term universe is applicable here. Therefore, reincarnation as a spiritual round trip becomes an acceptable philosophical subject and a possible fact of life, as it is in Oriental cultures.

Side-tracking of an issue can be easily noted if it begins with words like *suppose* or *what if* or *if we say*. The use of suppose, however is valid as an introduction to the stating of the basic conditions for an argument or debate. In such use, its purpose is to initiate an agreement regarding the issue being debated. Using *what if* may be pertinent or irrelevant. When used to present the opposite viewpoint for an analysis of both sides of an issue it is pertinent. However, introducing an illogical alternate like the zombie case, as an assumption is meaningless.

There are times when a well grounded philosophical argument degenerates into a contest of cleverness at solving seeming impossible situations. Such a case is that of the final resurrection for materialists. For Cartesian dualists this is no problem due to the continuity of an existence between death and resurrection. But for those materialists, like Christians who believe in resurrection, there are many problems to resolve. To do so requires conceptual thinking and even the assistance of God. For instance, how will the bodies that are cremated or decomposed be reconstituted? Will the precise atoms and neurons be brought together by God, even though they may have become parts of other bodies? According to St. Paul, the new bodies will be radically transformed. In that case what happens to the identity of the person since the new body is not the same as the old? Does God have to preserve some part of each human at death to retain the same identity? As you can see, these questions demand some clever answers. But in the final analysis, this effort is meaningless, except as consolation to the concerned materialists, that they have some answers. Such analyses without any knowledge of what happens after death are strictly philosophical exercises with no positive answers.

What should the skeptic look for in the philosophical arguments for or against dualism? The first question is whose theory of dualism is being advocated. Since there are conflicting versions, it is helpful to know the boundaries of the debate. The other question is which assumptions are used to initiate the argument. Terms and conditions may be introduced which could confuse the basic issue, even intentionally. Agreement to initial conditions can make the discussions meaningful.

Exposition of Double Standards

The detection of double standards alerts one to consider carefully what is being presented. However, detection often does not occur because the side being presented appears logical and, if based on facts, undeniable. Acceptance of a well-presented view is easy, but there are at least two sides to every issue. The presentation of only one side of an issue is a strong signal that it is meant to convince without argument. It is the skeptic who always introduces the opposite argument and finds the inconsistencies that reveal a conflicting standard.

An example of double standards is the neuroscientists' assertion that the soul does not exist although the Catholic Church (and others) claim that the soul is a reality. The scientists point out that the Catholic Church was grossly wrong in condemning the scientific claim of Galileo about the earth revolving about the sun instead of what was commonly thought at that time to be the reverse. The argument is correct to the extent that it is based on historic fact and, as hindsight, is undeniable. The Galileo incident refers to a debate about a scientific concept, which was outside the Church's philosophical domain. The Church eventually had to accept the facts of the earth/sun rotational relationship. The scientists claim that the same reasoning applies to the evidence of a soul. The deception that is overlooked and conveniently avoided is that although the Church was wrong in a matter outside its domain, the existence of the soul is in its domain, and more importantly, outside the domain of the scientific community. This is where detection of a double standard reveals the inconsistency of the scientific claim. The implication that the historical physical analogy applies to an unproven non-physical denial is completely invalid. A different standard for argument must apply. If the Church had to accept what was in the scientific domain, the scientific community will now be put to a similar test of admitting its error about the soul, which is in the Church's domain.

Another double standard is involved in the above example. If the scientists expect their mere statement without evidence of a mythical soul to be accepted by everyone, why don't they accept the mere statement without proof that it does exist? As I mentioned above, the detection of double standards is not obvious without some skeptical question-begging.

In conclusion, the above descriptions, with examples are intended to prepare you for what follows. In chapter 3, the evidence for both monism and dualism (in their various forms) is presented as applicable to the challenge that I am making to the scientific community. It is mostly physical evidence, although conceptual and unverified experiential evidence is included due to the possibility of its ultimate verification. It is crucial that the attitude in reading the next chapter is an objective one. Try to ignore any biases you may have—scientific, neuroscientific, cognitive, secular, philosophical, neurophilosophical, theological, atheistic, agnostic, or skeptical—to digest what is as far as I can determine, factual evidence. Do not don skeptic or agnostic garb until you have completed reading this all important chapter. This is evidence that has been available for decades but has not been recognized for unknown reasons. It should give you an appreciation for what the dualism debate should consist of and the direction that future research should take. Read carefully and apply all the rules that were discussed in this chapter not only to the material I introduce but also to my commentaries and conclusions.

3

Physical and Metaphysical Evidence

The Elephant has just walked in front of the scientists.
Will they see it or will they argue about what they see?

The material in this chapter has been anticipated for 350 years by Cartesian dualists and declared impossible by materialists. The conviction of impossibility has been the belief that physical evidence of a non-physical soul cannot be produced. But if it could be generated, what would it have to be to be accepted as valid by the skeptics and the materialists? Only the appearance of the actual soul would be the logical demand. But how could a non-physical soul appear in physical form? ***By reincarnation!*** *Reincarnation defines the soul and the soul allows reincarnation to occur.* The soul leaves one body at death and reenters another at birth and *subsequently identifies itself through positive references to the past life.* This phenomenon, dismissed by scientists and skeptics as paranormal occurrence unworthy of investigation, **has been proven scientifically** with complete scientific documentation and is a major contributor to explaining dualism and negating monism.

This chapter reveals the essential evidence, both monist and dualist, that calls for the rethinking of the dualism controversy. It is the basis for my challenge to the scientific community to recognize the evidence, acknowledge its credibility, and upon acceptance take appropriate action to redirect the analysis and definition of human nature. The conclusive evidence is reinforced by philosophical, conceptual, experimental, and historical information that adds credence to its validity through different viewpoints. But the evidence must be physical as required by scientific standards. *This requirement must apply equally to all physical and non-physical claims and evidence for or against the reality of the soul and to all supporting analyses, investigations, and conceptual explanations.*

For a comprehensive presentation of what constitutes confirmatory and deconfirmatory evidence, factors other than the evidence itself must be considered, that is, the standards for verification of evidence, the types of evidence, the distinction between belief and fact, the attitude requirements for acceptance of evidence, and the qualification requirements of the evidence generators. The mere presentation of the evidence without these considerations is an open invitation for unlimited skepticism about its generation and validity. There is too much unnecessary skepticism in the dualism debate as it is now argued. The challenge for a redirection in the thinking about dualism will, unfortunately, elicit much superfluous skepticism and resistance by its mere suggestion as a necessity.

In the debate, the need for the evidence, its nature, and its limitations, both physical and metaphysical, points out that there are two parts to the total problem of defining human nature. One part is proving the existence of a non-physical entity, the soul, in a human body, which must be shown before the second part can be adequately analyzed. The second part, as the soul does indeed exist, should be the analysis of how the two entities, the body and the soul, function together, that is, what are all the causes that make this human unit behave as it does. The second part is currently being addressed without a resolution of the first by a denial that the first exists. In such a situation, there never will be a satisfactory solution of the total problem because the complete structured *being* and its integrated functioning are not studied. This chapter is dedicated to solving the first part, which then becomes the challenge to proceed in solving the total problem. Suggested initial steps on how to approach the second part are given in the final chapter.

The dualism debate is full of claims: scientific, philosophical, theological, and psychological. Acceptable, supportive, pertinent evidence has not been presented by either side of the debate, resulting in the continuing non-acceptance by either side of the opposing views. There is a perception that the extensive physiological information about the human brain and nervous system is sufficiently reliable to support the neurophysical claims that physical causality is the sole initiator of human behavior. But the perception is misleading and unjustified because the detailed functional analyses of the neural system and the system elements are applicable *only to the explanation of the physical properties.* The knowledge base of the human and animal nervous systems is very detailed. Testing of these systems has been extensive to determine the functioning of these systems under various conditions and test plans. Surgical studies of damaged brain elements have produced consistent evidence of damage effects. Tests using test equipments have verified the analytical observations of brain functions. However, the extrapolation of this knowledge to explain the non-physical mental functions like memory, consciousness and free will with interactive potential is invalid because it has no supporting evidence. There is no, and I repeat no, supportable physical evidence for the explanation of *thinking.* (I can imagine the many arguments on this point based on the definition of "thinking". Only the mental process, not the brain's functional complicity is intended here.) The conceptual philosophizing about these functions is replete with admissions that the extrapolation of physical functions to non-physical capabilities is unproven but is being vigorously researched to bring closure *at some future time.* This is contrary to the scientific basic requirement for specific verification of conceptual theories before they can be accepted by the scientific community. Since this is the same requirement thrown at the dualists to provide evidence of a soul, it appears that both sides are at a standoff in current verifications. Physical evidence, acceptable under the same set of standards is an absolute necessity for mutual agreement.

Although these standards have not been followed in the denial of the soul's existence, physical evidence of the reality that does adhere to these standards has been found through rigorous investigations. Not only is it physical evidence but it also reveals the immaterial properties of the human being and sheds light on their misleading scientific explanations. The scientific quality of this evidence is assured by the generation, verification, and reporting by recognized professional investigators. Furthermore, with respect to the scientific requirement for repeatability, it expands the meaning of repeatability beyond the exact duplication of the same planned test through the consistent replication of similar evidence in other subjects.

In spite of the continuing scientific denials, the evidence of the soul's existence has been available for over four decades. The manner in which it was produced is as unique as the evidence itself, unique because test data have never been obtained in like manner. In a normal test of an organic material, compound, or human/animal reactive capability, a test plan is prepared with its specific goals, the test is run, the data recorded, interpreted and the results reported with conclusions as third person observations. The uniqueness of the soul evidence is that it is a FIRST PERSON, SPONTANEOUS, and UNPLANNED activity of the test specimen itself, vocally and with intent to be acknowledged with demands for reconnection to another environment. Although this evidence must be observed by a third person, it is the spontaneous act itself that is the evidence, subject to verification. How can this happen? It occurs because a reincarnated soul *speaks* through a new body about a previous body in its concurrent environment. The only possible procedure through which such a transformation can occur is for the person, the soul, to leave a deceased body and be reborn in a new body. Where this evidence clashes and disproves the materialist theory of physicalism is that the theory claims that memory is a physical property of the brain. Reincarnation demonstrates that the memory *is not* a physical property because if it were it should have remained in the deceased body instead of accompanying the person to the new body. (Note the use of the term *person* rather than *personality* in association with the soul. The eternal being is the person; the union with a physical body establishes a personality.) The use of the memory in the new body also verifies the existence of a non-physical, separate entity mind which directs the behavior of the reincarnated personality. The optimistic scientific hope of verification of the current concept of a physical memory will never be corroborated *someday* through continuing research because it does not represent the demonstrated contradiction.

Because the *first person* evidence occurs in real time, how can the scientific requirement for repeatability be complied with? In any activity we undertake, the action is completed and can never be repeated in exactly the same content under the same conditions. Consequently, the demands for the soul's spontaneous evidence to be repeatable in the same content, or as in a laboratory test, cannot be met but can and do repeat in the manner of continuing demands for recognition. The repeatability is of two types. The first is that the subject, the reincarnated personality, can repeat the same activity of remembering events of the previous life, which it does for several years, sometimes into the teens. The second type is the same activity in other individuals, thus showing consistency of similar occurrences. This repeatability has been confirmed by more than 2500 scientific reports of such cases. Scientifically, if a test can conclusively show the desired result in only one case, this is sufficient to declare that the result is positive. In the case of the reincarnated personality, one verified instance should be sufficient to establish the initial credibility. But the confirmation of 2500 cases far exceeds the demand for repeatability of the second type. Such physical evidence for the existence of the soul is presented later in this chapter. It has been available in the published literature since 1966. It is puzzling that this pertinent and related evidence has not been uncovered by the neuroscientific and neurophilosophic research communities as of the writing of this book (2009). Calling attention to this scientific apathy, and to other evidences with their confirmations of the soul's reality, supports the need for rethinking dualism.

Charles Darwin's theory, using the evolution of many animal species into changing forms as evidence of natural selection, was accepted scientifically. He did not have to explain *how* this happened (DNA had not been discovered). By the same scientific reasoning, the evolution of a human being *with a soul*, as now proven physically, without explaining *how* it came about should be accepted by the scientific community. Will it be? Or will there be denials of the need to recognize or evaluate the evidence on the basis of the generally insensitive regard for paranormal experience? Without evaluation of the evidence, such arguments are purely evasive and indicative of an arrogant and stubborn defense of a scientifically concocted *conceptual unreality*. Such arrogance has been displayed in scientific history by conceptual claims, only to be shattered with deconfirming evidence, which the concept originators refused to accept for various reasons. Examples are René Blondlot's N-Rays concept and the University of Utah's claims of hydrogen fusion in an apparatus no more complicated than a test tube.[8]

The order in which the material is presented below is: (a) definitions and descriptions of the types of evidence that follow, (b) the evidence supporting monism, and (c) the evidence supporting dualism.

Evidence is defined as the data on which a judgment or conclusion can be based. It must be prepared and verified by qualified personnel, documented objectively, and be specifically conclusive. What must its effects be? It must convince through detail, exhibit its pertinence, justify its use, and describe the limits of its application. In any testing for generating data, the test mechanism (procedure) must not favor the outcome of the test, such as limiting the conditions of the test to prevent questionable results. Data support either a single item or a theory. If the test involves a specific item rather than a theory, the results sought must be theory independent because theory dependence tends to slant the evidence to supporting the concept rather than the issue for its own confirmation. The inability to distinguish the test data of the single test item from its effects on a theory or system can lead to misinterpretation of the test results. (Obviously, when an investigation is intended to prove a theory, the evidence must support the conditions of the theory.)

The intent of any investigation is to produce a *yes* or *no* answer through the collected data. As an example, evidence of a soul's reality must prove that it is a distinct entity, not a mental phenomenon justifying the dualism claims. Only then can the effects of the soul's influence become part of the study of how the human being functions as an integrated unit of physical and non-physical interactions.

In a claim of valid evidence, what are the criteria for acceptance by everyone? How must it compare with all the other existing valid evidence? What other factors are involved in the acceptance of confirming or deconfirming evidence, like the ingrained attitude of researchers who refuse to consider any but favorable evidence? These are the questions, whose answers will shed light on the likelihood of the startling new evidence being accepted. So, let us review these ground rules which will help to evaluate the credibility of evidence and detect the possibility of misleading tactics.

Evaluation of any evidence must proceed with objective motivation by all concerned factions with objectivity defined as a search for facts rather than pursuit of desired verification. The conditions imposed on any scientist for producing verified evidence apply to any other type investigator, like a historic researcher, psychiatrist, psychologist, and medical researcher. First, there must be a thorough review of documentary validity. A scientific report must adhere to specific requirements for explanation and, more importantly, for verification of the stated results. Second, the professional background and experience of the investigator must be sufficient for the authentication of the resulting evidence. Third the disclosure of the evidence must be published in books and/or professional journals for public use and comment, with presentations to peer organizations. The comments, especially by the skeptics and any opposing factions, should become part of the evaluations. Lack of comment on a universal issue is rarely the case.

One of the scientific requirements for the generation of evidence is that it must be repeatable in a laboratory or comparable facility. This stringent requirement poses problems for planned cognitive tests since creating the *exact* situation repeatedly with constantly changing conscious reactions in the same test series is almost impossible. Similar results from repeated tests can give meaningful information but even this information is subject to interpretation. Although only the test results are meaningful as evidence, the test mechanisms used to standardize the conditions for achieving identical results help to verify those results.

How can the same scientific requirement for repeatability in any specific facility or location be applied in the search for evidence of a non-physical entity? How can a soul be invited to appear for testing at a given location in a manner that will be scientifically acceptable? What type of facility can fill the needs for unspecified soul tests? How can evidence be obtained about non-physical phenomena when non-physical entities have no substance to be handled? In other words, science demands physical evidence under normal laboratory conditions but there is no way to plan obtaining such evidence since there is nothing physical to test in compliance with test observation and verification requirements. It can only be done if the non-physical entity is identified and observed. There are no scientific requirements, principally because the problem is not a scientific one. But it was attempted, with recorded success as a planned experiment, by a psychiatrist, who anticipated the probable subsequent criticisms in setting up the tests (the tests are described later in the chapter). However, this inability to select a site, a time and the subject, emphasizes the importance of the spontaneity of the appearance of the evidence as described above in reincarnation. There are no requirements to adhere to. The evidence is observed wherever and whenever it appears by ever-present related observers, the parents and siblings of the subjects. This not only circumvents the normal scientific rules for testing procedures but introduces an incredibly unique and verifiable condition. It is not limited to any test location, test plan conditions, time of performance or specifically qualified observers. It is the *actual performance* by the subject without any inhibiting standards that confirms a scientific authenticity of the soul's transferred presence. The invitation to the soul for its testing is never required; the *soul invites itself* and initiates the testing opportunity that demands investigation for verification. A similar physical situation, which is never questioned, is that of a patient demanding treatment for an ailment that must be investigated and confirmed.

In searching for confirmatory evidence, certain tactics, which may not be readily detected, are used to circumvent the strict rules. Avoiding those tests that might result in dubious results is one of these. Another is changing some aspects of the theory to agree with the test results. The most obvious one is ignoring suspected anomalous data that might be deconfirming or might be difficult to explain. Evidence confirms, deconfirms, is questioned, or is ignored. How it is perceived depends on the attitude of the perceiver. In a legal jury trial, this is especially evident because the perception isn't always of the evidence alone but also on how it is presented, and more importantly, how the jury interprets this presentation. In scientific work, attitude should have no effect on the search for confirming evidence. The validity of the evidence should be dependent on the observed facts since the prerequisite is that competent personnel do the observing and that their recording of the facts can withstand the rigor of verification. This, however, is not always the reaction to factual data if it is of a deconfirming nature. If the evidence has the potential of changing existing theory or the expected results, these may be the reasons why it might be modified or ignored. What contributes to this potential for disregarding scientific rectitude? There are several factors among which are positivism, scientism and professional image.

Positivism, according to the Dictionary of Philosophy, is:

> The term 'positive' has here the sense that which is given as laid down, that which has to be accepted as we find it and is not further explicable; the word is intended to convey a warning against the attempts of theology and metaphysics to go beyond the world given to observation in order to enquire into first causes and ultimate ends. All genuine human knowledge is contained within the boundaries of science, that is the systematic study of phenomena and the explication of the laws embodied therein.[9]

In the dualism debate, positivism is a strong inducement to leave the materialistic monist theory alone. The soul's metaphysical character brings it under this rule as interfering with the scientific dictum: what "is" must not be questioned. But with the soul's reality proven, how does that fact overcome this positivist attitude, because the soul too becomes an "is" item? This is where the challenge for rethinking dualism becomes the clash between metaphysics and science. To respond to the challenge, science might be helpless for it does not possess in its vast knowledge base any of the characteristics needed for analyzing and formulating the soul's integration with the physical components of the nervous system. The response can be anticipated: "Do not tamper with the status quo of the previous 'is' by adding a new interpretation of all the analyses that have been funded to date. All this work cannot be thrown away." Rethinking, however, is not discarding but nursing the existing knowledge along a new path. Utilizing the physical base of existing knowledge can delineate what is required to expand the research to integrate the immaterial properties with the physical.

Another similar problem is scientism. Referring again to the Dictionary of Philosophy, it is:

> 1. The belief that the human sciences require no methods other than those of the natural. 2. In a more general sense, practices that pretend to be, but are not,

> science. In both cases the term is employed only by opponents.[10]

The phrase "used by opponents" indicates that the accusatory technique is used wherever convenient. When the dualists use philosophic arguments about the causes of mental states, that is not science because it is not supported by physical properties. When the scientists make unsupported assumptions about those same mental states, which are also not science, they are accepted as scientific technique *pending eventual verification.* The double standards in the ongoing debate expose the inadequacy of imposing exacting scientific rigor on the development of evidence for application to other than scientific matters. An argument can easily be made by scientists that evidence dealing with the non-physical is outside the natural laws and cannot be accepted by the physical sciences. But if the evidence is part of the study it cannot be ignored. Will documented evidence of the non-physical, obtained with "physical subjects", overcome these objections? The true nature of the debate will become evident with the reaction to this question.

Professional image is a driving force in the scientific community. Images that result from multi-year research projects and theoretical associations are to be preserved at all costs. Any new evidence that casts doubt on a theory, concept, or experience may endanger the careers of those professionals who are associated with those theories. However, when the evidence is conclusive through a thorough investigative effort, the impact of such deconfirming evidence *should* cause an emotional upheaval in the scientific community associated with the area of concern. This is inevitable due to the strong position intentionally built up to support the ongoing research. The resistance to the evidence could be overwhelming, considering the number of scientists involved and the accepted credibility of their scientific efforts.

The scientific view of the soul is that it is only a belief in the minds of the religious. But what is belief? For consistency in the philosophic vein, let's go to the Dictionary of Philosophy one more time.

> Belief. The epistemic attitude of holding a proposition p to be true where there is some degree of evidence, though not conclusive evidence, for the truth of p. Clearly

> related to "knowledge", belief may be characterized as stronger than mere ungrounded opinion but weaker than full knowledge. Importantly, while knowing p would generally be considered to entail, among other things, that p is true, believing p is consistent with the actual falsity of p.[11]

Because belief is between opinion and knowledge, it can interfere with the search for valid evidence. It can be responsible for finding reasons to interpret, ridicule, or even ignore, rather than simply accept what implies or is confirmed evidence. Belief can also be at fault in the generation of evidence by avoiding tests that could lead to deconfirmng evidence or even rationalizing against the need to perform certain tests. A strong belief can even develop into a bias, followed by skepticism, agnosticism, and finally denial in the extreme. However, a degree of this belief-based-skepticism has redeeming values, as for an author who subjects his developing text to such skeptical review as a test of true intended meaning.

By the above definition, a non-physical entity is a belief until proven according to scientific standards. But by the same standards, what are scientific arguments and concepts? Since those based on assumptions are not proven scientifically, they too are only beliefs. To merely state and assume that physical properties can perform non-physical behavior is a major assumption. On this level playing field, neither side has the advantage. The scientific evidence of the physical functions of the neural system is valid based on all the derived knowledge by cognitive science. But the credibility stops there when extrapolations of this information are carried on to the mental, the non-physical functions. Any evidence of the existence of a soul, obtained according to scientific standards, is then the equivalent of the physical evidence of the neural system. The reality of the soul then is not a belief but a fact, while the explanations based on assumptions are still beliefs. Soul evidence then becomes deconfirming evidence against the neuroscientific assumed solutions. When this happens in the presence of valid evidence, a review of the debate is called for under a new set of guidelines. Such a review is necessary since the mindset of the soul issue in the secular world is also at stake and acceptance of the evidence faces major hurdles under different conditions. As in many other issues, the worldview depends on science for verification of any new concepts involving natural law. This dependence on science also brings the secular side of the soul issue into the scientific field for acceptance. But science does not understand non-physical laws. The accuracy and veracity of the evidence may succumb to the adamantly uncompromising attitude of those unwilling to change that worldview. This is the situation I envision as the result of this book. Belief can indeed interfere with the search for the credibility of the evidence.

It may be difficult to distinguish between belief and attitude in one's feelings but there is a difference and a definable one. Although belief may be the motivating factor, attitude is the governing control of activity related to the belief. As an example, I may believe that I am being discriminated against in my professional standing, which I harbor internally. But my attitude, in response to comments or actions about my performance, results in external actions that may be slanted by my belief but have consequences favorable or unfavorable, depending on the strength of that belief.

Attitude can be very strong in the scientific community which considers itself the unopposed definer of natural laws. It can lead to a confirmed bias that is difficult, if not impossible, to change. Definitions of scientism and positivism (above) reflect this attitude. It is difficult to change because of fear of losing any accrued benefits. Acceptance of any opposing or contradictory solutions for application to ongoing research entails a possible loss of image, credibility, and even professional standing. It may even blemish the credibility of the scientific community. The resistance to such proposed changes is understandable as emotion-laden filters by human nature but it does little to promote the progress of research if the facts are applicable and valid. There are several well known examples of such an attitude of denial in scientific history. In a very interesting book relative to this subject (referred to earlier), *The Undergrowth of Science*, Professor Walter Gratzer of the Randall Institute, King's College London describes how various scientists refused to change their attitudes and thinking about original theories in spite of proven evidence that they were wrong.

I predict that the challenge posed by this book will encounter similar attitudes and denials. But attitudes can change since some are already in the doubtful stage. If attitude change has detrimental effects, it also has positive ones. It can lead to the pioneering and establishment of a new field of inquiry that may have more positive results, based on firmer credible grounds, than assumed possibilities.

3.1 Types of Evidence

Before going to the evidence disclosures for monism and dualism, we examine the types of evidence and the rules that should be followed in accepting or disallowing them.

Physical

The most available, reliable, and acceptable evidence is that of the physical makeup of the human body and the functions of most of the physical elements. There are still unknown properties and interactions of some brain components but their exact natures and functions are being studied. Until these components and functions are clearly defined, their functional applicability to other interactions cannot be considered valid evidence.

Cognitive science and neuroscience are uncovering much factual information about how the nervous system functions and how the brain's elements integrate the sensor stimulation from the external world. This is extremely important evidence in understanding how the brain controls the physical activity, not behavior, of the body. Non-physical mental activity is required for behavior.

Studies and testing of damaged brains are producing extensive data on brain operation under these conditions. The data are unquestionably correct and physical in nature. What the data represent, however, is arrived at through interpretation, which only explains the results of the specific damaged parts but not the total interactive damage to the rest of the brain. This information is acceptable as evidence but is limited in application to the total effect.

Laboratory testing of human and animal brain responses also provides conclusive results in terms of the type and location of observable activity in the brain through scanning and planned reaction responses. The observable physical data are again subject to interpretation for the explanation of the total brain intra-action.

Although physical evidence of the quantum theory has been found, the uncertainty of its occurrence and effects are still major issues to be resolved. The acceptance of any application of the data or the theory itself to human mental states in monism is still premature and not warranted as evidence.

Conceptual

Theory dependent evidence is composed of some physical evidence mixed with conceptual probabilities of conditions that appear to be factual in testing. An example is the concept of the brain creating mental states in which the physical processing of sensor stimuli is a physical certainty but their translation into a mental state is conceptual. The validity of such evidence has not been confirmed, although the claims may appear to be reasonable. The degree of acceptance depends on the number and type of assumptions that are made to arrive at the claimed results. Such conceptual evidence helps in understanding the issue involved, but the evidence will not be factual, as physical evidence is, until the assumptions are confirmed. Analogies used for understanding cannot be used with some physical facts as evidence.

If a concept cannot be tested in any conclusive test that might deconfirm it, then the concept originator can propose any view he desires without any fear of falsification. Such a concept cannot be used as evidence.

Computer Simulated

Computer simulations are major tools in understanding problems and supporting possible solutions. Their validity as evidence is based on the input to the simulations. If factual data are used and the solution mechanism is standard, the results are merely the result of using computer processing, not reasoning power, and are admissible evidence. If the input is based on assumed data, the results are informative but not factual, and consequently inadmissible as evidence.

There is much reliance on simulations in the computer-oriented world so that too often simulations are accepted as real solutions rather than merely the models for what they represent. As teaching and problem solving instruments, computers excel by far the abilities of the human being, but they can only produce reliable answers if they are given reliable data. As evidence, simulations have limited use due to the required verification of all the data that are used in the simulation.

One of the scientific goals is the simulation of the human mind (not brain) to exhibit intelligence comparable to that required for behavior. Although such a program called Artificial Intelligence (AI) would help in studying the mind, it would not be representative due to its inability to duplicate the mind's ability to vary its decisions in responding to the same repeated stimuli or to arbitrarily reverse its own decisions. The decisions are made through a designed selection program that cannot simulate the randomness of human volitional responses. An AI solution, therefore, is not credible evidence for proposed concepts.

The bottom line is that anything can be simulated in a computer analysis. Hurricanes can be predicted, the effects of antibiotics on an emerging virus whose properties are known can lead to drug remedies, traffic pattern improvements can be proposed, and even human behavior analyzed for social effects. It is with care that the selection of simulations should be accepted as evidence.

Experiential

Personal experience is a reality. Convincing anyone of its reality is the problem in using unusual experiences as evidence. If the experience is normal activity common to all people, it can be accepted as evidence. This is the type of evidence that juries are asked to accept in legal trials. Unusual experiences, though they may be real, are usually attacked immediately by skeptics with arguments that often use inconsequential or unreal faultfinding. However, if the action or experience is real, it should not be summarily dismissed. An example is when a prediction of a future event is recorded or documented before that event, the experience (and evidence) of prediction is real and should not be discounted since both were actual occurrences. Psychic information leading to the apprehension of criminals is an example of how paranormal ability can have physical effects but cannot be used as physical evidence in spite of the demonstrated performance. Although the effects of such experience and the environment in which it occurs are physical, the experience itself is the result of mental initiation.

Studies of extra sensory perception (ESP) have produced mixed results and do not, in general, qualify as valid evidence, although the studies indicate a potential in some humans. Again, the uncertainty of the source for this activity, brain or mind, prevents its use as physical evidence.

Several factors may contribute to establishing the credibility of experiential evidence, such as, the concurrent details at the time of the experience, the honesty of the claimant, or the number of similar experiences about the subject event. This is a case of third person evaluations of first person experiences, made difficult because the experiences were not obtained under intentional or preplanned test conditions. Nevertheless, such experiential evidence should be given appropriate recognition.

Historic

History is evidence if the subject matter is relevant. Philosophies, based on fact or opinions and through their repetitive applications, may add apparent historic credibility that is used as background evidence. A prime example is how Thomas Jefferson's philosophy of the limits of government has resulted in the principle of the separation of church and state. It is used in legal decisions although it is not the proper interpretation of the Constitution or the Bill of Rights. The freedom of religion has become a freedom from religion, not as Jefferson intended it. A court's decisions become legal precedent that may be used in subsequent cases, sometimes overriding the specific case evidence. Documentation, if obtained under oath, is valid evidence. However, if the documentation was not produced under oath in its own generation, it may still be acceptable if it was legally prepared as in transcripts of a trial. Other examples of historical evidence are government records, library collections, and collections of investigative reports. Written material enclosed by quotes is evidence that information was written or spoken in those precise words. Old documents, unless they are originals, are often questioned due to possible introduction of errors in transcription or addition of new material.

Archaeological findings are becoming a source of historical evidence of the development of the human species. If the relevancy of the evidence can be shown, it must be accepted, except that age determination methods may be questioned.

Witnessed and Experimental

Reports of events, activities and experimental tests should be valid evidence if the witnessing is credible, that is, if the facts stated by two or more witnesses are sufficiently clear and consistent.

Evidence obtained under oath should not be questioned. However, convenient violation of the sacredness of the oath by high government officials has left the validity of such evidence in question. This dereliction of compliance at high levels has had an effect on lower level compliance, leaving no positive assurance of truth under oath.

When tests are performed with video and audio taping, the validity of the evidence depends on the test setup, subject performance, and test result compliance with test objectives. The motivation for the tests should also be considered. As an example, a test setup designed to scientifically demonstrate a theory has more weight than a test designed for statistical data that will be used to support an opinion, like a political poll.

Medical scanning and imaging techniques provide documentary evidence of stated conditions. Credibility of their interpretation, however, is dependent on several factors, such as: reviewer qualification and state of medical or scientific knowledge with respect to the specific problem.

Spontaneous and Observed

Unlike witnessed evidence, as defined above, participation in or observation of spontaneous events is less convincing than planned testing but still in demand because the events cannot be duplicated. Investigations of such past events must also be included in this category because event information is based on memories of witnesses or participants whose retentive abilities may distort the facts.

It is important to distinguish between witnessing an incident and observing or participating in it. Witnessing is an unplanned third-person verification of a happening, whereas observation and participation in the development of a situation are seeing the increments of the action as they occur in real time. As an example, an auto accident can be witnessed by someone seeing two cars collide. Observed evidence of the same accident is the experience of the two drivers as they become aware that an accident is imminent and their subsequent descriptions of what happened prior to the impact.

When a person commits an act that is witnessed by two or more individuals, that information becomes evidence of its happening. If a person admits to performing an act, that too is evidence if it can be verified. The most effective evidence is the spontaneous (and continued) performance by the subject without any forced motivation. It is confirming evidence in the making. A criminal act is an example of such performance. Its validity lies in the first person generation of the evidence and its credibility is verified by any participating observers. It is much more powerful than a laboratory test of inorganic materials, animals or human beings because in such tests, there may be erroneous interpretation of the test results. However, in the actual spontaneous act as observed, reported and investigated, the evidence leaves no doubt that it occurred. In an unusual occurrence, the main questions may be why or how it could happen, but not that it happened. This is observed evidential experience. It is even more meaningful if it agrees with identical or similar performances by other subjects.

The above are types of evidence that will be described later in the chapter.

Confirming and Deconfirming

There is another categorization of evidence which states what the evidence does by its results: it either confirms or deconfirms a fact or theory. Whether the evidence is confirming or deconfirming depends on which side of the issue is being presented.

Confirmation depends on physical factual data. Philosophizing cannot confirm; it can only rationalize about the available or missing evidence and try to justify a position but it must state the limitations of that position. It cannot introduce additional facts.

Confirmation is desirable for its positive consequences. There may be some question-begging issues but these should be easily dispensed with if the confirming data are valid. But what if the data are deconfirming? Deconfirmation has more drastic effects. What are the probable reactions in a major effort in which the basic concept is deeply entrenched and cannot be changed to accommodate the evidence? In such a program, acceptance of the evidence and admission of the deconfirmation are the last items on the list of possible responses. The rationalized responses are an indication of the severity of the damage the deconfirming evidence has produced. Analyses of these responses will reveal the true weaknesses of the deconfirmed theory or concept and should be the basis for the opposition's question-begging. Will the program die or can it be redirected through incorporation of the deconfirming data? That is the question I pose in this book based on the deconfirming evidence about monism.

Having reviewed the various aspects of evidence and their applications, it is time to look at the evidence pertinent to the dualism debate. The seven categories described above are presented in the same order, if and as they apply. Exclusions or variations of the seven will be noted. Contributions of the evidence to the "big" picture will be noted.

3.2 Evidence for Monism

Monism is based on a theory that there is one, and only one, "substance". However, there are other versions that address the mind-brain problem, but not in a dualistic sense. In these versions the mind is a mental state that depends on conceptual combinations of properties, phenomena, neutral matter, or other convenient substitutes. In all such mental state explanations, however, only physical matter is causal. Mind as a separate non-physical entity does not exist.

Confirmation of monism requires: (a) physical evidence that the soul <u>does not</u> exist, (b) verification of the physical brain's ability to create the claimed mental states that control rationality, desire, volition, and emotion, (c) verification that the physical brain performs the mental function of creative thinking, (d) evidence that memory is a physical element of the brain, and (e) an explanation of how a researcher can support the absence of a thinking mind while creating the explanation with what functions like the very mind he/she is denying. If physical causality is the basic rationale for monism, the evidence must all be physical, as required by scientific guidelines for conclusive proof.

Conceptual solutions are not acceptable validation. Medical science has contributed much to the basic knowledge of the human anatomy and continues to expand this knowledge through research, experimentation, and surgery. Cognitive research has explained the basic functioning of the brain and the nervous system. The definition of the mind as a mental state dependent on the brain's functions lacks any supportive data. Explaining such a relationship can only be done conceptually or through interpretations of laboratory tests. Positive physical results obtained through laboratory tests on humans and animals with constantly improving test equipment are a promising source of valid evidence. Such tests provide much information but the weak link is that the data must be interpreted, leaving some doubt of their true meaning. Reliance on evolution as the answer to how an integrated thinking-brain developed through natural selection can satisfy some critics about its nature but due to the inability to postulate how it functions, it is only speculation. The relationships between the physical elements and consciousness must be explained before any conclusive evidence can emerge to justify the materialistic approach.

Regardless of these limiting requirements, let's look at what evidence can be used to support the monist approach.

Physical

There is no doubt about the anatomical composition of the human body. The brain's components have all been identified as well as their locations and functions in either the left or right half of the brain. The nervous system has been identified in its various elements (fibers and synapses), their structural composition (fiber layers and membranes), and the connections from the various elements of the brain down to the physical characteristics at the finger and toe tips. The functions of these various body elements are generally understood in their primary operations and interactions. This is incontrovertible evidence when used as the base for proceeding into the more complex definitions of how these elements result in human activity and, more importantly, behavior. The sensors that receive information from the external environment for use by the internal receptive system are well defined physically. The eyes, ears, tongue, nose, and nerve endings for receiving this information are understood in their physical makeup and receptive ability. Problems in seeing or hearing are easily analyzed and corrected with physical items such as contact lenses and hearing aids. The translation of the physical stimuli that these sensors receive into useable signals for processing and conversion into activation signals are not fully defined. The difference between knowledge and interpretation is difficult to detect in a technical explanation. As an example, there are only concepts on how the *awareness* of taste develops from the taste sensors in the tongue. How does the brain discriminate between sweet and sour or tangy? We are able to distinguish slight variations of seasoning, but never stop to consider how this happens. (A concept of taste is evaluated in a later chapter.)

There are, however, human properties associated with behavior that cannot be ascribed to any specific physical brain element. As cognitive research attempts to create this linkage, the information needed to reach conclusive answers is not available. Therefore, any concept attempting to explain the mental function as a physical operation is not admissible evidence. An example of such a function is memory. There is no identified brain component that is the storage place for acquired and stored information with a potential for future recall. The theoretical storage is in the synaptic areas. (This is also the subject of a later chapter.)

The human nervous system is a complex arrangement of billions of fibers and synaptic connections across which the neural signals are transmitted. Its physiological organization and functioning are well defined. References to its functioning and interactions with other physical elements are valid evidence; references to assumed functions are not. Motor nerves carry signals for activity to the end organs. Efferent nerves carry stimulant signals from the end organs to the brain. Thus when the brain sends or receives a signal concerning an activity at an end organ, it is either prompted by a physical environmental stimulus, like pain, or by a non-physical element (whatever it is called), for an intentional physical activity in which case the action is not an entirely physical one. The end actions in both cases constitute valid evidence because they are physical acts. But the initiating cause is only physical in the former case, not in the latter. As an example, when I accidentally touch a hot plate, the immediate withdrawal action is caused by a physical stimulus. However, when I'm working at the computer and decide to take off my glasses to rest my eyes, the computer is not the motivating stimulus; it is the non-physical I, the self, who calls for and initiates the action. This is the basic reasoning for determining how any evidence should be evaluated and used. This is the crux of the dualism debate.

There are other types of nerves that automatically (without voluntary initialization) control body functions, such as, breathing, blood circulation, digestion, etc. They are, therefore, totally physical elements and their functional performance is valid evidence. As an automatic control, this function does not rely on the conscious mental state, but operates within limits like the heart and breathing rates. It must, therefore, rely on some sort of memory. But memory is a mental state. Could it be that there are two memories, one for this subconscious activity and the other for experience, knowledge, skills, and willful storage of data? If so, the evidence of information related to each must be separated as physical or mental with appropriate validity. This unique and unprecedented idea of two memories is my contribution to the definition of human memory. (This duality is covered in more detail in the chapters on memory and the dualism system description.)

In summary, only knowledge of the body's anatomy and the evidence of physical activity that supports the properties of the "substance" in the materialist sense are valid evidence. Otherwise, the extrapolation of this evidence as physical properties of the mentally enhanced brain cannot be valid until it is proven according to scientific requirements, not by conceptual solutions. The attempts to bridge the gap between factual evidence and anticipated future verification are merely admissions that there is much to be learned about the complete human being. It is also puzzling that there is no effort at a scientific reevaluation of the dualism effect in the absence of so much confirmatory evidence.

Conceptual

Concepts, by definition, are merely assumed solutions to problems. They are, therefore, not valid evidence until they are supported by experimental data or demonstrated performance. This has not been done for monism's many concepts of physical/non-physical interactions. As conceptual solutions they are only *beliefs* that they will be confirmed at some future time.

There are three types of concepts being proposed to explain the monist approach. The first is the assigning of physical properties to non-physical *elements*, such as memory. It is an element, a thing, because it handles data, which are, presumably, physical signals that have been processed from physical stimuli. Conceptualizing how this happens and where the storage area is located depends on assignment of functions to parts of the physical system. Neither the storage location nor the information processing have been verified through physical evidence.

The second type seeking verification is an *action* rather than a thing. An example is free will, the ability to make a choice or to override an environmental cause. There is not even a question of its location in the brain since it is described as a mental state or totally denied as specific volition. Only the effect of its use can be observed, never the physical action that causes the effect. The explanations of free will are in the realm of neurophilosophic concepts.

The third type as a subject of conceptualization is *condition*, best exemplified by consciousness. It can neither be explained as a volitional condition nor an optional one. It is part of every human being, although it may vary in nature due to accidental or self imposed actions. It is not understood in spite of the intense research to identify it. Again, it is not a physical phenomenon, therefore, it will never produce physical evidence.

The attempts at explaining the above three types are many and varied. For an indication of the extent of this broad conceptual domain, I devote later chapters to the three subjects, memory, free will, and consciousness, for evaluation of the claims.

Computer Simulated

Computers are valuable tools in solving problems. They can be made to simulate almost any concept, theory, or problem and its solution, but these simulations are only as valid as the data that are fed into the computer. Since non-physical data of mental properties are not available, no simulation can provide a confirming solution to the materialistic brain problem. The use of computer simulations in medical research can provide many reliable paths to improving the mental conditions of patients. But the research information from studies of brain damage cannot be extrapolated to serve as evidence for conceptual analyses because of the lack of correlation between the dissimilar cases of brain damage. The resulting interpretations do not lead to consistent conclusions that can be used in the simulations for standardization of the various brain damage data.

One developing project that continues to simulate neurological conditions is Artificial Intelligence (AI). Its ultimate goal, as the name implies, is to duplicate the ability of the human brain to "think". Although the prospects appear promising, the basic problem that must be addressed is that which is the ball and chain of the dualism debate: the volitional ability of the human being. A computer must be given the conditions of thinking, which science does not understand. Basically, there will always be a computer operator, no matter how small the need for his control in any AI program.

Experiential

Personal experience is reality, in spite of any philosophical excursions to the contrary, but is it physical or mental? If the brain initiates the experience and completes it in the brain-caused mental activity, then it is all physically oriented. However, experiential evidence for the materialist cannot exist because it is subjective and explained in terms of mental properties, phenomena, and mental states that are claimed extrapolations of physical neural functions. All three are assumed conditions, with explanations that cannot be supported by factual data, therefore, they cannot qualify as valid evidence. A thought, which is an experience that does exist and can be verified, is not a physical item; it cannot be seen, heard or touched. Therefore, it cannot be a piece of evidence supporting the claim of being "substance". Any semantic effort to circumvent this view cannot comply with the rules of argument that since A and B are different, they cannot be the same. If the thought is recorded and accepted as documented reality, the relevance of its reality depends on other factors than just its content. Example: a thought that the earth is flat is a real thought, but its relevance depends on the available facts to verify its worth as evidence.

Since experiential evidence can only be supplied in a first person format, what can a person say that adds credibility to monism while demonstrating those qualities that imply dualism? Any inclusion of mind, awareness, free will, desire, etc. is an expression of dualism, not monism, since none of these are physical properties.

Historic

The materialistic monist theory has existed since Greek philosophers promoted it. However, until the logical attempt by Descartes to identify the dual nature of man, there was no call for evidence to prove or disprove the nature of man. With the rise of scientific demand for evidence of Nature's secrets, the same need was imposed for the explanation of dualism. It was the work of Charles Darwin that was seized upon to verify the evolution of man and his reasoning power. But Darwin's work only addressed the physical changes through natural selection. Such evidence was and is valid within the limitations of its physical implications. Some neuroscientists and neurophilosophers try to expand the evolutionary concept to include the natural development of reasoning power due to environmental and survival pressures. Since animals seem to have some reasoning ability, it would appear that the evolving process was and is similar, with Homo sapiens ahead in the game. However, this presumption of evolutionary applicability is only valid if the physical brain "thinks". Until there is physical evidence that the brain, not a mind is the rational element, the evolution theory has no basis for serious consideration.

Darwin did not have the benefit of the existence of DNA. This would have added some credence to his theorizing whether DNA could be considered Nature's way of evolutionary selection. But there is much more to the importance of DNA in the development of all living matter. DNA is the blueprint for the development of living things irrespective of the natural environment or the need for survival. These latter effects are complementary to the basic development of species due to their DNA. The evidence of DNA as the progressive developer of living matter has been established without any doubt and it is used as legal and scientific certainty, and evidence. But DNA has also opened new lines of thought about its ability to control behavior through the genes that vary in each individual. The question arises: Does DNA affect the reasoning power in the human being? Is there evidence that the physical DNA in some way integrates the mental state with the physical process? No evidence of this exists but the potential is apparent, although such potential circularly reverts to the impossibility of a physical element creating a mental "thing".

Witnessed and Experimental

Experimental tests are witnessed events. They are third person accounts of first person activities or organic tests. The data obtained can be used as evidence of what occurred. But the interpretation of the results must be verified because in some tests, the subjects, like animals, are unable to respond to questioning.

Evidence of cerebral behavior is obtained experimentally through the use of various imaging and scanning equipments: CAT, PET, fMRI, and EEG. CAT (computerized axial tomography) makes X-ray pictures of a predetermined plane by blurring out images of other planes. PET (positron emission topography) measures the relationship between neuronal synaptic activity and cerebral blood flow. When synaptic activity increases so does the blood flow. By administering short acting radioisotopes in the patient, the PET camera can image the activity in the specific area of the brain during the configured test. The advantage of using fMRI (functional magnetic resonance imaging) equipment is that there is no need for radiation exposure as the imaging of the variations in cerebral blood flow is done by magnetic field manipulation. The EEG (electroencephalograph) measures the currents within the brain by means of electrodes pasted to the outside of the skull. It records the various discharges of energy on a photographic plate as waves when the mind is relaxed or actively concentrating. It is used in diagnosing cerebral tumors, epilepsy, trauma, and degeneration of brain cells.

With all these equipments, the probability of evidential accuracy and correct interpretive analysis is high. However, this evidence is only physical in that it shows what is happening. It must be kept in mind that it does not show or help to determine how the actions happen from volition to accomplishment.

Testing of patients with abnormalities due to cranial injuries or traumatic emotional stresses, such as amnesia and dementia, produces information about the brain elements that are affected, their responses, and the conditions that cause these abnormalities. The evidence obtained through these methods is valid principally due to the consistency of the results. However, there is some doubt of total credibility due to the possibility that the investigator erred in the interpretation of the results.

Sometimes, unexpected findings in cognitive research require a redirection of thinking in how the human brain (or mind) functions.

When such findings produce measurable results, they become useable evidence in closing the gap between assumed explanations and factual certainty. Such is the case of tests performed by Benjamin Libet, Professor Emeritus of Physiology at the University of California, San Francisco, in researching the temporal factor in consciousness.

Normally when we experience a sensation due to an unintended sensory stimulus we assume that we are aware of it at the exact time of the occurrence. Experimental data indicate otherwise. The brain needs up to a half second (500 milliseconds) (ms) to cause awareness that the event occurred when the stimulus for a skin location is applied at the cortex. This was an unexpected result of the Libet experiments. Such a delay is for an unanticipated, involuntary stimulus. However, when the stimulus is applied at the skin (instead of the cortex), the response is within 10 to 40 ms depending on the distance from the skin location (finger or toe) to the brain. How can there be a 500 ms difference when the same pulse signal is applied to the same neural response destination? In other words, why does the brain take so long to react to a signal given within its own location, when the pulse reaction at the skin is much faster? This would indicate that there is awareness of the skin sensation before the brain is made aware of the same sensation.

This is a logical contradiction. The paradox led to a hypothesis that there could be a subjective referral of the timing for that experience back to the time of the primary response since neural timing is not the same as subjective timing. The awareness of the stimulus is antedated subjectively to the timing signal of the primary response.

An experiment was conducted to verify this apparent contradiction. It consisted of a comparison of the two different pulse signals: one at the cortex and one at a skin location. The cortex timing was the expected 500 ms but the response timing at the skin was within 30 ms from initiation, indicating that there was a difference by which the skin sensation arrived before the awareness time of the cortex pulse. Although the experiment seemed to prove the hypothesis, it gave no indication of how it happens. There appears to be no neural mechanism that could be viewed as directly accounting for the referral timing, which is of fundamental importance to the nature of the body-brain relationship.

How was this timing information obtained? Electrodes were taped to the cranial area which represented the sensitivity of the organ at which the stipulated test motion was performed. By activating the signal through the electrodes, the delay of signal transmission from the actual organ was eliminated. A threshold stimulus train was activated to determine the required time of buildup before a conscious sensation could be recorded. Thus the 500 ms time of the brain response to the activated signal was measured. This delay, establishing the basis for comparison of voluntary and involuntary sequences, was found to be consistent in all tests The somatic stimulation at the cortex also eliminated the possibility of varying times for stimuli travel by different paths from the intended organ.

A different set of tests than those for non-conscious testing were used for voluntary act tests. A specially constructed clock was designed to simulate a moving second hand by using a light rotating at the numerical positions. The rotation speed was increased to denote smaller increments of time. The subject would decide to make the prescribed motion, a flick of the wrist, at any desired time noting the position of the clock light when the action was taken. This position was reported at the end of the event for inclusion with other recorded test times. The results were found to be similar to the awareness times but longer when any preplanned activity was included. Preplanning did not affect the initiation time of brain activity because preplanning can go on for a long time with no resultant activity.

In planned or anticipated activity, there is an additional time up to 400 ms before the onset of the 500 ms delay. There are differences between unconscious and voluntary stimulations. By unconscious is meant those stimuli that are received without any intent to act. The voluntary intended acts initiate the process either with or without preplanning.

When we intend to perform an act, we think that awareness of the action and the initiation of the action are coincident. According to the Libet experiments, this coincidence is not true; there are three events involved before the intended act takes place (see Figure 3.1): (1) the onset of brain activity, the readiness potential (RP), (2) awareness of the intended activity (W), and (3) the act. What is most amazing is that the first indication that the action will occur is a signal of brain activity, the RP, 350 to 400 ms *before* any subject awareness of the desire to perform the act! The desired action then follows W by 200 ms. This is more surprising than the non-conscious situation because the decision is a conscious act. How can there be brain activity toward an intended act before the initiator of the act is aware that the act is commanded? This is a contradiction in logic.

Note that if the W time of -200 ms is combined with the 350-400 ms delay, the total time period may be as high as 100 ms (± 50) before the initiation of brain activity RP. In other words, there is a period of time before the brain begins to act (in this theory) that should be accounted for. The Libet explanation is that there may be another brain function contributing to this line effort.

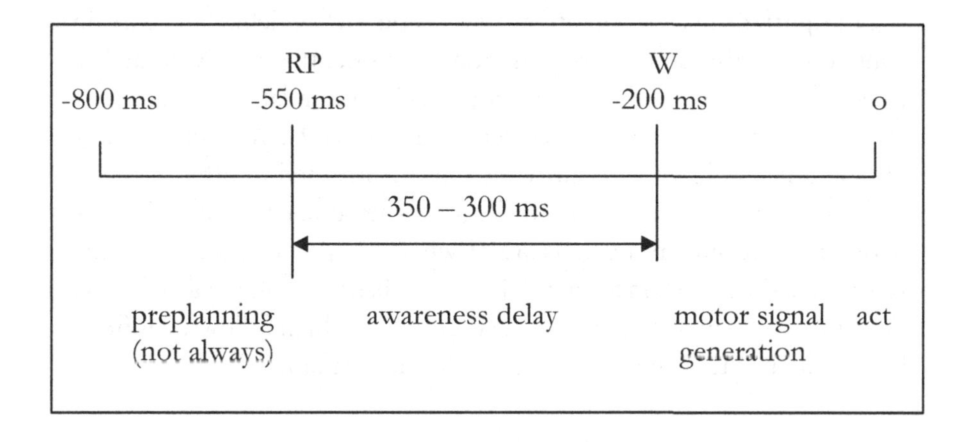

Figure 3.1 Timing for Voluntary Action

The sequence of events in an unexpected stimulus demanding an immediate reaction has an added element that Dr. Libet illustrates by a case of a boy running in front of an oncoming car. Although there is the 500 ms delay to awareness of the boy's appearance, the reaction is to apply the brake within the 100 to 150 ms before RP. (This is the same pre-RP interval I referred to in the previous paragraph. Here it is increased to 150ms.) The voluntary process is initiated unconsciously without awareness. It is interesting that a motor signal is given to the foot for brake activation before any indicated brain activity or awareness. So what activates the motor signal for the brake activation? This cannot be explained physically but it is a key point in applying dualist thinking because as will be shown later, the initial response to the sight of the boy *may not be a physical reaction.*

In this instance, however, the explanation is different from the one for the involuntary case since a definite reaction prior to the first brain activity must be explained. The hypothetical theory is that the delayed awareness is automatically, but subjectively, "referred back in time", an antedating as described above, through which the boy's appearance is seen immediately. I find this hypothesis difficult to understand, or believe, and consider it a good example of justifying a theory through reasons based on interpretation of experimental results; the conclusion being that the reaction is unconsciously automatic before awareness. Nevertheless, it is an important first step in attempting to physically define and explain consciousness through experimentation. I consider the experimental data to be valid evidence and will address this evidence in later chapters.

This sequence of events, unexpected as it is, has been experimentally obtained in the Libet laboratory tests and confirmed in other research projects. Dr Libet explains, at length, the apparent inconsistency between what our impressions are of what is happening and what the test results show. To add authenticity to this listing of physical evidence, Dr. Libet's interpretation of these experiments, which follows, confirms his confidence that the surprising results are verifiable.

> So what answer did we obtain to our original question, about relative timings for onset of brain activity (RP) versus conscious will to act? The clear answer was: the brain initiates the voluntary process first. The subject later becomes consciously aware of the urge or wish (W) to act, some 350 to 400 ms after the onset of the recorded RP produced by the brain. This was true for every series of forty trials with every one of the nine subjects.[12]

This straightforward statement of the timing confirms the contradictory nature of the theory. If you didn't catch the unintended contradictions it is because the jump from the physical brain activity to the non-physical consciousness is subtly glossed over due to the emphasis about the timing. The first contradiction is that a physical brain cannot initiate a voluntary activity, a decision by a non-physical state. This is verified by the second slip in that the "subject" later becomes aware. But it was the subject's physical brain that initiated the voluntary activity, the subject's functional responsibility. There is a basic flaw that needs explaining.

Although the timing is associated with consciousness, the results of the testing are theory independent in that they do not confirm the theory of consciousness, nor were they tested to confirm this theory. However, they do support a new theory of delayed awareness. The measured results were obtained for their values and sequencing, which can be used as evidence in either monist or dualist theory, although their full implications are not apparent. In his book, *Mind Time*, Professor Libet does propose his interpretation of how these results may affect consciousness and allow for free will.

The above experimental results show that measurements of brain activity with their effects on consciousness are a significant contribution, as factual evidence, to the potential ability of analyzing some of the undetermined functions of the mind-brain problem. The data can also enhance the explanation of other cognitive theories.

Spontaneous and Observed

Laboratory and clinical tests of brain functioning are observed, not spontaneous, when they are planned. The observers are participants in the testing rather than just witnesses. Test data are valid evidence for researching behavior due to cranial damage. Such data are more applicable to psychological or psychiatric application in remedial solutions than to proving the brain's creation of mental states.

According to the earlier definition of spontaneous evidence, for monism, everyone is the subject who observes own first person activity. This is pseudo-evidence for monism but invalid because it does not contribute any facts to how the only physical substance can transform physical stimuli into non-physical mental phenomena. No amount of own observation of own performance can supply that. We take it for granted and leave the generation of evidence to the scientists. And the scientists have not produced such evidence.

Laboratory or clinical tests on humans are not spontaneous but are observed because they are planned. The observers are participants in the testing, more than just witnesses.

Summary of Monism Evidence

The physical evidence for monism is extensive, as it should be because it is of a physical entity which can be seen, heard, and handled. It is constructive in that it defines the physical base of human nature which is necessary for understanding the complex functioning of the human being. Such evidence allows further excursions into the conceptual possibilities of those unknown functions that are non-physical. As valid evidence, however, only the verified physical evidence meets the criteria for acceptability. It is data, not theory, that must prevail in the search for the answer to the dualist debate. While much is known about the physical, the absence of information about non-physical properties hampers the definition of how the physical must duplicate the non-physical functions. That is why evidence such as the Professor Libet research on timing of consciousness is constructive in establishing a link between the physical and the non-physical.

The research on damage to brain elements is crucial to understanding the mental effects of such damage and the possible reversal of these effects. Testing of animals for these effects produces evidence but the interpretation is not as conclusive as that obtained from humans who can add first person information to help in the interpretation of test results.

The use of evolution theory to explain the purely physical development of the brain to contain a non-physical rational capability is inadequate even as a theory, much less as evidence; it only addresses the effect, giving no indication of the cause nor the incremental changes that led to the full physical-mental result. Nor does it explain why no animal of any species has been able to follow a similar path to the development of a rational capability. If evolution developed man, why has it not developed, by any process, at least one other similar rational animal like a human being?

As science is based on observation and experimentation, it must produce physical evidence to support its claims. If it cannot do so, relying on assumptions and promises of future verification, it is no longer science: it is scientific philosophy, like metaphysical philosophy, subject to criticism, discussion and continuous change.

The only evidence valid for the monist side of the debate is that which applies to the strictly physical "substance". In addition to the physiological data, a valid example of such physical evidence is the Libet experimentation about the timing in consciousness. Any concepts of the brain's mental capabilities are unproven as scientific reality, consequently invalid as evidence until proven in compliance with scientific standards.

3.3 Evidence for Dualism

What must the evidence for dualism prove? There are several issues. The most important is that the soul does exist. This must be physical evidence to satisfy scientific requirements for acceptance in spite of the seeming impossibility of developing physical evidence of a non-physical entity. Another issue is the confirmation of the separability of the two natures of the human being, the body and the soul. The evidence for the reality of reincarnation must be conclusive, for it is reincarnation that demonstrates this separable nature in the process of defining the soul. Another key issue is identifying the evidence of those properties of the soul that confirm dualism's replacement of monism.

As we progress to the description of evidence for dualism, we must adjust our thinking to accommodate the probable skeptical reactions to claims of immaterial reality. The disclosure of conclusive evidence of the soul's existence calls for a major attitude change in the secular and scientific mindsets about it. But is this possible within the established scientific position? If the evidence is scientifically presented, the scientific community, by its own admission, must evaluate it, and if found true, accept it. This section not only describes such evidence but also identifies the soul's properties that relate the evidence to the dualism debate. The divisive aspects between monism and dualism are postponed to the following chapter. As in the previous section on monist evidence, questions are raised that need to be addressed in a more appropriate format, limiting this section to establishing the critical base for rethinking dualism.

Although this most pertinent evidence has been available for several decades, the lack of its acceptance may be intentional or unintentional. Because it was developed in a different discipline, in medical research, for reasons that are theory independent of dualism, may explain its dualistic anonymity. It was a search for the reasons why birthmarks and birth defects appeared as results of reincarnation, replicating wounds received in the previous life. The role of reincarnation was thus established and became available for application to dualism.

Physical

The most dramatic and unique physical evidence of non-physical entities exists for the dualism case. It is an historic disclosure in that it has been awaited for centuries in the belief that it was impossible. How can physical evidence be produced of non-physical entities that have no physical qualities and occupy no physical space? This has been declared impossible by materialistic science. However, *evidence has been generated that living, physical, conscious human beings are conclusive proof of their dual nature through their verified activities that they are the reincarnated souls of deceased human beings in new bodies.*

To retain its personhood, a soul must exist between the occupancy of the two bodies. This evidence has been reported in detailed, scientific reports by Dr. Ian Stevenson, a medical doctor, who has personally investigated and verified the credibility of reincarnation. He has led the research effort and with other researchers has accumulated verified reports on over 2500 cases of reincarnation. Validated proof of only one case consisting of repeated recollections should be sufficient to establish the credibility of realism. But 2500 cases of similar occurrences should alert the scientific community of reincarnation's reality. In addition to investigating past reincarnation incidents, Dr. Stevenson has personally witnessed the generation of the evidence by reincarnated subjects. They described persons, articles, incidents, and locations in the previous life, which were subsequently verified as he accompanied the individuals in the verification process. These scientific-type investigations reported according to scientific standards conclusively prove that the human soul exists because it survives death and reappears in a different body with a memory of the previous existence. This recollection of the previous life demonstrates through verification of its details that the only possible connection between the deceased and the new personality is a soul with existential continuity.

This worldview-shattering evidence has been in published form since 1966. In that year it was published as Volume 26 of the Proceedings of the American Society for Psychical Research with the title *Twenty*

Cases Suggestive of Reincarnation. The evidence for dualism based on reincarnation in this chapter is taken from the 1974 revised edition of that book. Although Dr. Stevenson has published other books and material, I use this first book to emphasize the type of evidence that was available to the scientific community as early as 1966. It is amazing that this scientifically reported evidence with subsequent reports of continuing research has not been found and used to prove the case for dualism. The reluctance of the scientific community to even pay attention to such dramatic deconfirming evidence is understandable because it lays bare all the scientific claims of the soul's non-existence. The proponents of dualism, on the other hand, should have been following the developing mass of evidence and challenging the scientific community to respond to this scientifically conducted research.

The most incredible aspect of this evidence is that the *evidence itself is generating the conclusive data.* What do I mean by that? The subject of any other test must be observed by a third person to report the data. Whether the test involves organic matter, animal test specimens, or human beings, it is the *observation of the test subject*, the interpretation and recording of the data that constitute the evidence. But in the case of reincarnation, the *subject itself* supplies the data in a **first person generation** of information that establishes the evidence. Furthermore, this evidence is not a response to a test program or any test conditions. This is **spontaneous**, uncalled for information that is generated by the reincarnated entity to force a response to its demand for recognition as the personality in the previous life. Its demands are often ignored or efforts are made to stop the insistence. However, when they are believed and the information is pursued, *it is found to be true.*

The physical evidence for the dualism case is more specific, theory independent, conclusive, and challenging than the materialist's implied evidence of the physical aspects of the human mental state. Furthermore, this evidence negates all the conceptual explanations of human memory as a physically stored process completely contained in the physical body by demonstrating that memory is a distinct part of the soul. If it were physical in any sense, it would have remained in the deceased body, depriving the soul in the new body of any recollection of the past life. But recollecting the names of related persons, recognizing previously known people on first sight, remembering previous locations and environments and locations of stored items, clearly verify that *memory is an integral part of the soul.*

Due to the unprecedented method of generation, this unique first person evidence of the reality of the soul is theory-independent in that it does not rely on any theory to exist. The existence is a fact in itself: it stands alone as an occurrence because it would be true even if it were never shown to be true. The demonstrated existence explains the reality of reincarnation, through the unification of the same soul with each of two bodies. This repeated union of the two entities also establishes the reality of dualism. It proves that *the Cartesian dualistic nature of the human being exists before reincarnation as well as after*: proving through the many confirmed investigations that *such a combination exists in all human beings.*

The investigative effort of reincarnation was part of Dr. Stevenson's research of birth marks and birth defects for medical reasons. These investigations revealed definite relationships between injuries in the previous life and the subject's birth marks or defects, further pointing to the reality of reincarnation. These precise replications of previous life body wounds as reincarnated birth marks and birth defects are described in very detailed reports in Dr. Stevenson's voluminous book (monograph), *Reincarnation and Biology: A Contribution to the Etiology of Birthmarks and Birth Defects* (1997). For the readers who don't need the multitude of details involved in this research, Dr Stevenson prepared a book of abstracts of typical cases, which provides summaries with color photographs of the birthmarks and defects. This book of abstracts and associated comments, *Where Reincarnation and Biology Intersect,* was also published in 1997.

The reality of reincarnation is the main evidence produced by Dr. Stevenson. It refutes science's denial of the soul's existence. Using that fundamental reality, I show that the evidence of the soul's demonstrated properties in reincarnation negate all the neuroscientific and cognitive concepts of the brain's dominance in human behavior. The integration of the demonstrated mind's properties with the brain bridges the gaps of the current lack of understanding. This integrated argument is superior to the assumed and unsupported monistic concepts. The single demonstrated activity of the soul's remembering the existence in a past life with all the attendant circumstances alone proves that memory is part of the soul, not of the physical brain, If that were not enough, the demonstrated free will and new-body control by the reincarnated subject, which are also denied in the conceptual explanations, further attest to the futility of attempting to justify physical causality.

Descriptions of the activities involved in obtaining the evidence described above, samples of the reporting, cultural information, and the scope of their social acceptance are presented under the Spontaneous and Observing section below. Although this evidence is sufficient to validate dualism, other similar evidence is described, but not all of the strictly physical type. There is also experimental evidence based on brain scans with non-physical implications. This additional evidence is presented below under Experiential and Witnessed sections.

Conceptual

Evidence for current concepts of dualism has been just as invalid as for the monistic case for the same reasons. Actually, it was Descartes's concept of dualism based on logical and philosophic assumptions that started the debate due to the absence of justifying evidence. Lacking the evidence that is now available, he had to resort to philosophical argument, which was acceptable in his day. However, his concept of the free will and volitional control associated with dualism was defensible conceptually since theories of physical causality did not exist.

Even with the positive evidence of the soul's existence, the need for conceptualization remains. The proof of the soul's reality does not in any way explain how the integration with the physical brain occurs. If there is a new integrated research effort to combine the scientific with the philosophic attempts at defining the functional behavior of the human being, the potential for a final explanation will become evident. The conceptual effort will still be required until more facts of the union are defined.

Computer Simulated

It would seem that computer simulation as a scientific tool cannot study the soul's effects on human behavior since the "spiritual" laws of soul performance are not known. Since the existence of the soul has been denied, there have been no simulations of the effects of a soul on the functioning of the brain. Is there nothing on which to base future studies? I don't believe so. Any simulations of monistic brain functioning must use inputs based on observed behavior. Labeling them as brain generated or mind generated should not affect the study of behavior. The goal of Artificial Intelligence is to duplicate the working of an immaterial distinct mind, although that is not an admitted consideration. But the research in this area should be a starting point when the soul's presence with an independent mind is acknowledged.

Experiential

A personal experience is real to the person living through that experience. Proving that experience is difficult, if its nature is out of the ordinary. An example of such an experience is the near death experience (NDE) in which the soul leaves the body temporarily during which time the body is clinically dead. But the soul returns and the body recovers life. The immediate response to such a claim of body-soul separation is skepticism and an attempt to explain it as a psychological or hallucinatory occurrence. But how could it be hallucinatory when the brain was dead? To one having had the experience, it is neither. Anticipating such a reaction, many experiences of this type are never reported. Since the evidence of the soul's reality exists, the credibility of NDE experiences may assume greater importance in scientific research and the meaning of life. The subjects who experienced NDE show marked effects on their views of life and a reduced fear of death, so there is a residual effect as evidence of a paranormal event.

Not all scientists deny the validity of reported experiences like NDE. After more than thirty years of investigating conscious experience, Dr. Benjamin Libet (conscious timing experimenter above) became convinced that:

> Unless they can be convincingly affected or contradicted by other evidence, properly obtained introspective reports of conscious experience should be looked on like other kinds of objective evidence.[13]

A more general experience that neuroscientists are refusing to recognize is that of free will. To them, behavior is caused by physical reaction to external stimuli with no control over the resultant actions. But to anyone who experiences the ability to make a decision and then to reverse it or to intentionally override the effect of an external stimulus, the experience of a free will is real.

Consciousness is an experience of a different sort. It is not understood but it is experienced by every living being. Thus it can be a subject for argument because it is a first person reality for everyone, but no one can produce the physical evidence to prove it. Since it is not understood, it cannot even be described in precise terms for a meaningful debate. Only its effects can be discussed.

Since the argument about consciousness is in trying to explain what it is, could it be that it is the soul that has now been shown to be a reality? Since every living person has a soul, this could lead to the conclusion that the soul's presence is consciousness. More on this later.

Historic

Historically, the beliefs in the reality of a soul were prevalent in ancient times. They were recorded in Greek literature, the Old and New Testaments of the Bible, Eastern religious documents, and other cultural documentation.

The principal possible historical evidence for the existence of the soul is the documentation represented by the New Testament of the Bible and similar documents. The use of this documentation as evidence is limited for several reasons. The complete original manuscripts of these documents have not been found, leading to criticisms that existing texts may be altered versions, or even fabrications. The fragments that have been found are sparse and incomplete. The authenticity of the authors is also questioned. However, many archaeological findings have verified the accuracy of the persons, dates, social conditions and politics contained in these documents. The book *The Bible Is History* (1999) by historian Ian Wilson is an excellent description of the many archaeological verifications of Biblical text.

There is one case, however, that is unique and worthy of mention as authoritative evidence. This is the case of Jesus Christ as the proclaimed and claimed messenger of God, the Messiah, whose coming was predicted not only in the Jewish biblical prophesies but also in other ancient beliefs. I treat this as a historical and secular review of Christ's testimony about the existence of a soul without the religious connotations. This narrows the review strictly to the needs of this search for specific evidence.

Christ was a historical figure, not a legendary or mythical character. Another book by the historian, Ian Wilson, *Jesus: The Evidence,* is an objective review of the positive and negative facts and opinions concerning the life and death of Jesus, comparing archaeological evidence to the biblical narrative. Jesus' claim to be the messenger of God is the basis for the credibility of his testimony about the non-physical world. Since he claimed to be God's messenger, he came from the spiritual domain and became a material human (incarnation). He, therefore, knew of the spiritual universe: what it consists of, who inhabits it, what its spiritual laws are, and how they relate to physical laws. He was, therefore, able to give testimony about that non-physical environment. He supported his claims to know God and be the Son of God. He overcame the laws of nature by instantly healing all maladies and even raising the dead to life (all physical reversals overturning natural laws). Only the creator of a law has the power to reverse that law and Christ did this to prove his divine nature. Therefore, his activities, as recorded, justify his claims of divinity. Examples of our acceptance of this principle are: only a father can change a curfew law he has made for his teenagers; only a legislative body can change a previously enacted law.

As a visitor from the "other" world, he talked in terms of this other world. His testimony had and has as much credibility for acceptance as the testimony of one coming from a foreign land about that land. On an even playing field of acceptance, that must be valid. On several occasions during his three years of teaching, Christ referred to making room in this Kingdom for his followers (Matthew 19:28), and for one of the two men crucified with him when he said, "Amen, I say to thee, this day thou shallt be with me in paradise" (Luke 23:43). With his established credentials and credibility (there is no documented evidence of his lying or even evading a controversial question) his testimony is evidence that there must be a soul in each human body, because he knew that all souls survived earthly death. Recorded verbal testimony, as this was, should be treated as it would be in a court of law: credible evidence. Although such evidence, which can be contested, is not the critical evidence to support the need for a review of dualism, it does contribute supplementary and corroborative evidence, by a qualified "expert", to the total array of evidence for the existence of the soul.

Witnessed and Experimental

Witnessed tests most likely to withstand the rigors of scientific criticism are those designed and conducted by skeptical scientists. Such tests take into account the likely criticisms that will be directed at the methods used to obtain the results and the interpretation of the data. Further credibility is inherent in the tests if they are part of a scientific program at a research center. The following two sets of experiments were conducted to research communication with the spiritual "universe". The University of Arizona Tests,[14] were experiments in actual communication, while the University of Montreal Tests[15] were experiments of brain scanning during meditative activity.

University of Arizona Tests

The possibility of continued consciousness after death was the subject of the Arizona tests conducted by Professor Gary E. Schwartz. For there to be such consciousness, there had to be means of communication with such consciousness and, by implication, with entities in which such consciousness was implicit. In addition to the difficulty of designing the experiments, there was the need to identify and "invite" the entities to participate in those experiments. It was a challenging goal but Dr. Schwartz had to satisfy his skepticism about after-death consciousness and there was the possibility of some useable evidence or procedures for future work. He realized that the organization of the experiments would require unusual and controversial methods. In a preliminary experiment he obtained the participation of the University of Arizona through its Department of Psychology. The department head, Dr. Lynn Nagel, had previously agreed to offer a course on the psychology of religion and spirituality to be taught by Dr. Schwartz. The course was to pursue the discussion of the subject within a scientific framework. Dr. Nagel not only agreed to the precursor project, but also followed up by obtaining the cooperation of the Dean of the College of Social and Behavioral Science. Dean Holly Smith was also committed to promoting the dialogue between science and spirituality.

It took time and the assistance of specific people to organize the principal series of experiments. This included designing the "skeptically free" experiments, agreeing to a suitable arrangement with a sponsoring television station for filming, and most of all, the finding of reliable mediums to conduct the communications.

All claimed communication with the spiritual world has been done by mediums. However, due to the questionable (and sometimes fraudulent) performances of some mediums who exploit grieving and curious peoples' desires for communication with deceased relatives and friends, the reputation of the average medium is dubious, if not outright deceptive. This feeling of distrust is mutual. The mediums, especially the famous ones, consider scientists as disbelievers intent on proving all mediums to be frauds. Nor do they trust the media to report their findings favorably, instead, exploiting their work for sensational copy.

In spite of this general attitude toward mediums, Dr. Schwartz had to find reputable mediums, who would recognize the scientific nature of the program and agree to the reporting requirements: audio and video taping, and filming of the experiments. From personal experience and associated psychology programs he found five that had demonstrated credible communicative abilities.

Having found the communicators, the next critical task was to find "sitters". These were the subjects who had lost relatives and friends that would hopefully be contacted and who would disclose information that would be corroborated by the sitters. Two qualified sitters were found.

What I have described so far is the background of activity that had to occur before the experimental tests could even begin. This prelude to the description of the tests shows the deliberate efforts to assure the scientific nature of the project, which would enhance the credibility of the results, regardless of how they turned out. The results would be in the recorded data, not in a descriptive narration of the tests. Next, I will briefly describe the test setup and the testing procedure, followed by the detailed communications. The vital conclusive results are of interest for application to the evidence requirements for the dualism debate. (Anyone interested in the details of these and subsequent tests can find the information in the very detailed descriptions in Dr. Schwartz's book, *The Afterlife Experiments, Breakthrough Scientific Evidence of Life After Death* (2001), published by Pocket Books, a division of Simon and Schuster, Inc.)

The test setup consisted of an opaque screen with a chair on either side, one for the medium and one for the sitter. In separate tests, each of the five mediums would attempt to communicate with the deceased "souls" of a sitter's relatives or friends. The mediums and the sitters did not know each other, had never met and were not informed of each other by name or otherwise before sitting down on opposite sides of the screen. The sitter was only allowed to answer *yes* or *no* to the information being received and relayed to the sitter. The sitter and medium wore jellied electrode caps to record brain wave and heart data. The purpose of these data was to detect if the medium was receiving information from the sitter via telepathy. (These data were inconclusive.) The entire test scene was filmed, audio taped, and video taped.

In all the paired (medium and sitter) tests, there were communications from the spiritual side. The deceased subjects always responded to the mediums' calls and identified themselves with respect to the sitter who was only allowed to acknowledge or reject the identification or the information. The subject matter was always initiated by the spirits but the medium often asked for clarification and obtained it before asking for verification from the sitter.

As amazing as the communications and the acknowledged information appeared, they were all recorded <u>scientifically</u> and later scored by the sitters by studying the recorded data. The accuracy of the transmitted information was developed using a scoring method with a range of +3 for a completely correct piece of information to a -3 for a complete miss in the minds of the sitters. The information scores were based on accuracy of: names, initials, descriptions, historical facts, temperament, and opinions.

The results were an incredibly high average of 83% for the correct +3 items for one sitter (5 mediums) and individual scores of 90% and 64% (2 mediums) for the other sitter. To test the ability of correctly guessing answers for a comparison with these tests, Dr. Schwartz set up a control group of 64 students. They were asked to "guess" the answers to a set of questions. The overall result was 36% accuracy, which indicates that the mediums most likely could not be supplying random information to get their level of accuracy.

Of importance here as evidence is the fact that communication was achieved in all cases of this set of tests. There were subsequent tests that are also reported in Dr. Schwartz's book, but this series of tests is sufficient to prove that communication with the spiritual world is possible. The resultant, and more important, evidence is that there are souls who can communicate because they still exist, were part of a united body-soul, and who have retained their memories of the concluded life. *The memory resides in the soul, not in the body!*

This credible evidence, of both, tests and activities, were witnessed by Dr. Schwartz and others. The witnessing was reported scientifically with supporting evidence, such as video and audio tapes, photographs and written testimonials. With the availability of such evidence, the scientific community should have at least evaluated the data since it was developed by a scientific researcher and should have explained its continued unsupported denials of the soul's existence. This has not happened.

Testing for the existence of non-physical entities is, obviously, challenging since they cannot be seen, heard, handled, or their presence felt. Inviting souls to participate in any tests is restricted to mediums and their abilities and methods are not generally recognized as worthy of serious consideration. However, there are mediums who have demonstrated, as in these scientifically oriented tests that they do indeed perform as they claim. The evidence produced by these reliable mediums should be taken seriously.

The University of Montreal Tests

The University of Montreal experiments were performed by neuroscientist Mario Beauregard to test his hypothesis that the brain possesses a neurological mechanism for self-transcendence. Many scientists refuse to accept any hard evidence that conflicts with their materialistic prejudices. But Beauregard disagrees with the general concept that experiences can only be explained by physical causes. He believes that the mind is a non-physical entity and that in mystical experiences it remembers these experiences in the same manner that it remembers others. Therefore, these are not delusions or illusions which are forgotten; they are real. And because they are real, the meditative and contemplative states can be studied through neuroimaging techniques. He set up tests to determine whether the brain activity during these mystical experiences is located in the temporal lobes, as some researchers believe, and whether mystical experiences produce mental states that are different from normal activity states.

As subjects for the tests, he convinced fifteen Carmelite nuns from the Quebec area that the tests were scientific, not commercial, and informed them of the purposes of the tests. The tests were already funded by the John Templeton Foundation and would be performed at the University of Montreal Center for Research in Neuro-psychology and Cognition, with all its powerful neuroimaging equipment, There were objections to the tests from materialists and even the religious; the materialists wanting to head off any deconfirming metaphysical evidence and the religious fearing that the tests would distort religious meaning. In spite of these objections, two sets of studies were performed.

The objective of the first study was to determine whether there was a "God module" in the temporal lobes. The nuns would be scanned with functional magnetic resonance imaging (fMRI) while they recalled and relived their most significant experience (mystical condition) and their most intense state of union with another human (control condition). A third condition (baseline) was tested to establish a reference with a normal restful state. Immediately after the experiments, the nuns were asked to evaluate their experiences, to establish whether they were aware of any experience. They responded positively saying that they felt the presence of God and his unconditional love.

The results of the scanning showed that there is no "God module" - a communicating mechanism with God. Many areas of the brain were involved in the mystical experiences. The affected brain areas are described in the book, *The Spiritual Brain, A Neuroscientist's Case for the Existence of the Soul*, but are omitted here as both studies are too detailed for the purposes of these descriptions.

The second study was similar to the first in the use of the three conditions as the format for the tests but the testing was done with quantitative electroencephalography (QEEG) which measures brain wave patterns at the scalp that can be analyzed and translated into color maps. In these tests the nuns were located in a dark room, which was soundproofed and electromagnetically isolated, except for an infrared camera to observe the subjects.

The results expressed by the nuns were similar to the first study with some nuns describing a union with God and experiencing unconditional love and a feeling of self surrender. The neural results were more extensive with very detailed loci for the different conditions.

In summary, what do these experimental tests contribute to the evidence of the soul's reality? They are not physical, as I have maintained evidence must be, and the evidence, although produced scientifically, lacks that conclusive assertion that it leaves no doubt about the soul's existence. In the Arizona tests, the evidence appears conclusive but the communication with the spiritual world is not a scientifically acceptable procedure. In the Montreal Studies, the evidence is conclusive for what it defines: there are specific brain patterns for meditative states. The extrapolation of that data to an existence of the soul, as claimed in the subtitle of the book, is not clear although there are other more pertinent subjects. The experiment may demonstrate a meditative union with a Supreme Being but does not prove the existence of either a non-physical communicating entity nor of God. It only defines the probability of action rather than an entity. So why did I include them? As activities involving a metaphysical transcendence they contribute to the verification of a spiritual world that must be defined because the soul exists. As will become evident before the end of the book, if there is to be an acceptance of the soul's reality, it will be necessary to innovate experimental procedures to reveal some of the non-physical laws that govern the interfacing between the body and the soul. The two sets of tests described here are proof that there are methods for probing the metaphysical as well as for getting physical results that will comply with the scientific standards for evidence.

Spontaneous and Observed

The evidence of reincarnation is conclusive through scientific reporting and observation by Dr. Stevenson and his associates. I don't hedge by calling reincarnation by a more scientifically acceptable term. The evidence speaks for itself. Although Dr. Stevenson has documented his work in several books, the evidence I use here is taken from only one book, *Twenty Cases Suggestive of Reincarnation* (1966). Twenty cases describing different situations in different parts of the world are sufficient to establish the consistency of the phenomenon, which should have attracted the attention of some scientists and the media. But why didn't it? One fully documented and verified case should be sufficient to arouse scientific interest.

It is strange that such extensive and revealing information has been ignored by the mainstream scientists and medical professionals who were and are the doctor's peers. He hoped that other physicians would pay attention to his findings because they might shed light on the causes of phobias, birth defects, and birth marks that have no explanations. His commitment to the investigations of the medical effects of reincarnation obviously precluded any expansion to different effects in other disciplines, perhaps not realizing how his work refuted the research approaches of the new neuroscientific and cognitive disciplines. For reincarnation is contrary to the very doctrine of neuroscience. It had to be "outsiders" who could see the many interdisciplinary dots that had to be connected to create a more encompassing picture of human nature.

In addition to Dr Stevenson's reporting, there is also corroborative evidence by a journalist Tom Shroder, who accompanied Dr. Stevenson on his final investigative trips to India and Lebanon. This witnessing of not only Dr Stevenson's investigative methods but also of the actual spontaneous subject performances and related interviews are the subjects of Shroder's book, *Old Souls, The Scientific Evidence for Past Lives* (1999). Summaries from it are also given below. A more recent book, *Life Before Life* (2005) by Jim B. Tucker, M.D., who accompanied Dr. Stevenson on his investigative trips, sheds more light on the investigative methods and cases of reincarnation.

There have also been other investigators who have reached similar conclusions about reincarnation. In fact, there is a Guide to the evidence for past-life experiences, *Exploring Reincarnation* (2003) by Hans Ten Dam. It includes a 24 page bibliography of works in English, Dutch, German, French, Italian, Spanish, and Portuguese, including three pages on related subjects. This extensive bibliography indicates the extent of interest and exploration of reincarnation. It is not a neglected subject.

In presenting this evidence, I do not interrupt with any justifying remarks or speculate on the many scientific and skeptical arguments that will be directed at this evidence. That is material for the next chapter. The only questioning of the possibility of other explanations for the evidence is Dr. Stevenson's own "devil's advocate" approach in analyzing each case.

Dr. Stevenson's Investigations

As mentioned above, the evidence presented here is from Dr. Stevenson's first book[16], which was published in 1966, and updated in 1974. Although he wrote several books later about additional investigations, only the information from this book is used here to show what was available to the scientific community as early as 1966. To the date of this writing, 2009, most scientists, especially neuroscientists, and neurophilosophers still deny the existence of the soul claiming there is ABSOLUTELY NO EVIDENCE of its existence. Some scientists have doubts about such denials but they too have not bothered to find, or if found, to disclose this information.

Evidence, as collected by Dr. Stevenson, is unique in that it was produced spontaneously by the subject, not by a third person's reporting of a planned test. The spontaneity of such evidence is crucial to its importance because it demonstrates the *desire* of the subject to be heard and recognized for who he or she had been! It is done regardless of anyone else's concern or prompted motivation. Evidence cannot be more credible than when it is generated spontaneously by the subject, WHO IS THE EVIDENCE. The subject makes the statements about the past life, statements that are subsequently verified by visits to the locations and recognition of claimed relatives, friends and, at times, chance meetings with local residents. *For this to happen, the memory of all these past places and people had to be transferred from the previous body to the new body by an entity of which memory is but a part.*

The 20 investigations described in the book were carried out in several countries: 7 in India, 3 in Ceylon, 2 in Brazil, 7 in Alaska, and 1 in Lebanon. Reincarnation is an accepted fact in the Eastern cultures because of its demonstrated reality. It has an effect on human behavior there because of the belief that the only way of avoiding reincarnation is to lead a morally good life. Such a belief would raise havoc in life's meaning in the Western cultures which are more concerned with accumulation of material wealth and attainment of personal accolades in the current earthly life than any mythical life after life. Perhaps that is the reason why any revelations of a returned existence are ignored or squelched with the result that reincarnation evidence in Western countries like the United States is far less prevalent than in the Eastern cultures. Another reason is the fear of being ridiculed for reporting paranormal events.

Although the investigations of some of the cases were performed some time after the occurrences, they were based on witnessed information which was verified by other sources. Searches of official records and newspaper accounts produced corroborative historical material. Some of the investigations, however, were the result of real-time observations by Dr. Stevenson and his assistants.

That reincarnation was the transfer "agent" of the appearance of birthmarks and birth defects on the reincarnated body from a previous body was of major interest and importance to Dr. Stevenson. As a medically oriented investigator, his goal was to relate the appearance of the birthmarks on the reincarnated body to the previous body with respect to the cause of the initial event. Obviously, how this happened was the ultimate medical goal. For instance, there was a birthmark across the throat of a boy in the new body in exactly the same location and shape as was the cause of the murder in the previous body. This similarity of birthmarks in other cases between the two bodies occurred frequently. Another example, this one of a birth defect, was the birth of a girl without a left leg. When she could talk, she claimed to be someone else who had been killed in a train accident in which *her left leg was dismembered.* When her story was investigated, it was found to be true in the details of the accident and the identities of the related people. There are no medical conclusions that can be drawn about how or why birthmarks and birth defects occur in reincarnation. Dr. Stevenson comments on the various possibilities in the book, *Where Reincarnation and Biology Intersect.*

Two representative cases are given here to demonstrate the scientific quality and detail of the reporting. As should become obvious from the length of these two examples, they must suffice to invite the reader to read Dr. Stevenson's books for more extensive coverage of the subject. The purpose here is to show what the evidence is and the extent of its coverage. The investigative reports in both cases, summarized here except for the Tables, have the same format as follows:

(A) Summary of the case and its investigation.

(B) Relevant facts of the geography and possible normal means of communication between the two families.

(C) Persons interviewed during the investigation.

(D) Statements and recognitions made by the subject.

(E) Relevant reports and observations of the behavior of the people concerned.

(F) Comments on the evidence of paranormal knowledge on the part of the subject.

(G) Comments on long term observations in this case.

(H) The later developments of the subject.

Note the amount of identified information in each report from the above listing of the coverage of all aspects of the case—from its beginning with all the analyses of related material to the follow-up of the long-term effects on the reincarnated person. As I have mentioned before, the scope of reporting is detailed to include the possibilities that there could be explanations other than that of a reincarnation. In other words, Dr. Stevenson assumes the role of a skeptic and comments on the possibilities, or lack of, of other explanations. This emphasizes his objective (scientific) approach in the investigations. The 2500 cases that were investigated in this manner are those with positive conclusions. Whenever an investigation demonstrated a lack of positive information, he terminated the investigation.

The reporting length of the cases as described by Dr. Stevenson, precludes my presenting them in total content. I summarize his descriptions and comments about each of the two cases but I show the entire table of evidential events and verifications as the gist of the investigated evidence with its witnesses, verifiers, and pertinent comments. This will allow the reader to judge whether the evidence merits the recognition and acceptance as scientifically valid reporting.

The Case of Ravi Shankar (India)

(A) On January 19, 1951, Ashok Kumar, familiarly called Munna, the six year old and only son of Jageshwar Prasad, a barber of the Chhipatti District of Kanauj, was enticed from his play by two neighbors and brutally murdered with a knife or razor. The motive for the crime seems to have been the wish to dispose of the heir so that one of the murderers, a relative, might inherit his property. One of the murderers was a barber, the other a washerman. They were seen walking off with the boy, but after the murder they were acquitted because there were no actual witnesses of the murder. A few years later, word reached Sri Jageshwar Prasad that a boy born in July 1951 in another district had described himself as the son of Jageshwar, a barber of Chhipatti District and had given details of "his" murder, naming the murderers, the place of the crime, and other circumstances of the life and death of Munna. The boy, named Ravi Shankar, was the son of Sri Babu Ram Gupta. According to his mother and older sister, when he was between two and three years old, he kept asking for the toys that he had in his previous life. When he was about six years old, he talked to his schoolteacher about the murder.

When Sri Jageshwar Prasad heard about the boy's statements, he visited the home of Sri Babu Ram Gupta to obtain more information. The latter was annoyed by this intrusion and fearing that his son might be taken away refused to give any information, especially since the boy had talked about "his" toys. Ravi Shankar's father continued to oppose discussions of the case and even beat the boy severely to make him stop talking about his previous life. The effects of these beatings were noted by the boy's schoolteacher.

However, Sri Jageshwar Prasad arranged to talk to Ravi Shankar's mother and to the boy, who after a while recognized his previous father and talked about his previous life as Munna and about the murder. This occurred when the boy was about four years old.

Ravi Shankar's mother testified that he had a linear mark across the neck resembling the scar of a long knife wound. She first noticed this mark when the boy was about three months old.

In 1956, Professor B. L. Atreya corresponded with Sri Jageshwar Prasad about the case and collected considerable testimony from some of the other witnesses, although he did not personally interview them. He made the documents available to Dr. Stevenson, who thought that personal interviews should be made. This was done in 1962 by Dr. Jamuna Prasad and two others and the documentation was made available to Dr. Stevenson. In 1964, Dr. Stevenson personally visited the site of the case and interviewed the witnesses. Sri Jageshwar Prasad, with whom Dr. Stevenson had corresponded earlier, was not available at this time.

(B) From the testimony of witnesses, the two families involved in this case had very little personal acquaintance prior to the attempts at verification of Ravi Shankar's statements. Their houses were located about a half a mile from each other with an irregular street path between them. Ravi Shankar's mother stated that Sri Jageshwar had never visited her before that verification visit. The Ravi Shankar family had heard of Munna's murder, as did most of the residents of the city and may have come to offer condolences. Sri Jageshwar had tried to bring the murderers to justice but it was impossible to get witnesses who did not want to get involved. Is it possible that Ravi Shankar could have obtained information about Munna from some of Munna's family members or acquaintances? Considering that he began making the pertinent statements before the age of three, it is unlikely that he was allowed away from the house unaccompanied at that age to wander to the other house a half mile away. Nor could he have acquired knowledge of Munna's toys since they had been put away by the parents after the murder.

(C) A list of Munna's family and neighbors.

A list of Ravi Shankar's family and neighbors

Schoolteacher and classmate.

Written depositions furnished by Professor B. L. Atreya

(D) In 1962, when Ravi Shankar was eleven years old, he had forgotten the events of his earlier life as Dr. Jamuna Prasad talked to him. He did say, however, that when he saw either of the two murderers he experienced fear but did not know why.

Table 3.1 consists of the Ravi Shankar Tabulation of events, verifying data with witnesses, and comments. According to witnesses, at least sixteen of the items occurred before any members of the two families had met, the rest after the first meeting. This Tabulation, pages 96-100, is taken from Dr. Stevenson's book, first published in 1966, to illustrate the information that was available regarding reincarnation. Note that the pronouncements by the reincarnated child were spontaneous and generated without any prompting, and in spite of the father's beatings. Note also the straightforward presentation of the report with comments about items and terms that may be locally clear but strange to a foreign evaluation.

Table 3.1 Statements Regarding the Case of Ravi Shankar

Tabulation

Summary of Statements and Recognitions Made by Ravi Shankar

Item	Informants	Verification	Comments
1. He was the son of Jageshwar and was killed by having his throat slit.	Maheswari, oldest sister of Ravi Shankar Raj Kumar Rathor neighbor of Ravi Shankar's family	Jageshwar Prasad, father of Munna Kishori Lal Verma, neighbor of Jageshwar Prasad Confession of Chaturi, alleged murderer as reported by Jageshwar Prasad	Sri Jageshwar Prasad had a son, Munna six years old who was murdered on January 10, 1951.
2. His father was a barber.	Raj Kumar Rathor	Jageshwar Prasad	
3. His father lived in the Chhipatti District of Kanaui of Kanaui	Maheswari Raj Kumar Rathor Uma Shankar, older brother of Ravi Shankar	Jageshwar Prasad	

4. His murderers were named Chaturi and Jawahar.	Maheswari	Jageshwar Prasad Confesions of Chaturi as reported by Jageshwar Prasad.	In addition to the confession by Chaturi, some pieces of shoes owned by Jawahar were found near the clothing and body of the boy.
5. They were a washerman and a barber.	Shriram Mishra, Ravi Shankar's school; teacher Raj Kumar Rathor	Jageshwar Prasad	The alleged murderers were a washerman (Chaturi) and a barber (Jawahar).
6. He had been eating guavas before he was murdered.	Maheswari	Mano Rama, mother of Munna.	Munna had taken some guavas just before he left the house to play and it was while he was playing that the murderers had induced him to accompany them.

7. He had been enticed by the Murderers with an invitation to play Geri.	Jageshwar Prasad Uma Shankar	Mano Rama	Sri Uma Shankar was a secondhand witness of this statement of Ravi Shankar. Geri is a game which Munna often played with Chaturi and Jawahar, so it is likely they would have invited him to play it in order to lead him away from the neighborhood. No one actually heard the alleged murderers invite Munna to play the game that day. This is probably correct, but not verified.
8. He was taken by the murder ers to the riverside.	Raj Kumar Rathor Kali Charan Tandon	Jageshwar Prasad Kishori Lal Verma	Munna's body and clothes were found near the river.

9. He was killed in an orchard.	Shriram Mishra	Swaroop Rajput, neighbor of Jageshwar Prasad	Probably not completely accurate, but the route from Munna's house to the site where the body was found traversed several orchards. The site where the body was found may not have been the place where the child was murdered, but presumably they would be in the same area.
10. He was murdered near Chintamini Temple.	Raj Kumar Rathor	Kishori Lal Verma	The head of the murdered boy was found about 250 yards from Chintamini Temple. It was thought that the murder was committed in this area.

11. The murderers cut his neck.	Raj Kumar Rathor Shriram Mishra Kali Charan Tandon	Asharfi Lal Rajput, neighbor of Jagesgwar Prasad. Kishori Lal Verma	The murdered child's head was found severed from his body. In his (retracted) confession, Chaturi had said they killed the boy with a razor.
12. The murderers buried him in the sand.	Raj Kumar Rathor	Kishori Lal Verma	At least part of the body was found buried.
13. He had a patti (wooden slate) at his house.	RamdulariRam Gupta, mother of Ravi Shankar Jageshwar Prasad	Mano Rama	According to Sri Jageshwar Prasad, Ravi Shankar also correctly stated that this slate was in the almirah (large closet) of their house. It is perhaps noteworthy that Ravi Shankar used to say that his slate and toys (see succeeding items) "had been kept." He seemed certain that they had been preserved so that he could

			have them again if only his parents would get them for him. In fact, his mother had carefully, almost reverentially, preserved many of Munna's belongings, including his toys.
14. He had a bag for his books at his home.	Raj Kumar Rathor	Mano Rama	The school bag of Munna has been preserved by his family and was shown to me in 1964.
15. He had an ink pot.	Ramdulari Ram Gupta	Mano Rama	
16. He had a toy pistol at his home.	Maheswari Ramdulari Ram Gupta Jageshwar Prasad	Jageshwar Prasad Mano Rama	Munna was particularly fond of toy pistols. Ravi Shankar had no toy pistol. Poor people cannot usually afford to purchase toys for their children, but as Sri Jageshwar Prasar had only

			one son, Munna, he could afford to buy him toys. The toy pistol had been kept and was shown to me in 1964.
17. He had a wooden elephant at his home.	Jageshwar Prasad	Jageshwar Prasad	The toy elephant of Munna had been kept and was shown to me in 1964.
18. He had a toy of Lord Krishna at his home.	Raj Kumar Rathor	Verified by me in 1964.	Munna's toy statuette of Lord Krishna had been preserved and was shown to me in1964.
19. He had a ball attached to an elastic string at his home.	Raj Kumar Rathor	Verified by me in 1964.	This toy of Munna's had also been preserved and was shown to me in 1964.
20. He had a watch in his home.	Raj Kumar Rathor	Mano Rama Jageshwar Prasad	Munna's watch had been kept and was shown to me in 1964.

21. He had a ring given to him by his father which was in his desk.	Raj Kumar Rathor Jageshwar Prasad	Jageshwar Prasad	Ravi Shankar told Sri Jageshwar Prasad "The ring which you got for me is in my desk. Have you not sold it?" Munna's father replied: "Your ring is safe. Would you recognize it?" To this, Ravi Shankar replied, "Yes." Srimato Mano Rama stated discrepantly that the ring was not in the desk at the time of Munna's death.
22. Recognition of Chaturi, al-Leged murderer of Munna.	Ramdulari Ram Gupta	Ramdulari Ram Gupta	Chaturi was the alleged murderer who confessed to the crime. He was unknown to the family of Ravi Shankar when the boy noticed him in a group of people at a religious ceremony. Ravi Shankar told his host's

		son that he would revenge himself on Chaturi. Srimati Babu Ram Gupta lived in purdah and so cannot not have known a man like Chaturi from outside her family and another district. When Ravi Shankar showed this reaction of fear on seeing Chaturi, his mother inquired as to the identity of the man her son pointed out and learned who he was.
23. Recognition of Sri Jagesh war Prasad.	Jageshwar Prasad	In a lecture to me of July 9, 1963, Sri Jageshwar Prasad described the recognition as follows: "I sat down at the door [of Ravi Shankar's house]. Ten or fifteen women asembled. The boy, whose name is now Ravi

Shankar was called. He stood at a distance of about one and a half feet and looked at me quietly. First of all, I addressed him. 'Dear boy, come here. What is your name? Do you know me?' I repeated these words twice or thrice, but he did not speak and became shy, as if he were going to weep. I again said, 'Oh, dear boy! Do not be afraid. Did you forget that you used to take money from me?' After twenty to twenty five minutes he drew close and sat in my lap. Then he said to me, 'Father, I used to read in Chhi-patti school and my wooden slate is in the almirah…' "

24. He had attended the pri mary school of Chhipatti District.	Kali Charan Tandon Jageshwar Prasad	Jageshwar Prasad	
25. Recognition of watch own ed by Munna.	Jageshwar Prasad	Jageshwar Prasad	Sri Jageshwar Prasad had put on Munna's wrist watch and was wearing it when he met Ravi Shankar. During their talk, Ravi said: "It is my watch." Munna's father had brought the watch for him from Bombay.
26. Recognition of Munna's maternal grandmother.	Jageshwar Prasad	Jageshwar Prasad	Sri Jageshwar Prasad was not himself present at this recognition. He wrote: "My mother-in-law went to some other person's house and a boy was sent to call him [Ravi Shankar]. He was chewing sugar cane. When he came the women asked

who had come there.
For a while he looked
downward and then said:
"Grandmother [mother's
mother] has come. She has
come from Kanpur' ."

Events regarding the case of Ravi Shankar

Dr. Stevenson mentions and discusses one discrepancy in a witness report that he has been unable to resolve. Two neighbor reports state that Ravi Shankar was taken by his father to the residence of Sri Jageshwar Prada. The discrepancies are that it is not clear that it was the father and it could also be mistaken with a visit to the grandmother's house. Why would the father do this? It was possible that in the absence of Sri Jageshwar Prada, the father may have wanted to verify the son's statements.

Description of the birthmark in 1964. Dr. Stevenson interviewed Ravi Shankar in 1964 and examined the birthmark on his neck. It was smaller than in earlier years but still resembled a scar of a knife wound. This is a medical appraisal of the appearance of the birthmark.

(E) The testimony of several witnesses justifies the conclusion that Ravi Shankar had fully identified himself with Munna. His mother testified to his extreme fear when he first saw the two murderers but by 1962 he still felt fear in seeing them but did not know why. By 1964, he no longer felt fear or anger and did not recognize one of the two.

Munna's mother became mentally ill after the loss of her son. In 1964, during Dr. Stevenson's interview, she showed signs of depression and agitation, and several times burst into tears at the mention of her son. As a further sign of her imbalance, according to one witness, a neighbor, she at times reproached her husband as the murderer of her son, an accusation that probably caused him much pain.

(F) In the present case, the verification came entirely from the family of the deceased Munna. The family of Ravi Shankar took no steps toward verification. Instead the father actively opposed such steps. Their resistance stemmed from their fear that their son would return to the home of his previous life. There was also the fear that there might be reprisals by the murderers after Ravi Shankar openly accused them of the murder. Such opposition makes it extremely unlikely that the case could have been worked up for fraudulent purposes.

With respect to the location of the houses of the two families, the possibility of communication by persons familiar with the case with Ravi Shanklar would have to consider the means of communication. To assume that it could be telepathic communication is unrealistic when considering the extent of specific detail that would have to be communicated.

In this case, as in others, the evaluation of birthmarks that reflect events or similarities in a previous life cannot be separated from the evaluation of informational and behavioral features. The story of the previous life cannot alone explain the birthmark, but when the two are so specifically related to the case as to suggest they were caused by experiences of the previous life, such cases become of major interest in the analyses suggestive of reincarnation.

(G) Not included in this case.

(H) In 1969, Dr. Stevenson met Ravi Shankar, who was then eighteen years old and in college. He had lost all the phobias of earlier life and had forgotten most of his memories of a past life but was often reminded of them in hearing people talk about them. He no longer wanted revenge against the murderers of Munna. In that same year he saw Munna's father and expressed pleasure at having met him.

In 1969, Dr. Stevenson also met with Sri Jageshwar Prasad but did not learn any additional important information. Dr. Stevenson has sent him a copy of the book that this material is taken from. Since Munna's father spoke no English, a translator read the report on Ravi Shankar and he responded that all the facts were correct. In fact, he wanted to use the report as evidence for reopening the case against the murderers but found that the courts would not accept the evidence. Munna's mother was still calling for the return of her child, but at other times wanted the entire matter forgotten.

Dr. Stevenson again met Ravi Shankar in 1971. In both visits, 1969 and 1971, he examined the birthmark resembling the knife cut. Since 1964, the location had changed. Whereas originally it had been on the neck just below the chin, now it was under the chin and near the point. It was still clearly visible as a distinct line of dark pigmentation about 3 millimeters wide.

The Case of Paulo Lorenz (Brazil)

(A) The case of Paulo Lorenz occurred in the same family: the personality reincarnating as Paulo was that of his deceased sister, Emilia. From an investigative standpoint, this case has a weakness due to the possibility of normal communication of information between the present personality and the older people who knew the previous personality. Despite this weakness, the case is presented because it illustrates: (a) a difference in sex between the two personalities, (b) a highly developed personation by the second personality of the first, and (c) the expression in the second personality of a talent for sewing which, although not unusual in itself, was in this family most highly and almost specifically developed by these two children and no other child in a family of thirteen children.

Emilia was the second child and eldest daughter of F. V. and Ida Lorenz. She was born on February 4, 1902. From all accounts, she was extremely unhappy all her short life. She felt constrained as a girl and a few years before her death, she told her brothers and sisters, but not her parents, that if there was such a thing as reincarnation, she would return as a man. She also said she would die single and never married although she had several proposals. Several times she tried to commit suicide unsuccessfully, but finally succeeded by taking cyanide. She died on October 12, 1921.

Paulo was born on February 3, 1923. Up to about the age of five, Paulo refused to wear boy's clothing and preferred to play with dolls and with other girls. He exhibited several traits that were Emilia's and made statements identifying himself with her. At about the age of five, he was given a pair of trousers that were made for him from one of the skirts formerly worn by Emilia. This seemed to appeal to him and he wore boy's clothing from then on and he began to shift his sexual orientation toward masculinity. However, a strong feminine orientation remained with him into the teens.

Some time after Emilia's death, her mother attended some meetings of amateur spiritualists at which she, as a communicator, received messages on three occasions from a spirit who claimed to be Emilia. Emilia said she wanted to come back as her mother's son. Ida Lorenz and her husband doubted the authenticity of the messages, especially the desire for the change in sex (remember, she never told this to the parents when still alive). Ida Lorenz did not expect to have any more children after having twelve, but she did have one more child, a boy she named Paulo.

(B) Not applicable in this case due to the in-family event.

(C) All informants were family members: Paulo, 6 older sisters, brother, and parents.

(D) Behavior and statements of Emilia and Paulo indicative of Paulo's identification with Emilia. In the tabulation, items of both Paulo and Emilia are listed together to show similarities and behavior. The informants all had personal knowledge of the items so the column for verification of witnesses has been omitted.

As in Table 3.1, pages 206 thru 209 listing the events and related data are presented here in their entirety. Noteworthy are the lengthy comments which help understand the family relationships and Brazilian culture.

Table 3.2 Statements Regarding the Case of Paulo Lorenz

Tabulation

Summary of Behavior and Statements of Emilia and Paulo Indicative of Paulo's Identification with Emilia

Item	Informants	Comments
1. Statements by Emilia before her death that she wanted to return as a man if she reincarnated.	Ema Bieszczad, older sister of Paulo Lola Moreira, older sister of Paulo Ana Arginiro, older sister of Paulo Ema Moreira, older foster sister of Paulo W. Lorenz, brother Paulo	Not told to the parents by the children who heard Emilia make these statements. Lola Moreira stated that the children did not have sufficient familiarity with the parents to tell them something of this kind. W. Lorenz recalled that as an adult he told his father about Emilia's distaste for being a woman and his father showed surprise, not having heard that before. Emilia was better at making men's and boys' clothing than at making feminine clothes.

2. Interest of Emilia and Paulo in traveling.	Ana Arginiro	Apparently one reason why Emilia wanted to be a man. As a woman in Brazil in the earlier twentieth century she could not travel easily. Paulo, according to W. Lorenz, with whom he lived in 1962, was particularly fond of travel and occupied his vacations with it.
3. Unusual competence of Emilia and Paulo in sewing.	Ema Moreira Ana Arginiro Lola Moreira W. Lorenz Florzinha Santos Menezes, older sister of Paulo Marta Lorenz Huber, sister of Paulo Rma Bieszczad	Emilia had exhibited great skill in sewing and owned the only sewing machine in the family. Several witnesses testified to the precocious competence of Paulo in sewing. Lola Moreira recalled that when Paulo was "extremely small" and a servant of the family was trying clumsily to work the sewing machine, he pushed her aside, showed her how to work the machine,

and made a small sack with it. W. Lorenz and Florzinha Menezes both recalled that once when Paulo was about four years old, she (Florzinha) was having difficulty in threading the machine and Paulo showed her how to do it. Marta Lorenz Huber and Lola Moreira recalled that once Marta left the sewing machine with some unfinished embroidery on it; in her absence Paulo finished the work she had left. All the above three episodes occurred before Paulo had any lessons in sewing. Ema Brieszczad rerecalled seeing Paulo working Emilia's sewing machine before he had any lessons. She stated that once when someone asked Paulo about

		how he could sew without lessons, he replied: "I knew already how to sew." Ema Moreira also recalled Paulo's ability to work the sewing machine when he was four and before he received any lessons. Ana Arginiro also recalled that Paulo could sew very well before he had instruction and resisted having instruction, saying he knew already. In addition to describing Paulo's talent for sewing, several informants mentioned his liking for it. He would frequently go to the sewing machine and work it by himself despite prohibitions from his elder sisters.
4. Unsuccessful attempts of Emilia and Paulo to play the violin.	Ema Bieszczad	Both Emilia and Paulo wanted to play the violin, tried to do so, but lacked competence.

5. Preference of both Emilia and Paulo for Lola among the brothers and sisters of the family.	Marta Lorenz Huber	Lola was the favorite sister of Emilia and also of Paulo, who expressed a wish to move out of W. Lorenz' house and live with Lola, who was a widow in 1962.
6. Weak interest in cooking on the part of Emilia and Paulo.	Marta Lorenz Huber	Marta stated: "He [Paulo] was not much interested in cooking and neither was Emilia."
7. The first words spoken by Paulo at the age of three and a half, on seeing another child put something in his mouth were, "Take care. Children should not put things in their mouths. It may be dangerous."	Ema Bieszczad	Paulo delayed speaking so long that doubts were entertained about his ability to do so or that he could hear. Some children (they are often younger children whose needs are met by others) do not speak until three or four years old and then begin in full sentences and Paulo seems to have given an example of this behavior. After two suicidal attempts with swallowing poison,

		the second successful, a surviving Emilia might have become cautious about putting things in her mouth.
8. Emilia and Paulo each had a habit of breaking off corners of new loaves of bread.	Ema Bieszczad	This habit seems to have been uniquely possessed by Emilia and Paulo in the family.
9. Refusal of Paulo to wear boy's clothing before the age of four or five.	Marta Lorenz Huber Lola Moreira Ema Moreira	Florzinha Menezes and Ana Arginiro recalled that Paulo liked women's clothes. They did not mention any actual refusal on his part to wear boys' clothes.
10. Statements by Paulo about being a girl.	Marta Lorenz Huber Ana Bieszczad Ema Moreira	To Marta, Paulo once said: "Am I not beautiful? I am going to walk like a girl." To Ema Bieszczad he used to say: "I am a girl." Ema Moreira also recalled that he said he was a girl.
11. Preference of Paulo for playing with girls.	Marta Lorrenz Huber Lola Moreira Ema Moreira	

12. Claim by Paulo that he had been in the house of Dona Elena; accurate description of the house of Dona Elena.	Ema Bieszczad	W. Lorenz said that Emilia had taken sewing lessons from Dona Elena.
13. Statement by Paulo that he had taken sewing lessons from Dona Elena.	Marta Lorenz Huber	Emilia had taken sewing lessons from Dona Elena. as quoted by W. Lorenz who did not himself hear this statement directly.
14. Dislike of Paulo for milk.	W. Lorenz Lola Moreira	Upon the occasion of her earlier unsuccessful suicidal attempt with arsenic, Emilia was forced to drink large quantities of milk. The phobia of Paulo for milk (his dislike was intense enough to justify this word) may have been related to this episode. W. Lorenz could not recall whether Emilia had such a dislike of milk during the interval between this unsuccessful suidal attempt and her later successful

one. He enquired of another older sister, Augusta Praxedes (born June 18, 1905, and not otherwise a witness for this case report) with regard to the occurrence of the milk phobia in Emilia. She recalled that Emilia had taken milk with pleasure as a small girl and had developed a milk phobia in adulthood. It is therefore not known exactly when Emilia developed her phobia of milk, but it seems reasonable to infer that it came on after the use of milk in the treatment of one (and possibly another) suicidal attempt. At any rate a phobia for milk was observed in Emilia in adult-hood and in Paulo at a very early age. W.. Lorenz stated that Paulo had disliked milk "all his life."

15. Recognition by Paulo of Emilia's sewing machine.	Marta Lorenz Huber Ema Moreira Lola Moreira	Upon the occasion of completing the embroidery of Marta Lorenz Huber which she had left unfinished on the sewing machine (item 3, Comments), Paulo said that the machine was his and he used to work it. On the occasion (see item 3, above) when Paulo pushed a servant aside in order to show her how to work the sewing machine, Lola Moreira asked Paulo: "How is it you know how to do this?" and Paulo rereplied: "This machine was mine and I have already sewed a lot with it." Ema Moreira also recalled that Paulo said that the sewing machine was his. He said: "This machine was mine. I am going to sew. The sewing machine had in fact belonged to Emilia.

16. Recognition by Paulo of Emilia's grave and concern for it on his part.	Marta Lorenz Huber Lola Moreira	Marta Lorenz Huber took Paulo to visit the cemetery. Instead of going around to see various graves, Paulo stood during the entire time of the visit on the grave of Emilia. He said: " I am looking after my tomb." Lola Moreira recalled that Paulo stood a long time on Emilia's grave. Once he took a flower from another grave, put it on Emilia's grave and smiled. Florzinha Menezes recalled that when she visited the cemetery, Paulo gave her flowers to put on Emilia's grave.
17. Recognition by Paulo of a dress that had belonged to Emilia.	Marta Lorenz Huber	Material from a discarded skirt of Emilia was made into trousers for Paulo. He recognized the material and said: "Who would have said that after using

this material in a skirt, I would later use it for trousers?" He was particularly fond of these trousers and preferred them to others. According to Lola Moreira, after having these trousers at age four or five he overcame his reluctance to wear boys' clothing.

Events Regarding the Case of Paulo Lorens

(E) Of specific importance was the sewing ability of Paulo. Emilia was very proficient in sewing. Because she had shown such ability, a sewing machine was bought for her, which she used constantly. After she died, the other sisters were taught to sew but none of them showed the skill of Emilia. In contrast, Paulo showed a skill in sewing before the age of five, and that without any instruction.

After he took a turn to a more masculine development, his interest in sewing waned. Some feminine traits remained with him. He never married nor showed any inclination to do so. He never had much to do with women except his sisters.

(F) Was it possible that Paulo could have been influenced by the family members into believing that he was Emilia? The children did not tell the parents about Emilia's statements of wanting to return as a boy. Conversely, the parents did not tell the children of the spiritual messages between Emilia and her mother. So there was no consistency of any effort to influence Paulo into a feminine orientation or into a conviction that he was Emilia. It was Paulo who clearly considered his life a continuation of the life of Emilia. The skill in sewing and the dislike for cooking in both personalities argue against such influence. There were also other similar interests and dislikes that associated the two personalities.

Even if there were a possibility of influence it could not explain the skill in sewing. There is a difference between having an interest in acquiring a skill, or an aptitude toward acquiring the skill and actually being able to use the skill without any instruction.

(G) Omitted in this case. Dr. Stevenson investigated this case in 1962 and did not see Paulo after that and did not contact the family members until 1972.

(H) In 1967, Dr. Stevenson received a letter from Paulo's brother saying that Paulo had committed suicide in 1966. In 1972, the doctor visited Paulo's brother and learned the details of Paulo's life. Paulo had spent time in the Brazilian Army and retired early due to ill health. He spent some years in convalescence. From 1951 on, he was employed by the Department of Highways. In later life he took part in some political activities. After the 1963 insurrection the military tightened their controls on political activity. Paulo was taken for questioning and was beaten during the interrogation. Afterwards he was in constant fear of further intimidation. Finally he committed suicide, presumably from this fear, although there may have been other motives.

Summary of the two cases

These two cases are examples of the type of reports that Dr. Stevenson prepared for each investigation. Due to the lengthy details of the reports, I have summarized the descriptive portions to indicate the type of material that Dr. Stevenson provides in addition to the main item, the tabulation of events and statements with the corroborative witnessing and verifications. The Comments add explanatory information pertinent to each event or statement to clarify what might otherwise be confusing about the reason for the item and its supplementary details. The reports indicate that the doctor, as a psychiatric medical professional, questions some of the reported items, but then goes on to show why the skeptical aspects do not pertain to the event or statement.

A key element in these reports is the repeatability of the activities that confirm the reality of reincarnation. Time after time it is the *spontaneous demand* by the reincarnated soul through actions or verbal comments to identify itself with the past life. This is repeatability as demanded by scientific standards for verification and acceptance. It is not repeatability controlled by a test controller; it is much more authentic because it is repeatability initiated by the "test item" itself. This is far in excess of the usual repeatability requirements, especially since it has also complied with those requirements in 2500 verified cases.

For anyone desiring to get all the details supporting the tabulations, as well as the other eighteen similar cases, I strongly recommend reading this 400 page book of the initial investigations by Dr. Stevenson that was first published in 1966. Even more important is the follow-up of other cases in Dr. Stevenson's other books. Before a skeptic attacks what I have presented here, it is mandatory that he/she read the complete reports of these and the other eighteen cases, as well as the later books by Dr. Stevenson, to realize that these are authentic and credible scientific reports by a medical professional. Knee-jerk reaction to paranormal phenomena is an indication of a slavish submission to an unsupported rejection of reality by semantic irrelevancy.

Reincarnation proves the reality of the soul. In the typical demand by the scientists and philosophers for physical evidence of the existence of the soul, it is equally demanding that they produce similar physical evidence of the non-existence. Not only is there no such evidence but what is most disturbing is that the unsupported denials are used as the basic assumptions for all the ongoing scientific work in support of the materialistic approach. This is against the basic requirements of scientific analyses: the major underlying assumptions must be stated and explained.

Dr. Stevenson's reports are far superior in the details and related comments than a sterile compendium of laboratory test data. In his reports can be detected a determination to cover all the circumstances surrounding each case that might be questioned by reviewers of the reports. In raising some of the questions himself, Dr. Stevenson provided the rational information that supported the answers. This was professional coverage with rebuttal details in anticipation of skeptical reviews rather than speculative interpretation. He presented the investigations as new scientific information for use wherever it could be applied, not because he believed it but because that was what he found. He did this in each case report.

In conclusion, it is mandatory to accept that Dr. Stevenson's work of forty years has produced the undeniable evidence that reincarnation is a factual occurrence. This compilation of irrefutable facts proves that the only possible means for that transfer of personalities to occur is through the existence of a soul with a memory of the past life.

A Skeptical Journalist's Evidence

The following evidence is not so much of detailed individual events and reincarnated personalities as it is of the confirmation of Dr. Stevenson's investigative methods through actual journalistic observation. Not only was the investigative procedure witnessed but the journalist, Tom Shroder, actually became a participant in the investigations by questioning the subjects and evaluating their behavior during the interviews. The strength of this evidence is that a skeptical journalist set out to convince himself that his skeptical views of the paranormal and of Dr. Stevenson's work would find flaws and contradictions in the investigations (as any skeptic would). Instead he became a believer. The entire venture, consisting of two trips of investigations, is described by the journalist in his book, *Old Souls*.[17]

The first trip of three weeks to Lebanon included interviews with the subjects, their families, and witnesses of specific events.

There was the case of a young man who had been killed in an auto accident. The reincarnated boy identified himself as the dead victim with details about the accident. Later, before a witness, he recognized and identified his past brother who had been the driver of the accident car.

A college student, who at the age of 23 still remembered the details of how she was killed by four armed men who stole the jewels that she had on her person. Tom Shroder questioned the student.

A woman, seeking treatment for a terminal illness in an American hospital, could not communicate with her daughter by phone just before she died. Her early actions as a reincarnated child were to pick up the phone and ask for her previous daughter by name. Later she talked of her past husband and other children. In Dr. Stevenson's investigation in 1997, accompanied by Tom Shroder, this young woman still remembered her past life, met with her previous family, and often called her past husband, who had remarried. Tom Schroder participated in the questioning of this subject and was impressed by the matter-of-fact answers.

A 11 year old boy who still remembered how he died of a shrapnel wound. Although alive when he was left for dead, he still remembered how his car was robbed. Tom Shroder questioned him.

The second trip was to India where many current cases were brought to Dr. Stevenson's attention since reincarnation is an accepted reality. His agenda forced him to spend little time with these. The following were among his planned interviews.

A seven year old girl remembered being killed by a car as she ran across a road. She recalled falling from a height. There were conflicting reports by witnesses: a confirmation of her being thrown in the air by the impact and another report saying she was dragged by the car. In such cases of conflicting information, Dr. Stevenson disregards it as inconclusive verification of the girl's recall. Other statements and recognitions, however, were confirmed.

A 5 year old girl remembered being her cousin. (Such in-family cases are not uncommon in India.) She remembered being burned to death while working at making bangles. Although in-family cases are suspect because of the availability of past information to the claiming reincarnated personality, Tom Shroder noted that it was the behavior and frankness in responding to the interview questions that established the credibility of the answers about the past life.

Part of the time was spent in tracing down confirmatory documents of past incidents like traffic accidents and death certificates for past cases. Shroder experienced the frustration and determination that accompanies such searches by Dr. Stevenson to verify the historic accuracy of his reports.

In summary, throughout the investigative interviews, Tom Schroder continually maintained his skeptical attitude, asking himself if there were any flaws in the procedures, any indications of fraud, or any doubts about the credibility of the information uncovered in the interviews. He was forced to admit that he could not detect any such faults. Although the specific information supplied by the subjects was revealing, it was the manner in which it was presented that convinced him that it did indeed support the only logical conclusion that reincarnation is real. It still left unanswered questions about how this could be possible but it did not dispel the fact that reincarnation did occur in those investigations he had witnessed and in which he participated. What most impressed him was the certainty with which the reincarnated children insisted that they were someone else.

These observed investigations are corroborative evidence, not so much of the actual details of each case, which would appear in Dr. Stevenson's reports, as the overall uninhibited disclosures by the subjects. As mentioned previously, this evidence is much more powerful because it is not a matter of interpreting the data of a test, as much as witnessing the actual generation of the evidential data *by the subject.*

3.4 Applicability of Reincarnation to the Dualism Debate

The evidence presented above conclusively proves through reincarnation that there is a transfer of personality from one human to another in two separate periods of time and two separate physical bodies. A skeptic could pose questions about the relativity of the evidence to dualism since there is no obvious connection between the physical medical findings and any neural investigations. The appearance of birthmarks and defects through reincarnation as a physical phenomenon was sufficiently intriguing as a research program without expansion into the dualistic effects of the soul. This was a theory oriented study of how often it happened and the how and why of the happening. In the process, however, the carrier of the information, the migrant spirit in reincarnation, was identified as a theory independent reality. Thus these two aspects of the Stevenson research are separate issues, each verified scientifically. Defining the duality effects relies on the demonstrated properties of the reincarnated person in the new personality. These properties of mind, memory, emotion, desire, and free will transcend the times and distances between the lives of the two personalities—they are the same in both, surviving the death of the previous body. Because they survive bodily death, they are immaterial properties.

The demonstrations of these immaterial properties are in direct conflict with the substituted materialistic explanations of physical causation or non-existence. Thus they are directly applicable to the research of the mind-brain problem and duality. To argue otherwise is to ignore the 2500 conclusively investigated cases of reincarnation.

The elaboration of how this evidence applies to the verification of dualism is given in the following chapter on arguments and philosophy. A combination of scientific, philosophic, and engineering logics is necessary to understand the complex union of the physical with the metaphysical.

Not only is the application of the reincarnation effects pertinent to the mind-brain problem, but it also applies to the duality of every individual, although in a different sense. The certainty of a continuing eternal afterlife should be included in one's understanding of the meaning of life. As I stated in my previous book, there are many goals in life: prosperity, personal relationships, professional and artistic achievements, acquisition of desirable possessions, etc. But these are goals "in" earthly life. With the knowledge that there is an afterlife due to the soul's eternal permanence, there is only one goal "of" life, that being the goal each individual sets depending on the belief of what that hereafter entails. This application of the soul's reality needs no scientific contribution because it is out of the scientific domain of conceptualizing.

There are questions about the effects of reincarnation on the dual human nature that only new research with philosophic participation will (or may never) be able to answer. Typical questions are:

Are all souls reincarnated after some residence in a spiritual location?

Is reincarnation a matter of personal choice or of assignment?

Is reincarnation repeatable for the same person?

Why are there only sporadic reports of reincarnated children?

Is there a reason for reincarnation like another opportunity to live a full life since so many of the investigated cases were the results of accidents, homicides, or war casualties?

Is reincarnation responsible for homosexuality as demonstrated in the second case of Emilia becoming Paulo?

Is reincarnation responsible for psychological or psychiatric cases of dual personalities?

Is reincarnation a substitute for the concept of purgatory? It is that in the Eastern cultures in which lives are lived morally to forestall the penalty of reincarnation.

These typical questions indicate that the research field will be very broad with the reality of reincarnation applied to the dualistic nature of the human being. It should create many alliances between the disciplines that so far have protected their specific areas of interest from invasion. The paranormal domain is limited in adequate answers since the laws governing its operation are not known. Only the results of the laws' operations are observed.

Unconfirmed Applicable Evidence

There is additional evidence that lacks confirmation because it deals with the non-physical phenomena as witnessed and conceptualized by third person interpretations. Dr. Melvin Morse, a practicing pediatrician, has written several books describing his investigations of children who have had NDE's and memories of past lives. I distinguish this evidence from Dr. Stevenson's investigative work in that Dr. Morse does not give the detailed scientific type documentation of the cases he discusses.

One of the consistent effects of the NDE on the children is their behavioral response to the experience. It affects their lives through tranquility about death and a desire to help others in need. In some cases they acquire unique skills and abilities, like extra sensory perception (ESP) and the ability to predict future events.

Although the evidence is first person generated, it cannot be used as physical evidence because it cannot be verified in its content, as Dr. Stevenson's cases were. For this reason criticisms on any grounds are impossible to refute. But the consistency of the reports given by children is the determining factor of their credibility. Children, at different ages and in different environments, have no reason to fabricate similar stories of the traumatic NDE experience except that they are real. The unusual happenings cannot be fabrications when the out-of-body descriptions agree with the actual observed scenes during the "absence".

Dr. Morse believes that there is a portion of the brain that is the communication link with the spiritual world. It is the right temporal lobe of the brain. Through this link, psychic powers and spiritual contacts are experienced. In his book, *Where God Lives*, he states that in 1997, neuroscientists from the University of San Diego claimed that they had found an area of the brain that was hardwired to be the communication link. He did not supply any verifiable evidence that supported this neuroscientific analysis. The acceptance by Dr. Morse, a medical professional, of this concept, which he uses in all his explanations, may be one of those cases that scientists claim will be substantiated in the future. For the purposes of this book, I mention it as an alternative to the Descartes version that the point of contact was the pineal gland, which also was not supported by any evidence. However, as noted in the University of Montreal tests described above, no evidence could be found of a "God module".

3.5 Key Conclusions

After a lead-in of the types of evidence that are valid and the conditions that affect the acceptance of evidence, the evidence for both monism and dualism was presented. The evidence for dualism based on the reality of reincarnation is more credible than the assumptive type for monism that lacks verification. The inability to explain how reincarnation occurs through unknown "spiritual" laws does not negate the fact that it does occur as physical reality. This contrasts with the lack of understanding of how conceptual solutions can be implemented through known natural laws.

The evidence for the soul's existence is conclusive. Confirmation of reincarnation as a paranormal reality has been demonstrated in a scientific manner through the spontaneous *first person* demand by the human soul for recognition as a personality in a previous life. The demonstration has also proved that memory is a property of the soul, a non-physical property, which refutes the neuroscientific and cognitive conceptual claims of a materialistic memory.

Physical evidence is the only valid evidence for confirmation or deconfirmation. There is evidence for both monism and dualism with varying degrees of credibility due to claims, demonstrations and experimentation. The evidences for monism and dualism were separated to allow a distinct evaluation of each.

The evidence for the materialistic concept only explains the physical attributes of the human body. Any conceptual explanations of how physical stimuli transform into mental states are inconclusive because they are based on assumptions with the proviso that they will be validated through some future successes in ongoing research. But this is **invalid** evidence until the hoped for substantiation occurs. In view of the dramatic evidence for the existence of the soul, this substantiation will never occur because it cannot. The only meaningful physical research which attempts to explain a non-physical property is the Libet timing data associated with consciousness. That evidence is not only in physical measured format but is also indicative of an unknown activity in the first 150 ms, which may be the link to understanding dualism.

The evidence for dualism is not only dramatic but also overwhelmingly conclusive. Were the evidence of reincarnation only proven in one case, that would have been sufficient to overturn the scientific denials of the soul's existence. However, the scientific investigative reports of 2500 positive cases by the same medical investigator and his associates magnify the confirmatory importance of the evidence to the undeniable level. The scientific community must accept this evidence in the same manner it accepts other scientific data. The investigations of the reincarnation effect are theory-dependent because their objective is to validate the occurrence of that act which can correlate the appearance of birthmarks with incidents in past lives. However, the definition of the soul as the "performer" in reincarnation is a theory independent finding because it becomes a defined entity of all human beings irrespective of reincarnation. The element that makes this possible is the continuing memory of the "traveling" entity. The existence of memory within that entity is the defining link between the two living events. Also, the ability of the reincarnated personality to control the new inexperienced brain to act in accordance with its desire to establish the interrelated connection between the two existences shows that the brain alone does not dictate the totality of human activity and behavior. It should be obvious that the mind needs the brain in the performance of human physical activity.

In addition to the confirming superiority of the reincarnation evidence of the soul's existence over the assumptive beliefs of the scientific community, there is other confirmatory evidence. Although not as impressive, the evidence of afterlife reality was obtained under scientific testing guidelines, further demonstrating that there is a non-physical entity, a person, who lives on after the death of the physical body. There is also the experiential evidence of out-of-body events, such as the near death experiences, that confirm through the many similar corroborative occurrences that since there is a separation of a non-physical entity from a physical body, there must be two individual parts to the human being. Finally, there is the historic, but credible, testimony of the soul's existence by the foreigner from the other spiritual universe, Jesus Christ, who had first person knowledge of that foreign locality and the nature of its inhabitants.

That these types of evidence can be obtained about a non-physical entity completely contradicts the naïve unsupported scientific claim that there is absolutely no evidence of the existence of a soul. The term soul must be used rather than an ambiguous term because it has been shown to possess a memory, which is a vital part of a mind. And science accepts the fact that there is a mental state in a human being although it may refuse to identify it as a separate entity. There is no semantic escape from these facts.

In contrast to this array of understandable and believable evidences, there is not one bit of verifiable evidence of the transformation of physical stimuli into non-physical mental activity, like desire, decision, free will and mostly consciousness. The concepts may be believable and understandable but they are not supported by any physical or non-physical evidence. The conceptual explanations are only assumptive claims, the kind that are denied to anyone making opposing claims. The criticism of belief as "folk psychology" is obviously also applicable to the scientific beliefs of physicalism.

The evidence for the soul's existence is undeniable according to the scientifically performed investigations. For the *immediate-response* skeptics, the only reasonable advice is, "Read the reports, realize, then criticize".

If the scientific community, which demands positive evidence in scientific research, admits that, given credible physical evidence of an immaterial soul, it may reverse its position of denial, it now has the opportunity to do so.

It is obvious from this overwhelming evidence in favor of the soul's existence that the worldview of the composition of the human being is due for major rethinking.

The evidence Descartes sought to justify his philosophy of dualism has been found!

PART II

The Closing Arguments

Having stated the Case and the Evidence in Part I, the opposing sides are given a chance to present their arguments. The arguments have been going on for centuries, in the Cartesian mode, and even longer in the broader philosophical domain. The principal relevant argument for the challenge made by this book is how the evidence of the soul's reality presented in the previous chapter negates the materialistically mandated conceptual mental solutions. The solutions attempt to substitute physical causes for non-physical mental functions. As the materialists have no factual evidence about any mind-brain relationships, their arguments are philosophic projections. Therefore, Part II must address the philosophic nature of the problem as well as the effects of the physical evidence. With a soul in the system the functional analysis must include the soul's demonstrated properties. Explaining the immaterial laws is a philosophic challenge. The impact of the soul's evidence on neuroscientific research demands rethinking of the basic analytical organization for future logical coherence. Metaphysics cannot be excluded at this point. Neuroscience and metaphysics must be partners.

To relate this new viewpoint to the current thinking, materialistic concepts and arguments are used in Part II augmented by comments about the soul's effects. Thus it is possible to envision the difference the new evidence can make in a before-and-after comparison. This necessary plunge into the direct conflict between the two ideologies paves the way for Part III, *The Deliberation and Verdict*, which will complete the challenge by defining a holistic approach and suggesting a course of action.

The Chapters that illustrate the conflicting nature of the debate are:

Chapter 4 Arguments and Philosophy
Chapter 5 Conflicts and Contradictions
Chapter 6 Memory
Chapter 7 Free Will
Chapter 8 Consciousness

4

Arguments and Philosophy

This is the debate about the Elephant,
its existence, nature, and behavior.

The importance of the disclosed evidence of the soul's reality cannot be overstated. Although the scientific evidence has been in published form for several decades, its surprising but straightforward disclosure should have disturbed many individuals and disciplines. Because it didn't (for whatever reasons) it is imperative that the arguments for its relevance to dualism be clearly stated for general understanding. The philosophical differences between metaphysics and science that result from this disclosure must be addressed to show the need for a closer alliance between the two. The effects on the meaning of life for all human beings must not be overlooked. This chapter, therefore, addresses these objectives and in a historic sense, concludes René Descartes' argument for the soul's reality.

In the philosophical discussions of the meaning of life, the soul was always accepted factually by the religious and, to some extent, the secular communities. In the past fifty years, however, this acceptance has been moderated by scientific research of the ability of the human being to apparently function in a completely physical mode without participation by an immaterial entity. With the development of techniques to test and study the human brain and nervous system (and of animals), impressive brain research has added credibility to this theory of physical causality. But in the investigations of the complexities of the neural functions, it became obvious that not all functions are of a physical nature. Something is missing. Since scientists can only understand the real world in terms of atoms, molecules, waves, and other particles, the neural research expanded to explaining the non-physical phenomena in physical terms. The specter of dualism reappeared to mar the positive explanations of physicalism. This was easily circumvented through science's ability to convince most of the skeptics through a simple denial of the existence of any soul stuff since there was no physical evidence of its reality. The soul is called a myth and ridiculed as solely a belief, pure folk psychology. It was and remains a scientific goal to disprove any otherworldly existence. This is obvious from the constant reminders of the mythical soul, in fear that a non-physical entity might invade the restricted domain of conceptual scientific thought.

In an amazing neglect of scientific procedures, research was, and is being, pursued without any analyses of the possible effects of an immaterial entity. The entire effort relies on the accuracy of the unsupported, unverified assumption of no "soul stuff". Not only is the soul to be ignored, but its effects, if any, have to be explained in physical terms. Physical causality is the objective of cognitive scientific research. Neuroscientists supported by neurophilosophers have embarked on this mission realizing that there were unknowns even about the physical aspects of the brain and the nervous system. Assumptions have to be made about the unknown properties with the proviso that these unknowns will *eventually* be explained. Evidently the extent of the analytical vacuum created by the research "dead ends" that had to be given "new life" has not penetrated the scientific minds to realize that there is a major flaw in the fundamental approach of physical causality. It is based on an incomplete human network. Consequently, it is crucial that in disclosing the evidence that the soul does really exist, the relevance of that evidence to scientific research must be made clear. The historic reversal of scientific theories is about to happen again. The scientific evidence of the soul's reality as described in the previous chapter is sufficient to show the error of scientific thinking. But when it is made clear, there is the incumbent obligation of the scientific community to acknowledge that relevance and its credibility.

Why is it, then, that cognitive science has not been aware of, or has intentionally ignored, the research done by Dr. Stevenson? Is it because the research was done in a discipline removed from that of neuroscience? But since there is a link through psychiatry and psychology between cognition and medicine, an interest should have existed in both disciplines in any work involving non-physical research. Another reason could have been that the linkage of the medical research with reincarnation, considered by skeptics as paranormal illusion, suffered the loss of recognition by scientific peers. The very concept of reincarnation is contrary to neuroscientific discipline. It is regrettable that such detailed scientific research has not been accorded the recognition it merits, not only for the credibility of reincarnation, but also for the knowledge of possible medical reasons for abnormalities other than birthmarks and deformities that may be due to reincarnation. The necessary condition of reincarnation within which the medical phenomena were verified, had the awesome secondary result of proving the existence of the soul, which had been sought for centuries. Is it possible that the inability, or the mere absence of interest, to relate the two has been the reason for missing the impact of the investigations on neuroscientific research?

For whatever reason this awareness did not occur, the result is that the question will most surely arise about how reincarnation evidence can be used to challenge and negate the neuroscientific claims of physical causality. Even if cognitive scientists were aware of the reincarnation evidence and disregarded it as paranormal phenomena unworthy of scientific involvement, such disregard was uncalled for when this paranormal evidence is physical evidence reported in a scientific manner, putting it within science's ability to evaluate it.

I am building a bridge to connect the soul's presence in reincarnation to the cognitive conceptual claims. But I cannot complete the bridge because there is no abutment of supporting evidence on the current scientific side. My side has the support of physical evidence. The admission by the scientists that their abutment is not in place but is being built and will eventually be there is not adequate because due to the soul's evidence it can never be completed. Why is this so? Simply stated, it is because the evidence shows that two, not just one, entities make up the human being and the physical side merely performs physical activity while the non-physical mental side controls the operation. How can that concise statement be illustrated simply without technical jargon or idealistic philosophy? The answer occurred to me as I watched a TV program on (what else but) the brain.

The brain's functioning was being explained in the only manner that it could be, that is, as a strictly physical activity with the unknowns glossed over as temporary gaps that were being researched. The brain was the total independent element for human behavior. In other words that brain was ME described in a third person observation. The subject of neuroplasticity and its effects was introduced describing its possibilities for overcoming some of the brain's inability to cope with psycho-physiologic conditions. *Through neuroplasticity I can change how the ME brain might overcome these deficiencies.* Did you catch the gross contradiction between the use of the capitalized words, ME and I? If you did, you understand dualism. How can I change an independent brain that is ME unless I am also a ME entity that can control the other entity? Figure 4.1 illustrates the lack of clarity in the TV explanation.

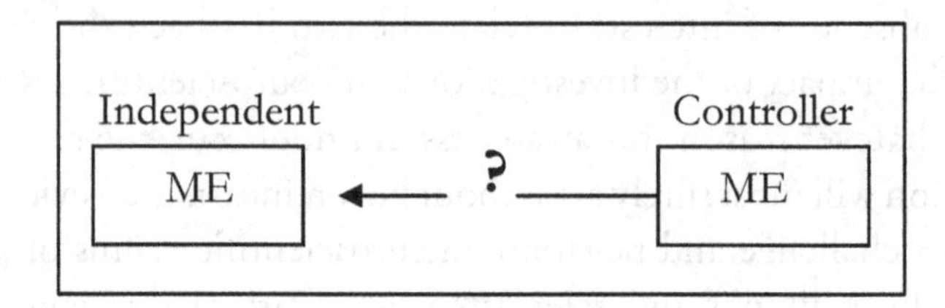

Figure 4.1 Which is ME?

To clarify the above statements, I will elaborate. As the brain was explained, if there is no other entity than the brain, then that is ME with all activity controlled by this brain. This is what neuroscience claims since a mind does not exist; the brain creates the mental states. This is a third person description of ME. But the next statement saying that I can modify the brain admits that I as first person controller can alter that independent brain because I am something else than the ME brain. For something to be changed it must be done by an external cause. ***There are two entities involved!*** The conclusion is that I do it to my brain, but who am I to override the brain that is supposed to be in control? This is dualism featuring a receptive brain (because it can be changed) and a controlling mind that can do the changing.

The mindset of the scientific community is that it can *explain* the human being as an entity that is driven solely by physical causes without realizing that it is demonstrating dualism subconsciously. By using the creative thinking process that does not depend on physical causes they develop the explanations for arguing away the very mental agent they are using. If the explanations of physical causality were valid, what physical stimuli would develop the concept of a brain that replaces a mind? That is philosophic circularity as demonstrated by the following set of questions and answers about the mind which are the same as the circularity concerning the existence of the soul.

Statement: You say the mind (soul) does not exist.

Q. How do you know the mind does not exist?

A. There is no physical evidence of a mind.

Q. How do you know there is no physical evidence of a mind?

A. Because a mind is not a physical entity, therefore, it cannot be proven physically.

Q. Can you support your statement in any way?

A. I don't have to because it is obvious.

That is the simple scientific rationale for the non-existence of the soul and with it the mind. The simplicity is replaced by complexity with only one more question.

Q. What will you do if there is physical evidence of a mind (soul)?

A. ?

What is the rationale for a mindless brain function that depends on external stimuli for activation? A current version proposes that the brain receives representations of the external world through the senses. It extracts the information from these representations for processing the motor signals needed for activity and behavior. What representation does the brain use for the activities of a mind? There are no representations of a mind because it is not a physical element, therefore there is no mind representation to process. Consequently, the conclusion is that there can be no mind. But a physical brain cannot create rational thoughts independent of physical stimuli. Such a cognitive omission is one of the gaps in a series of dots that needed connecting for a meaningful explanation of a total system. This I have done using the analytical approach with emphasis on logic. But why wasn't this done with scientific or philosophic logic since both disciplines are based on logic? The reason is that a different logic was needed: engineering logic. Philosophic logic is used to convince that an explanation is reasonable through rhetoric or semantics. Scientific logic is used to verify a theory or concept through the application of physical natural laws. But when a problem has to be solved, it is engineering logic that seeks the unknown parts, material and innovative connective ideas, which will make the integrated unit perform according to the required specifications. In this sense, knowing the specifications for human performance, it is engineering logic that can define a human system based on the knowledge that there must be two identifiable parts of the system, the physical body for physical activity and the spiritual soul for rational and emotional behavior. Defining how the system works is now the problem seeking a solution

The following points based on the scientific investigations of reincarnation can be made:

- Reincarnation as an event has been proven a reality through scientific research and verification.
- For reincarnation to occur there must be a separation of an immaterial entity (soul) from the physical entity (body) at earthly death. The ability to separate establishes the dual nature of the human being before and after reincarnation.
- The soul leaves the physical body at the termination of life in the body. It isn't clear whether the soul's departure causes death or the reverse.
- The soul continues to live in the spiritual world until it enters a new body.
- The soul retains its previous cognitive and volitional properties during migration because they are still identified as the same after migration.

- The soul reveals itself in the new body by demanding to be recognized as the previous personality. This *physically* expressed presence and associated activities establish the soul's reality.

- Reincarnation, through its reality, defines a migrating spirit, the soul. The aoul, through its continuing life, makes reincarnation possible.

- Reincarnation, therefore, refutes science's denial of the soul's existence. The results are the negation of the conceptual materialistic mind-brain solutions and of the complete concept of monism.

- The demonstrated presence of the reincarnated soul in the new body *with verified experiential memory* of a previous life confirms that a soul's personhood remains intact in the acquisition of a new personality.

- The demonstrated ability of the soul to remember the experiences of the previous life proves the location of memory is in the migrating soul.

- The demonstrated ability to understand and respond to questioning about the previous life proves the soul's ability to think rationally about the previous life. The new brain cannot do this not having participated in that life.

- The demonstrated ability to desire to retain the previous personality and to return to that locality proves the existence and migration of "desire".

- The demonstrated ability to choose activities related to the previous experiences in spite of parental objections proves the existence and possession of free will by the soul.

- The reincarnated soul's control of the new brain is demonstrated by its reference to the past life and initiation of activities related to that past life. The new brain does not and cannot resist this control and cannot contribute to the mental content, having had no connection with that life.
- The reality of the soul should be the base for rethinking dualism.
- Analyses of the soul's confirmed independence and union with the physical body require revised philosophies, metaphysical and neuroscicentific, based on fact rather than idealistic optimism.
- **A new program philosophy must be defined for the areas of influence by science and philosophy before any revised research is initiated.**

All the above dots are a logical extension of the meaning of reincarnation if pursued in an analytical process sans a biased mental obstacle. There is some duplication in the statements due to slightly varying interactions, but the path is a clear one leading to the confirmation of dualism and the need for a new research effort.

In addition to the effects the soul will have on the reality of dualism, there is an even broader effect, the effect on the worldview of the meaning of life with the knowledge of a confirmed soul. The religious community is comfortable with the idea of a soul. In general, however, the secular community is not because it involves acceptance of responsibility for transgression of moral values, if a life after earthly life is a reality. Introducing an afterlife into the secular mindset will cause much debate about this claimed reality, and the criticisms will probably be of the paranormal type disregarding the scientific evidence. Acceptance of this dual union and its effects is inevitable.

In connecting the logical relationships between the issues of the above dots, a complex allocation of physical and metaphysical responsibilities, which takes precedence, argument or philosophy, since logic plays a role in both? Can the basic arguments benefit from philosophic justification, or is it the reverse with philosophy setting the stage for a logically confirming fill-in?

How does one present an argument against a concept or theory that is based on assumptions? The concept can be shown to be technically or intuitively wrong, but the net effect is that as a tool to reach understanding, it leaves a lingering impression of a possible solution. In some arguments, after constant usage the concept assumes a credibility it lacked in the initial explanation. Countering conceptual arguments consists of *repeated reminders* that they are only conceptual and noting the assumptions, if possible, upon which they are based.

For philosophy, the challenge is to expand on the basic issue of *being*. The new soul's evidence changes the explanation from pure ideology to a factually supported conclusion. This may be possible in metaphysics but not in scientific philosophy which will have to deal with justification of non-physical reality and *being* without knowing the non-physical laws that determine the soul's behavior.

Argument, however, is challenged to establish the validity of the evidence of a soul to dualism, a connection that is not obvious merely from the presence of the soul. What are the relationships of the soul's properties to those of the physical brain and body? It is this need for the definition of the *what* and *how*, versus the *why*, that in my opinion justifies the presentation of the arguments first. As a secondary reason, this order postpones the use of the troublesome *if's* and *assume's* of philosophic semantics without tainting the analytical progress with doubt.

4.1 ARGUMENT

What does the soul's evidence prove by its nature and what does it confirm by its existence? What is the meaning of the evidence and what are its effects? The conclusive nature of the evidence, such as the spontaneous first person cry for acknowledgement, was covered in detail earlier and will not be repeated here. It is valid evidence on its own merits. The present objective is to construct the bridge over the gap between the reincarnation event, with the resulting certainty of the soul, and the argument for dualism. The soul's existence puts to rest those questions of life after earthly life and justifies the acceptance of the soul's control of the body. This is the evidence Descartes needed to justify his philosophy.

The bridging of the gap between reincarnation and dualism has some inescapable and devastating consequences. The evidence of the soul's reality pulls the rug from under the basic scientific assumption upon which science is building its case for physical causality as the solution to the body-soul and mind-brain problems. But this is science's own fault in closing the door so completely by not analyzing the possibility of a soul's participation in a holistic system. The following arguments are intended not only to point to the conflicts between monistic and dualistic versions of the human being but also to point to new areas of research that may lead to a better understanding of the new complexity. Consequences and effects are included wherever appropriate.

1 Reincarnation is a demonstrated physical/non-physical phenomenon as proven by scientific investigation and acknowledged by Buddhists and the Eastern cultures that experience it. It is not an illusion or paranormal oddity as skeptics try to dispute it using various reasons that themselves are definitely unproven practices, such as ESP, clairvoyance, and mind reading. Reincarnation lacks scientific approval because prior to Dr. Stevenson's extensive medical research and meticulous reporting, accounts of such phenomena were fragmentary, unscientific, and often merely hearsay evidence, hardly worth the effort of scientific verification. Furthermore, reincarnation is contrary to neuroscientific discipline precluding any voluntary acceptance of its possibilities. But Dr. Stevenson's contribution was valid research with a medical objective that was verified. The positive correlation of birth marks and defects in the newborn bodies to incidents in past lives was established through the verified descriptions of the reincarnated victims.. This transmission of physical defects by spiritual entities exposed reincarnation as the process of transfer. The appearance of a birthmark in the precise location on the new body corresponding to its cause in the previous life also links the same personality to the two lives. There are no explanations for this phenomenon, but it does happen and it confirms the linkage between the two bodies of the single occupant.

The incidents of the past life affect the behavior of the reincarnated personality, some of them causing fear of past relationships and even the instruments or objects that were involved in the end of that life. These factors are direct links with the past that can only be introduced into the new life by the person from the past life. This information cannot be communicated to the reincarnated child by sources in the past life because such sources are not aware of the newly identified person or of the location of the new union.

The number of confirmed cases demonstrate repeatability, a scientific requirement for verification of test or theory. Had only one case been verified, it should have been sufficient to prove the phenomenon. Although the research results were disclosed to the scientific community in 1966, scientists either ignore these researched reports or intentionally refuse to acknowledge them. This is shown by the following quotation from the book *Brain Wise* by neurophilosopher Patricia Churchland which was published in **2002!** After describing a case of reincarnation and calling it a fable, she continues:

> This fable illustrates the selectivity in considering evidence, and it is something to which we are all prone. Consequently, we have to work hard to be as tough-minded with respect to hypotheses we hope are true as we are with respect to those we fear are true. Whether all accounts of reincarnation share the weaknesses illustrated in the fable <u>is not known</u>, but because so many that have been studied <u>do</u>, and because one does not want to be gullible, we need to exercise careful scrutiny, <u>case by individual case</u>.[18] (underlining by author, not in original)

"Selectivity in considering evidence" is certainly obvious here. Note that only the weaknesses are stressed although there is uncertainty about the other accounts. Does the uncertainty imply that some were found to be true? If so, why isn't a "tough-minded" statement made that some were confirmed. *Such an admission would be fatal to the neuroscientific discipline because even one confirmation is positive proof.* The 2500 confirmed cases of Dr. Stevenson and his associates, a major contribution of consistent "scrutiny, case by individual case" isn't even mentioned. Such a major research effort certainly should have received the attention of anyone searching the literature for evidence rather than fables.

"Careful scrutiny" was exercised "case by individual case" in the 20 cases first published in 1966 by Dr. Stevenson with details of the scientific verification. Such evidence (as included in Chapter 3) was sufficient at the time to call attention to the work that was uncovering facts about reincarnation. The results were also published in a scientific journal over forty years before the publication of the book from which the above quotation is taken. The other 2500 cases were verified in a similar scientific manner by the year 2000! The evidence was available for anyone who *wanted* to find it.

With reincarnation revealing the identity of a soul through a scientific pursuit, why was this fact not recognized by the neuro and cognitive scientists and philosophers for its effect on materialistic monism? Was it ignorance or arrogance? It could have been the lack of interest or communication between the disciplines but with thousands of scientists involved, this was one of the best kept secrets in the scientific community. Regardless of the history of scientific inter-discipline secrecy, the argument for the reality of reincarnation based on the scientific verification is conclusive.

Science is incontrovertibly committed to factual reality as proven through scientific investigation. Since that is the procedure for proving *its* theories and accepting the reports of the investigations as verification, it is bound to accept all reports of scientific research in other disciplines if prepared with the same professional methodology. The investigations performed by Dr. Stevenson and his associates in this scientific manner prove that reincarnation is a reality not a paranormal myth or illusion. The verification of so many cases more than fulfills the requirement for repeatability not only by each subject but also in the number of cases.

In the 1920's, a philosophy called logical positivism was introduced by the European Vienna Circle that addressed verification of statements. It restricted the truth or falsehood of a statement to the degree to which it could be verified by empirical observation or the form of the words, that is, through scientific study, logic, or mathematical analyses. Ethical, metaphysical, religious, and aesthetic statements were meaningless because they could not be verified empirically. They were neither valid nor invalid. By the 1950's the philosophy had lost its appeal due to its many detailed errors. I include this bit of philosophic history because Dr. Stevensons's scientific study did precisely what the discarded philosophy stated could not be done: he verified the truth of reincarnation, a metaphysical phenomenon, through empirical observation. The exposure of metaphysical properties through such observation is an example of the potential effects of this method on existing concepts, practices and experimental planning.

Interesting questions are raised through the certainty of reincarnation. Due to this certainty, are all souls reincarnated? Why don't all children have memories of past lives? Perhaps they do but they don't become obvious because they are not encouraged to disclose them and forget them with age. How often do we see and hear young children talking to their unseen friends and pass that off as childhood fantasy? With a distant past relationship or no interest in returning to any locality, near or far, there may be no reasons for referring to a past life. But there are cases of recognition of specific locations and identification of equipments at these locations without any prior visits there. The apathy, if not downright disbelief, by the Western culture to reincarnation precludes any major effort at pursuing these intriguing cases of past life memories. But psychiatrists find them in troubled patients.

2. There is life after earthly death. For reincarnation to occur there must be a survival of an entity from a previous body to a new body, with memorable continuity of the experiences involving activities, acquaintances, and locations from the previous life. This between-lives existence is a non-physical state because there is no physical matter retained after death. Spiritual life must be an eternal life with no spiritual death as believed in Eastern and Western religions and as demonstrated by some reincarnation cases in which the subjects exhibited retentions from more than one migration.

The Afterlife Experiments by Dr. Gary Schwartz at the University of Arizona argue for a life after earthly life. The confirming facts in the experiments about an afterlife were the accuracies in the communicated information. As reincarnation evidence has shown, information is retained by the soul, not left in the dead body. But the communications consisted of not only remembered information but also of real time intentional messages like apologies for past behavior or questions about current happenings. The latter type of information indicates a continued desire to converse rather than merely to relate to past memories. Such conversation demonstrates rationality in progress, rather than just storage of information in a memory, indicating that other properties are dynamically present. The ability to respond to a medium's specific call indicates retention of behavioral control on a spiritual level; free will to agree to communicate; and rationality to comprehend and respond in a personal manner. Only through results can these arguments be made; there is no explanation for why such communications are possible.

The realization that part of the human being survives earthly death poses a problem for the scientist in forcing an admission of an error in claiming afterlife to be a myth and to come to grips with the dualistic implications. For all humans, the redefinition of life as a continuing process after death forces a reevaluation of the meaning of the earthly sojourn. For the religious believer, the argument is a confirming consolation. However, for the atheist who denies the idea of a hereafter, reincarnation and the Arizona experiments pose a warning signal that his entire philosophy is on shaky grounds because a soul and a hereafter lead to a theistic conclusion.

3. The non-physical entity, the soul, does exist. What type of physical evidence of a non-physical entity would satisfy the materialists that the soul exists? What would force them to renounce their persistent denials of soul existence? The demand for the most impossible evidence would be for the soul to appear and identify itself. As impossible as this demand would appear, it ***has*** manifested itself in reincarnation. If the evidence were based only on third person observations, it could be subject to skeptical disbelief and criticism. But it is *unprecedented in having its occurrence verified as evidence itself,* especially in not being the result of planned tests, which is the normal scientific method for verifying theories, physical properties, and performance. The term test is not applicable because the soul's demonstration of its own existence is a SPONTANEOUS FIRST PERSON DEMAND for recognition as a previous personality. It is not the result of any prearranged situation. **The soul proclaims its own existence! No test item ever produced**

evidence of itself spontaneously in a rational manner. It is the same as any physical human claiming own existence vocally. The impossible situation that science mistakenly assumed could not be produced, the appearance of physical evidence of a non-physical entity, *has been demonstrated and verified*!!!

The other scientific requirement of test repeatability is improved on because the recurrence of the soul's reality happens under many different conditions, always with the same confirming result, the spontaneous and rational plea for acceptance, which never happens in planned tests. The spontaneity occurs repeatedly in the same subject, and in like manner in all cases of reincarnation. Spontaneity is the soul's continuing insistence on being recognized. Even in cultures that accept reincarnations as routine occurrences and plan on them, some souls are denied this recognition for fear of the social consequences in conflicts within families and with associates of the previous life.

Because the soul's identification with a past personality and relationships could be, and was, investigated and corroborated, there can be no doubt regarding its authenticity. This association and revelation by the reincarnated soul happens mostly in childhood when the memories are most distinct, for they fade with time as memories normally do.

4. The soul with a mind is a separate entity from the body. The reality of reincarnation proves the separability of body and soul. Even if the soul's existence is acknowledged, there are skeptics who will argue that it is part of the body, an integrated element, but not separable from the body. Assigning the mental state to the soul but having it responsive only to the brain functions would still leave the soul concept alive, but it would have the soul die with the body. This conceptual and semantic ploy has no connection with the reincarnation evidence. It shows how conceptual innovation can distort factual information to advance a preconceived objective.

A concept of a physical brain having to create a non physical mental state because a mind does not exist due to the absence of a soul is not only illogical but also unscientific. It is illogical because the concept admits the need for a function that a mind would enable and yet excludes it because it is committed to ideological materialism.

It is unscientific because, while science admits the lack of supporting evidence for the materialistic approach, it continues and expands the misleading research. This is contrary to the scientific standards for verification before expanding a concept. It is incredible that such explanations by scientific personnel without credible evidence can be accepted by the scientific community, especially in expanding research to validate such illusory and unverified concepts.

Reincarnation refutes such practices because it supplies demonstrated evidence that both, a soul and its mind, exist and should be taken into account to fill the void of unexplained concepts. The soul migrates with all its properties from a dead body to a new body. One of its elements is the mind that the scientists are trying to explain physically or eliminate. Had the brain in the old body been responsible for the mental actions defined as mind, the reincarnated soul could not function in the new body because its functioning element was still in the dead body. The brain is the tool of the soul through the mind. Obviously the mature mind of the old body acts as a mature mind in the new body assisting in the development of the inexperienced new brain. In reincarnation, the mind with its memories of past experiences expresses the desire to associate with the environment of the past life, or discusses it, a recollection that the new brain is unaware of. Thus the new brain is incapable of directing the mature mind in its quest for recognition.

Since the soul is a reality, what is it? Is it a spirit with components and properties like a body with its components, or is it a unit with no distinctive parts? If it is such an integrated unit then the soul, mind, and memory are one. Since its own functioning and integration with the body are not known, it is impossible to segregate the elements or the functions. As an example, is the memory a separate entity and how are memories stored? What is the interface with the body? Since it is impossible to have two thoughts at the same time, is thinking a prime function of the mind, although it can be accompanied by emotion, or be channeled into making a decision? These are examples of the research projects that the inclusion of a soul, and mind, should initiate.

If memory is part of the mind, as was shown above, then the mind and the soul are one entity that does not change after earthly death. There are no other distinct parts, only activity potential, like free will, and conditional response, as with pain and emotion. Consciousness is a property of the soul but its influence on the body and other integrated activities is not clear. All these are missing in a dead body. This is a radically different approach from the assumed physicalism of scientists. As it is based on reality rather than optimistic assumptions, but lacking in specifics, a tentative conclusion is that the soul and the mind are one. There could be differences such as the soul being life and the mind being the functioning part of the human person that interfaces with the brain.

5. Memory resides in the soul not in the body. This is the bridge that allows crossing the gap between the evidence of reincarnation and dualism. The neuroscientific explanation of memory is incredibly naïve in stating that non-physical information (experiential and learning) is stored in a physical memory because it has been transformed into physical bits of information by neurons that have no intelligence to interpret such information. What is the process that translates intelligent thought into storable bits of physical data? What is the physical/non-physical interface? Assuming that such unexplained functions exist is symbolic of the liberties scientists take in their conceptual explanations to appear knowledgeable; and do it without supporting evidence. The memory of events, experiences, and associated activities is not stored in weighted configurations of the physical synapses. If it were, this physical storage would still be resident in the dead body. However, the memory is alive and exhibiting itself by the recollections of the past life. How could the soul be recalling that information from a decomposed brain when it is resident in another body with another brain? For the memory recalling the past life has survived as part of the soul, not as part of the body. It is alive and integrated with the new body as it was in the old. **This completely negates the neuroscientific claim that the memory of experience, learning, and skills is a physical attribute of the nervous system, storing information in synaptic configurations. This contradiction, also applicable to free will and emotion, is a major contribution to the refutation of the entire concept of materialistic causality. The devastating negation, along with the refutation of the scientific denial of the soul's existence challenges the neuroscientific community to rethink dualism and reevaluate the effects of the soul on the brain's physical systems and human behavior!**

The above comments apply to the non-physical experiential information. The storage of information in the synaptic configurations as claimed by scientific conceptual explanation may still be valid for physical instructions to control the automatic functions like breathing, heart action, digestion, and sensory operation. These functions require no conscious effort by the mind so they must rely on the stored operating instructions in a physical body component. The obvious solution to this duality is two memories: one for the physical body and one for the non-physical soul. I propose this unique but logical explanation of two memories as a key contribution to the rethinking of human nature with a soul.

6. Free will is a function of the mind not the brain. Free will can only be free if it is unconstrained by external causes. In the materialistic concepts there is no free will because there is no mind and the brain is like a machine with deterministic behavior having no capacity for making decisions. With the evidence of a non-physical mind, these materialistic concepts are baseless. Even when there are external pressures attempting to force a decision, the mind can still refuse to be controlled by them. This override ability refutes the claims that the brain is motivated and controlled by external stimuli in a deterministic manner. A decision to comply with the external demand is still a willed decision.

Volition is simply the mental activity to accept or reject an order or request, or to initiate a decision for a specific activity. It does not include the subsequent chain of events to perform the activity. If I am offered a hand for a handshake, I can accept or refuse as an act of free will. If I accept, I will my arm to move out with extended hand but I do not consciously generate the motor commands to move my arm. I direct how far to reach out, but how are the motor commands generated for the extension? The procedure for extension may be stored in the physical memory, relieving the mind's memory to concentrate on the purpose and follow-on of the handshake. This is another example of an interface problem created by the inclusion of the soul in the human network and seeking future resolution.

Free will was demonstrated as a property (not an element) of the mind in the reincarnation evidence. If it were part of the brain's deterministic subjection, as the materialistic concepts claim, what external stimulus would cause the brain to demand reconciliation with the previous life in which it did not participate? The determination and demand to return to that previous situation continued in spite of parental objections, which demonstrated willed and deliberate acts.

7. Learning is a function of the mind not the brain. Learning requires storage of information in memory. As argued above, memory is part of the mind for non-physical information. Learning skills and language is, therefore, a mental operation with a mind-brain interaction, which is impossible to explain without knowledge of the involved interface(s). Converting skill knowledge into physical manipulation requires communication between mind and brain for the generation of the accurate motor signals. Is this a case where a second physical memory can contain the physical movement memory after an initial storage with reinforcement through continued usage? An example would be a musician playing a memorized musical composition which has been practiced at great length to overcome the technical difficulties. The physical movements of the hands and fingers take on an automatic quality while the mind concentrates on the emotional input to the performance. The situation is even more complicated in speaking when the mind is developing the thought train as the vocal cords respond almost automatically, except in cases of dubious pronunciation.

In the Lorenzo case of reincarnation, the retention of the information by the mind's memory was dramatically demonstrated. The retention of sewing skills by the mind is not as difficult to accept as the translation of this retained information into motor commands by the inexperienced brain to perform the physical manipulation of the sewing machine. It is even more difficult to explain how a child with no lingual instruction could carry on a conversation coherently in a foreign language with acquaintances from the previous life. In another case, a reincarnated girl danced a foreign dance and sang in the foreign tongue without instruction or even understanding the meaning of the words she was singing. The memory and the ability to perform in like manner remained with her into her teen years. (These two cases are described in Dr. Stevenson's book.)

The argument here is for the learning to be part of the mind as mental information, including the ability of the mind to direct the translation of stored information into physical activity. How this can occur is definitely a challenging research project for the rest of the century.

8. Consciousness defies definition. Consciousness is not a "thing" but a condition. As such it affects the physical and mental states. It cannot be argued against because it is experienced. Is it physical through brain initiation or is it a non-physical quality due to the presence of the soul? Or is it the soul itself? It is retained by the soul in reincarnation but the body loses it when it separates from the soul at earthly death. With the introduction of the soul into the debate, this argument, which was metaphysical, now becomes a joint effort with both metaphysics and science participating. In Chapter 10, I return to this subject as an issue in systems analysis. Thinking is a conscious activity of the mind not the brain. A thought is purely non-physical and cannot be initiated by physical components of the brain. Nor can it be stopped willfully indicating a relationship with consciousness. (I maintain that dreaming is thinking non-consciously while asleep.)

A conscious thought, other than day-dreaming, which is a form of mental relaxation, cannot be seen, heard or touched. It is a volitional activity, with selectivity, that can utilize external sensor experience but cannot be created by it. Can thoughts be created without a brain? That question cannot be answered unless we take into consideration paranormal experiences such as NDE. Refusal to legitimize NDE evidence shuts out a source of valuable information for defining the laws of non-physical potentiality.

If the cognitive scientist rejects the mind as the functional element for producing thoughts, what substitute does he/she propose and what is the role of consciousness in such an arrangement? According to the mandated policy of materialism, it must be a physical element, which of course is the brain, but consciousness is not physical. Obviously the next question is how does the physical brain produce non-physical thoughts, motivations and rationalization? Is it because the brain also produces consciousness? At this point it is only a half step into the realm of assumptive generality with the theory of representations. What is a representation, what does it do and how does it do it? Supposedly, the brain uses ideas to represent objects, like features of the environment, rather than through direct apprehension of them. (But ideas are non-physical.) Since the brain is compared to a computer, it performs operations on these representations to extract pertinent information, stores it in memory, makes decisions, and directs activity. This is not even logical commonsense. Any offhand concept would do as well without any analysis and it would have as much credibility as this naïve nonsense since neither would have supporting scientific evidence. If the mind is brought back into consideration, the picture begins to take shape because first person experience, which is a reliable source, can validate explanation through immediate demonstration.

It is amazing that an intelligent scientist can argue against the existence of a mind as a third person occurrence, while *consciously* performing a first person mental activity in describing the act; arguing against the existence of a mind while using one. In other words, how can one use thinking to deny the use of the ability to think? Which physical sensory inputs contribute to thinking when the eyes are closed, there is no sound, and the body is inactive, or when authoring a book?

9. The soul controls the body. Control is not a property; it is an activity that can be exercised by only one entity. Body control by the soul's mind, as demonstrated through reincarnation, is the argument against the materialistic concept of physical deterministic control. Since the new brain has no experience at birth, the reincarnated mature soul takes control of the pristine, unconfused physical brain. According to materialistic explanation the new brain begins creating mental states based on developing experience, a constant process of developing the synaptic configurations. But the evidence shows that it is the mind's desires that are accommodated by the body's reactions. The brain generates the motor signals demanded by the mind to be taken to the location of the previous existence; even to attempt to go to that location without any assistance. In the reincarnation experiences, this control is also demonstrated uniquely in the application of skills from the past life. Demonstrating the skills without any instruction is conclusive evidence of the ability of the soul to utilize its memory of skills to direct the generation of physical motor commands for execution. The child's

conversing in a foreign language can only be explained by the soul's latent mental ability to control the neuronal direction for that linguistic expression while it is occupied with the content of the conversation. The trained mind is in charge of the operating system. There can be only one controller of the mental activity and the motor activities processed in the somatosensory cortex. Are there interfaced functions involved in the control function? Yes, and they are the subjects of the needed, redirected thinking..

10. The mind, the soul and the person are one. Once the soul is separated from the body, the person and the personality both continue as residuals of the past life. As the person in the new body is given a new name, with new parents, the old personality begins to lose its association with the past, as demonstrated in the investigated cases when the memory of that personality fades. In some cases, it rejects its new name and living requirements, demanding the old ones. This is clear evidence that the personality is the same, with its old desires, identifications and needs, like the need for the food eaten by the social caste in the old life that is not the food of the social class in the new life. To exhibit its uniqueness, personality depends on the characteristics associated with the mind's functioning, the memory's retention of the past, and its gradual loss of past life experiences in the awareness or consciousness of the present. But these are all properties of the soul migrating from one body to another. The soul then is the person that lives on into eternity despite the two or more bodies (and personalities) that he/she may have occupied. Actually, as reincarnation has demonstrated, the soul is genderless with gender determined by the body which the person (soul) occupies.

From a first person understandable awareness, as you and I experience it, there appears to be no distinction between the general term soul, the identification of the personality, and a mind that is aware of what is happening as well as what its behavior should or will be. It is one functional and identifiable person. Perhaps this incorporates the meaning of "self".

11. Monism is dead. The above arguments lead to the major conclusion that monism is no longer a viable explanation for human

life. Dualism must be rethought regardless of the problems that it will raise in the scientific community and in the philosophic, secular, and theological sectors. *The explanations of human behavior must change from a third person viewpoint of deterministic causality to a first person awareness of the experienced motivation for volitional activity.*

The public sector may lose confidence in science's ability to explain "everything" due to its insistence on denying for 350 years what has been accepted philosophically and theologically throughout that period. The recent scientific fault is the failure to address the evidence that has been in published form for over four decades. On the other hand, the secular opinion may be slow in reacting due to the ingrained anti-religious bias that has been growing in the liberalized version of how life should be defined, preferably without an afterlife. Beliefs are difficult to change. A change of this nature may be a historical event and I may be the historical accident that made it happen for which I must take full responsibility (or credit). However, a complete picture with the effects of the deconfirming evidence must be generated along with a holistic system description of the dualistic human nature for better understanding and agreement.

12. Radical change in basic reasoning. The final argument is why must the current direction of neuroscientific and cognitive research be changed? If you're driving on a road that leads to a dead end, you don't continue on that road when you see a sign telling you it is a dead end. You turn immediately to find another road that leads to where you want to go. The above eleven points argue that a radical change in basic reasoning is required to lead to a definition of the human system based on the two known facts about the human makeup. The first is the physiological and biochemical information that has been researched about the human physical system. The second is the new fact that there is a soul interacting with the physical elements to control human behavior. As Descartes philosophized; if you want the truth you must throw away all previous information and begin with the basic facts. This calls for not only a new approach but also a new science, one based on the combined efforts of both scientists and traditional philosophers. They must overcome past interdisciplinary biases about science, religion, philosophy, and physicalism in the search for the laws

that determine or do not apply to a dual human system. There is no room for preconceived notions that must be verified with reasonable but optimistic explanations. The basic facts are now available to initiate the new science

Continuing the research based on physicalism is a waste of talent and resources that could be reapplied to a new science. A reevaluation of the admitted gaps in current understanding of the mental–physical interactions would be a promising beginning because the effects of the unexplained gaps are known. With the new evidence of soul properties, some of the gaps can be explained immediately.

The above logical points are facts, not conjectures or hearsay evidence. They point to the conclusion that the neuroscientific conceptual solutions are unsupported and will never be proven scientifically. The reliance on future scientific discoveries can no longer be a technical crutch to lean on for unsolved mind-brain and body-soul problems. The soul's presence should demand the rethinking of human behavior as a neuro-spiritual activity with social responsibilities. Dealing with non-physical laws, properties, characteristics, and their physical interfaces is the direction of future research.

The scientific redirection is my view of what must be done to get the redefinition of Homo sapiens back on the right track. I am neither a scientist nor a recognized philosopher (we are all philosophers of some importance). I have not developed the evidence that deconfirms the scientific conceptual solutions of the mind-body problem. So I ask: Why am I the first to propose this new scientific approach based on the evidence that was available since 1966? As an observer of the misdirected and illogical research rather than a contributor, I found the inconsistencies and functional gaps that my engineering experience would not accept. I had to find more credible answers. Or was it to complete an unproven argument. Since reincarnation has been shown to be real, is it possible that I have returned to validate my philosophy? As improbable as that may seem, there is no valid argument for or against; it is strictly an unknown but under the circumstances, appropriate.

The situation is clear now and needs development. As a start, in the final chapter of this book, I propose a logical approach for what I believe is a long adventure in the search for the truth by a unique combined scientific, philosophic, and theological team. The evidence, therefore, should not be questioned for what it does conclusively: prove the existence of the human soul through its migration from one body to another.

Two unanswered questions remain: Why wasn't this linkage to dualism comprehended earlier? and Why should there be resistance to it before it is finally accepted? The mission is obvious but the missionaries must be identified.

4.2 PHILOSOPHY and SCIENCE

The dualism debate is a philosophical debate in spite of the scientific applications of physical neurological evidence to create an apparent certainty of materialistic monism. The claims of physicality in the creation of mental states are offset by caveats of pending solutions for verification of how this creation is accomplished. Such uncertainties, subject to being categorized as beliefs, are immersed in optimistic philosophical and scientific rhetoric promising eventual confirmation by the new and expanding efforts in neuroscientific and cognitive research. The debate, therefore, will continue to be a philosophical one until confirming evidence for these uncertainties is produced. The probability of such confirmations is in serious doubt by the appearance of physical evidence that negates materialist claims and instead supports dualism. Although the evidence is physical, it confirms a non-physical entity which creates a major philosophical paradox: how can a philosopher armed with positive confirming evidence explain the physical/non-physical interactions without knowing the non-physical laws that unite the two entities? On the other hand, how can science accept a non-physical reality with no clues to how its effects are imposed on the physical body? Philosophers, scientific and metaphysical, are adept at explaining conceptual situations as long as they are not encumbered with conclusive evidence that places strict boundaries on their rhetorical acumen. It may not be obvious to scientists and philosophers that with this unique evidence, there will be a need for a humbling and compromising union of metaphysical and scientific philosophers to shed previous biases and prejudices and to search for the direction in which the scientific effort must proceed. It will not be sudden or well directed in this scientific world of materialistic commitment and the initial efforts at defining the path must be governed by patience and innovative vision.

Generally, the relationship between science and philosophy is not a discernable one. There are scientific philosophers and there is a philosophy of science. Defining the boundaries of the two domains is difficult. The French philosopher Jacques Maritain clarifies this relationship in his book *The Degrees of Knowledge*.[19] If there is to be a new direction in the research of the body-soul and brain-mind interactions, an awareness of the roles science and philosophy can play will help in appreciating the problems that will be encountered. By summarizing some of Maritain's philosophic views with interspersed quotes and comments, I attempt to support my statements of the previous paragraph that the dualism debate is a philosophic, not a scientifically controlled, encounter.

Although it may happen that the material object of philosophy and science is the same—for example, the world of bodies—nevertheless, the formal object is essentially different in each case and it is this that determines the specific nature of intellectual disciplines.

The scientist studies the world of objects by linking one observable with another observable in the structure of matter by either laws of the phenomena or conceptual entities which determine how the material complex is constructed in space and time. To the philosopher, the spatial or spatio-temporal reconstruction of the matter in molecules, atoms, electrons or linking in wave systems is irrelevant. He seeks to learn the nature of corporal substance as intelligible being.

The scientist proceeds from the visible to the visible, the observable to observable, or at least indirectly observable but not imaginative. But in the atomic order, when continuous observation is eliminated, he passes from a world of imaginatively representable to a world without imaginable form.

The philosopher proceeds from the visible to the invisible, to what is of itself outside the observable, because he deals with pure objects of understanding, not of sensible apprehension or imaginative representation.

> Since they [science and philosophy] have utterly different formal objects, other principles of explanation, diverse conceptual instruments, and, on the part of the knowing subject himself, quite distinct intellectual virtues or discriminating lights, the domain proper to philosophy and the domain proper to the sciences do not overlap. No explanation in the scientific order will ever be able to displace or replace an explanation belonging to the philosophical order, and vice versa. One would have to be very naïve to imagine that recognizing, on the one hand, an immaterial soul in man, and studying, on the other hand, the glycogenic function of the liver, or the relations between idea and image, are two explanations that both belong to the same field and run counter to each other.(page 50)

(I interrupt this explanation of the differences between science and philosophy to point out that the last quotation has been compromised by the incredible physical evidence of the soul's reality. What Maritain refers to as naiveté has become reality. Both science and philosophy are faced with the problem of domain overlap which was non-existent in the earlier explanation. The soul is to be recognized in man while studying the physical brain which is influenced by the mind, a functional element of the soul. How grave is this reversal in the roles of both disciplines? It is expressed in the opening quotation of the Introduction to this book. If a non-physical entity is found, it will require a new field of research to study the effects of that non-physical entity on the physical components. The future roles of both science and philosophy may be radically changed in the dualism debate, which will also become a monologue by the exclusion of monism.)

The second explanation of the relations between idea and image (in the quotation) refers to the handling of each of the mental identities with respect to a "thing". The *idea* of a thing is not bounded by any restraints of size, dimensions, or material composition, leaving the conceptualization to be defined. It may be non-physical like the soul which has only ethereal qualities difficult to describe. The creation of an *image*, however, is in the scientific domain because it involves the experience of visualizing a dimensionally physical and existent item. The ultimate form may be different but its formulation begins with an intuition of an existent item. Thus idea and image belong in different domains due to the doctrine applied in addressing them.

There is a certain dependence by science on philosophy in its attempt to explain the reasons for "being" (entity and existence). Finding difficulty in the attempt, it seeks the answers in a higher knowledge. Not that it is dependent on philosophy in defining its basic principles for its activities but it is in the philosophic domain that it justifies these principles. Perhaps scientists do not admit this dependency for they feel no need to express it; yet if they reflect carefully on their activities they will recognize the presence of a whole philosophy in a pragmatic but subtle participation. When scientists exercise experimental methods and critiques of their results, that constitutes applied or lived logic, a property of philosophic discipline. They raise questions of the value of principles, especially the principle of *causality,* with the world of experience, that is, the adequacy of the change to explain itself by itself by itself......

Philosophy is handicapped, in a sense, by its modus operandi. Its tools of conviction are words whose interpretations are often at variance with the intent. It has not been able to use experiment to verify its conclusions. Traditional philosophy deals with abstractions, such as *being, essence, universal reality*, that invite limitless interpretations. Semantic battles are endless conflicts in which the war is never won. Conceptual situations that cannot be verified are desirable topics for lively discussion in which conclusive agreements are not anticipated. How will this change, when the philosopher is suddenly confronted with a reality that has verifiable, and possibly conflicting, aspects on both sides of the being's physical and non-physical interface? The need for the semantics will persist in the search for the unknown properties but the physical evidence will anchor the base of the argument in logical concrete. Such philosophic ambivalence may not be easily acquired, or tolerated, in a debate with an uncompromising opponent.

Science is similarly involved in abstractions but with different objectives. The abstractions are theories about natural law which need formalization and confirmation. When only the natural laws are involved, research including experimentation usually suffices to confirm the theory. However, when sufficient evidence is not found, the specter of abstraction begins to conflict with physical certainty as, for example, in quantum mechanics. Abstraction prevails in the absence of certitude. Additional theories are proposed but if resolution is not imminent, the level of activity may or may not be reduced until a breakthrough is achieved. This is not the case, however in the current dualism debate. The neurophilosophic doctrine is to propose a concept of a solution in the absence of a verified solution with reliance on a future verification. This tactic may be acceptable occasionally but its constant use for the many unsolved mind-brain problems emphasizes the shaky foundations upon which the materialistic house of monism is being built. Such promissory vindication is tolerated by the scientific community in its own applications but is disallowed for opposing arguments. Professionally it is in direct violation of scientific and ethical standards for the continuation of a project that cannot relate the many unverified component solutions to a holistic mind-brain solution. Consequently, proposing the concept of physical causality as the logical solution to the mind-brain problem is only a philosophical (conceptual) product to date (2009).

There are two types of causality in the philosophic explanations of the mind-brain problem. As the scientist applies the causality principle to the scientific picture of the world, there is a causality that is responsible for the mental activity and a different one for the observed subject. The difference is the first person experience of the former and the third person view of the latter. The former is agent causality, produced by the agent through volitional intent. The latter is mechanistic causation which is a phenomenon causing another phenomenon as when sensory information produces neural reactions.

When arguments concerning metaphysical entities are initiated, philosophy enters the discussion without waiting for an invitation. This is especially true when science is a participant, for there is little physical evidence of a non-physical nature involved, and science is limited to relying on physical evidence in any argument. So it is constrained to be an argument of conceptual thoughts sans factual reality. But the neurophilosophers involved in the dualism debate introduce extrapolations of physiological facts that tend to skew the credibility of the debate in favor of materialistic ideals. They argue that there is no physical evidence to support the metaphysical arguments. Philosophically the advantage lies with the neurophilosophers because they can refer to physical test data that become a base for their extrapolations. The traditional philosophers are thus at a disadvantage since they cannot counter by discrediting the assumed concepts since they are only concepts, not facts. But what if they are suddenly thrust into a position of factual advantage by having physical evidence of the soul? How does metaphysical philosophy change in this unprecedented role, for it has always been a position of arguing without the evidence? The next section explores the requirements for a new philosophy in an unprecedented research effort in which the analytical roles are reversed: science as the conceptualizer and philosophy with its base in factual evidence.

4.3 THE NEW PHILOSOPHY

Describing a philosophy to satisfy envisioned requirements is an exercise in pure speculation. But there must be some perceived direction if the general problem is defined. The following perspective has never been attempted since the problem was never broached as a possibility.

The inclusion of the soul as an essential participant in the human configuration calls for an acceptance by science of human duality which it now ignores as being outside its domain of strictly materialistic involvement. Metaphysical philosophy (as distinguished from neurophilosophy) is called upon to explain the effects of the soul's characteristics on the physical elements of the body. Obviously both domains overlap as never before, leading hopefully to an unprecedented alliance between the two. As pointed out in the previous section, these two domains, although dependent on each other, never overlapped. Since the issue has never been forced into any serious consideration, the resolution will not be easy or expedited.

What are the current philosophies and how may they change within the new direction? As the soul's reality is the principal positive evidence, compared to the neuroscientific promised but unsupported solutions, metaphysics may inherit the dominant role in explaining human nature and behavior. Prior to any such change, science has the advantage by relying on physical and experimental data in promoting its philosophy of physical causality. In contrast, metaphysics relies on the Cartesian philosophy of dualism with no attempt at explaining the how of the union of the two entities. It is merely concerned with the *what is it* and the *why is it* of the soul. There is competition between the two approaches for the dominance of human behavior: is it deterministic by environmental conditions or is it a volitional manifestation of a spiritual entity. The chasm between the two cannot be bridged without compromises by both sides. The required result appears to be a new philosophical direction with a new identifying name like metascience or neuromet. It must be a distinctly separate domain from the traditional ones so that allocation of the required degree of control to the separate spheres of influence will be possible.

In contrast to the current idealistic scientific arguments for the brain's supremacy in mental states expressed in semantic innuendoes, the new evidence now gives the philosophers a major advantage over those scientific beliefs. The mind, replaced by brain functions in the conceptual solutions, is back in control. It can no longer be ignored as an unnecessary entity. Along with its intrusion into the scientific search for the solution to the mental state problem it introduces the extremely complex issues of integration and interfaces. The responsibility, and involvement, in defining the course of action in this complex problem cannot be unilateral: philosophy and science must collaborate, not only in the definition but also in the ensuing research.

In addition to the research for resolving the dualism issue, there are issues that are not so obvious in the effects brought on by the dual human nature. What is life as existence in either the physical or non-physical states? What is the self of human nature, the person that has now been shown to be the soul, with the dual nature of man being personality? In other words, is a person a living entity in an eternal life while a personality is a physical experience for one earthly lifetime with residual memories and after-effects into the next personality? What is the meaning of human life, now that the existence of a hereafter has been demonstrated, and what is the moral effect on the value of human life with eternal consequences? Is there a duality in other aspects, such as memory, consciousness, sequential/independent brain-mind operation, and distribution of activity control? These are considerations that will require the joint participation of science and philosophy to define a broader than current worldview of the human being for social, secular, legal, professional, and even governmental communities. The religious communities are already oriented in some of these issues.

In the pursuit of philosophic dominance in future research due to the acquired factual base in competitive brain-mind physical evidence, what can neurophilosophy contribute to facilitate the birth of a new science? With the semantic experience of promoting physical causality, it could as a meaningful start explain how the conceptual solutions could be restudied and restructured to close the pending gaps by the injection of the soul's demonstrated properties. But rethinking and restudying depends on a submission to the philosophy that there are two entities responsible for human behavior. Is there sufficient willingness in the neuroscientific community to do this? Due to the unexpected nature of such philosophic demands for prescribing the direction of future research, the fundamental practices of the scientific researchers and philosophers will probably change gradually. A materialist researcher cannot be expected to suddenly become a philosophically oriented experimenter searching for a non-physical link between the brain and the mind. Explaining the mental functions as a physical brain activity has been an experience in analytical dancing and the addition of a fundamental fact may be the life saver in this sea of unresolved issues. However, science and philosophy may be reluctant to have their domains overlap with requirements for broader and possibly conflicting methodologies. This might restrain a researcher from barging ahead with innovative searching for the answers to the complex issues.

How can traditional philosophy contribute to the generation of new conceptual thinking since such thinking must introduce useful metaphysical information? Developing that information will be difficult but necessary in rationalizing the gaps between the physical and metaphysical problems of control. The neurophilosophers have already done this by studying the interactions in their conceptual approaches. Although the soul is factual evidence, the non-physical laws governing its operations with the physical body laws are unknown, creating a major chasm that will pose an intellectual challenge in a new science.

This redirection of effort could use much of the current analyses about the functions of the brain and nervous system.

Because it has never happened before, the motivation to initiate such a joint effort may take time to penetrate the indoctrinated rigidity of formal rules of logic. The pursuit of this radical philosophical change must be left to the surprised philosophers who never expected to be given this absolutely inspiring gift of factual evidence for bolstering their tenuous metaphysical positions in the current debate. The dualistic philosophy now has more credibility than the materialistic tenuous one. When physical evidence is introduced, the side introducing it should have the philosophical edge in the argument. Without valid physical evidence to back the scientific side for physical causality, this should be a slam-dunk for the soul philosophers. But it will take time for the world, scientific and secular, to admit and adjust to the soul's permanent presence.

This is the role the dualist philosophers inherit by the soul's reality. It is not even remotely apparent what the outcome will be. Will the evidence be evaluated and acknowledged as promised by the neuroscientific community or will the deleterious effects on their research cause them to renege on those promises? The non-question-begging facts cannot be skeptically attacked by criticisms of ESP, clairvoyance, or telepathy because these are not only irrelevant but also unproven. Simple opinionated disbelief is not an argument although it may hamper the activities in some research areas.

In the new philosophy there must be a systems approach for defining the research requirements. Apparently the current research was begun as a bottom-up synthesis, rather than a top-down analysis. Instead of a preliminary analysis of the total system needs, the advances in the understanding of the functioning of the components promised the greatest reward. There is no holistic system philosophy or concept which may have controlled the direction of the total effort. The only prevailing analytical motivation appears to be the necessity for assuring that physical causality dominates the picture. Due to this attitude, the need for a totally new research philosophy without the biases of the past efforts is apparent to complete the work that Descartes initiated. But before any research is initiated, there must be a notion of what the objective research must accomplish. Through abstract thought, ideas independent of past experience can be generated, so independent that one can generalize them analytically and set them free from the intuitive structure in which they were first conceived. This is what the new metascience demands.

4.4 INTRA-DISCIPLINE DIVERSITY

There have been many arguments about Descartes's dualism since he wrote his philosophy in the Meditations. During the Meditation debates, the philosophical nature of the arguments was the only method of debate since the physical properties of the human neural structure were not known, therefore, not an issue. Extensive cognitive research in the last century has supplied physical evidence that has been construed as explanatory of the workings of the human mind. As scientists are only concerned with physical matters, this increase in scientific neural data has spurred the investigative effort to justify the concept of physical causality as the explanation of human behavior. Not all scientists have agreed with this approach. A few examples expressing pragmatic variance with the materialistic approach are given here to illustrate this dissent. The title of the book from which the statements are extracted is given after the quote.

The importance of the first example is that it is made by a neuroscientist of long standing. His acceptance of the need for a non-physical explanation for the subjective human properties is an unqualified rejection of physical causality.

> However, the nonphysical nature of subjective awareness, including the feelings of spirituality, creativity, conscious will, and imagination, is not describable or explainable directly by the physical evidence alone. As a neuroscientist investigating these issues for more than thirty years, I can say that these subjective phenomena are not predictable by knowledge of neuronal function.
>
> Ref. *Mind Time*, 2004, by Professor Benjamin Libet, page 5.

The second example is a strong statement by a materialist for the inclusion of a mind in the brain-mind problem. The acknowledgement of the underlying reliance on the concept of a mind argues against neuroscience's elimination of the mind as a viable entity.

> We are then driven back once more on mind itself, for only mind itself, in each of us, can determine what the functions of our mind can and cannot do. And since even those of us who are materialists constantly make determinations that this or that is truly the case, it seems that all of us acknowledge, at some level of discourse, a reality-attaining competence that belongs to mind itself.
>
> Ref. *Mind Regained*, 1998, by Edward Pols, Keenan Professor of Philosophy and The Humanities Emeritus at Bowdoin College, page 135.

The objection to physicalism is obvious in the next example. The definition of physicalism referred to here is the principle that all phenomena, if determined to be factual, must be described in physical terms. Such a limitation prevents explaining subjective behavior in non-physical terms.

> Psychologists sometimes feel pressure to employ the vocabulary and the methods of the physical sciences, as a matter of principle—often with the result that they depart from psychological explanation and move toward neuroscience. In acknowledging the importance of perspectival instrumental constraints on explanation, we can highlight the importance of the distinctive goals of psychological explanation—goals that call for a level of explanation different than that of neuroscience. As a result, we can see that psychological explanation faces no threat, as a matter of principle, from standards for explanation set by the physical sciences. The physical

> sciences have, in the end, no monopoly on the explanation of human behavior.
>
> Ref. *Philosophy After Objectivity*, 1993, by Paul K. Moser, Professor and Chairperson of Philosophy at Loyola University of Chicago, page 226.

The following statement of how science eliminated free will and the mind as a related non-essential is given as a professional explanation, not as an acceptance of its validity. The theme of the book argues against the validity of the exclusion.

> Freud elevated unconscious processes to the throne of the mind, imbuing them with the power to guide our every thought and deed, and to a significant extent writing free will out of the picture. Decades later, neuroscience has linked genetic mechanisms to neuronal circuits coursing with a multiplicity of neurotransmitters to argue that the brain is a machine whose behavior is predestined, or at least determined, in such a way as seemingly to leave no room for the will. It is not merely that the will is not free, in the modern scientific view; not merely that it is constrained, a captive of material force. It is, more radically, that the will, a manifestation of mind, does not even exist, because a mind independent of brain does not exist.
>
> Ref. *The Mind and the Brain*, 2002, by Jeffrey M. Schwartz, M.D., Research Professor of Psychiatry, UCLA School of Medicine, page 8.

The next example points out the philosophical objection to the scientific attempt to explain the brain's ability to create mental states that control conscious behavior.

> Though we are fond of presenting the puzzle of conscious experience as a matter of conflict between materialism and our subjective, intuitive conception of our experience, it isn't really physicality that presents the problem. The point is that merely positing a new kind of property—call it a basic mental property—doesn't really shed any light on how to understand conscious awareness.
>
> Ref. *The Purple Haze.* 2001, by Joseph Levine, Professor of Philosophy at Ohio State University, page 177.

The final example illustrates the critical opinion of the materialistic approach by this eminent philosopher not only about its subconscious motivations but also science's credibility.

> How is it that so many philosophers and cognitive scientists can say so many things that, to me at least, seem obviously false? Extreme views in philosophy are almost never unintelligible; there are generally very deep and powerful reasons why they are held. I believe one of the unstated assumptions behind the current batch of views is that they represent the only scientifically acceptable alternatives to the antiscientism that went with traditional dualism, the belief in the immortality of the soul, spiritualism, and so on. Acceptance of the current views is motivated not so much by an independent conviction of their truth as by a terror of what are apparently the only alternatives.
>
> Ref. *The Rediscovery of the Mind,* 1992, John R. Searle, Mills Professor of Philosophy and Language at the University of California, Berkeley, page 3.

The above quotations illustrate the lack of conformity with the current direction of cognitive science's promulgation of physicalism. I concluded with John Searle's statements because I believe they not only summarize the views of all the other dissenters of the current approaches but also express the underlying fear of the consequences of what this book is presenting. The mere thought that man is responsible to a Supreme Being in an accountability that takes place after earthly life frightens a secular society that prefers to think of life as an indulgent existence with an end devoid of residual effects. Evidence of a soul's reality is a devastating imposition on the hedonistic outlook on life's meaning.

This is especially true in the argument that Descartes would have had with the modern philosophers in supporting his philosophy. The arguments in his period were philosophical tests of the power of persuasion. Currently, this is not so; it is a contest of supremacy between cultural relativism, scientific positivism, philosophic diversity and theological certainty. It has social, moral and relativistic overtones in addition to the credibility of one scientific concept of neuronal control versus another. The credibility of free will is more at stake than any analytical justification of neural paths and connections. For this reason, the arguments are more diverse than strictly scientific, extending into the non-physical aspects of free will and consciousness (as did Descartes). These demand rationality that science lacks because science can only vouch for physical facts. With these thoughts in mind, the arguments and philosophies in this chapter are representative, as were the above quotations, of the state of the current debate of dualism (and monism).

4.5 CONTINUING RELATED ARGUMENTS

There are some arguments based on modern philosophy and scientific research that represent current thinking which I believe need rethinking in light of the soul's reality. The following three are representative arguments: support of the concept of physical causality in claiming evolutionary development of the mind-brain complex, use of the effects of brain damage to deny the existence of a mind, and the derision of dualism as folk psychology. The evolution argument is a speculative and suitable one since there is no evidence to support it. The inclusion of the soul into the argument, however, introduces more questions like when in the human developmental history did the soul enter with its properties of rationality and free will? The brain damage argument relies completely on a materialistic application of medical, surgical and psychological data. The folk psychology argument is an ironic one in that it can be applied to either side of the soul argument although that option has not been exercised by the dualism victims in labeling conceptual studies as similar folk psychology.

When philosophers expound on their conceptual solutions, the number of *ifs, supposes, and assumes* that are included in the elaboration reveal the factual shallowness of the material. It is pointless to question the conceptual credibility because they are merely beliefs, proving only that the vexing problems have been addressed with some results. As Paul Moser states at the beginning of his book *Philosophy After Objectivity:*

> Controversy over proposed theories, of one explanatory level or another, is thus a hallmark of philosophy. A familiar reaction from observers outside philosophy proper yields a simple injunction: Beware of philosophers bearing theories! The only problem with that injunction is its narrowness. Impartiality suggests another dictate: Beware of anyone bearing theories![20]

As a retired engineer, I consider myself such an observer outside of formal philosophy and I relate to that advice. Impartiality, however, was not possible in the three examples I selected because of the new evidence. Instead, I can show how the philosophical theories and arguments lack the credibility they intend to generate.

In the following arguments, each topic begins with a brief commentary about pertinence of the topic to dualism. Points of view or claims by individuals are added to give diversity to the topic as well as to bring out the merits or inconsistencies of these views. The reader may find openings for making additional arguments, and using the views for conversational inquiry. These are live topics brought about by the soul's reality.

Specific arguments of individuals are identified by their names and the books in which these arguments appear.

Evolution

Evolution is a theory of a biologically physical process through which the human being, as well as all living matter, developed. It has nothing to do with creation in the context of the primordial protoplasmic beginning of cellular life. It does have a stronger argument for species development than that of the proponents of the theory that species were created in their final forms. Intelligent Design (ID), which also appears in these arguments, has a more logical application in the beginning of life than in the design and creation of all living matter in fully functional forms.

Evolution explains the development of the physical nature of Homo sapiens. But how does the human brain develop with an immaterial mind? Darwinian evolution cannot explain this union. So how do philosophers explain it? Through Darwinian application, evolution has become a sort of biological theory of everything. But evolution must have something to evolve from. There is no creation of new species, only changes from existing types. The three actions that the theory of evolution attempts to explain are: natural selection, random mutation, and common descent. At the observable level, these three seem to support the theory. At the microscopic level, they are not as obvious and suffer from severe limitations and contradictions.

Through natural selection, the fittest organism of a species will produce more surviving offspring than the less fit, thus assuring the continuation of the superior line. Random mutation is a chance deviation of an organism's biological makeup that may or may not be reproduced in the offspring. It may lead to the resistance of an invading germ or create a debilitating condition. An example is the mutation that was responsible for malaria. But its effects are limited by climate. In all the time that malaria has been a killing disease, it has been unable to mutate further to allow it to spread from the hot climates to more temperate areas. If mutation makes an organism stronger or more capable of performing its functions, natural selection takes over. Common descent tries to identify the genealogical development of a species. These changes and improvements can only be physical since they occur in the physical structure of living things.

However, the evolution picture is different at the cellular level, which is more relevant to the creation of species based on information that was not available to Darwin. Are behavioral responses to the external environment caused by random mutational change at the cellular protein level or at the organism's system level? Genetic regulatory networks and detailed genetic plans are necessary for building animal bodies. There is no evidence that random mutation leads to changes in these networks. The networks are constructed; they are not a product of mutational modification.[21] Therefore, mutation has no effect on the manner in which the physical bodies are assembled.

In the natural selection process, somewhere between the random mutational effects and the appearance of the changes in the physical compilation, there is the system adaptation to account for evolutionary change. But there is no system level description of this process, only the cellular structural changes. These cellular changes result in actions of the genes, which are portions of the DNA that code for the enzymes. And gene complexity is mind-boggling. But that is all physical. There is no explanation of the reasons why or how behavioral variations are influenced by the physical selections. Nor is there an explanation of how the basic elements of the DNA and RNA developed to be ready for the integrations as the DNA. How did proteins evolve, or amino acids, or nucleic acids, or nucleotides?

This brings us to the explanation of the evolution of the brain. All higher order animals and man have brains. There is no doubt that these brains were products of biological evolution, although they differ in degrees of development. With the human, however, a problem arises when the mind is included in this biological development. It isn't a problem when stated as Patricia Churchland does in her book *Brain-Wise*. But a little skepticism reveals that it is a deceptive piece of philosophical doublespeak.

> The more we understand about brains, their evolutionary development, and how they learn about their world, the more plausible that the pragmatists are on the right track concerning the scope and limits of metaphysics. The explanation is quite simple. We reason and think with our brains, but our brains are as they are—hence our cognitive faculties are as *they* are—because our brains are the products of biological evolution. Our cognitive

> capacities have been shaped by evolutionary pressures and bear the stamp of our long evolutionary history.[22]

These statements are a prime example of not only biased and confused thinking but also of the method of controlling the thinking of the reader through the mixing of facts and conceptual conclusions. (Remember the warnings in Chapter 2 about reading carefully?) The following statements may not be in the order of importance due to the mish mash of thinking in the quotation.

1. It is not a fact that the brain does the thinking. This is a preconceived notion demanded by the materialistic view. It is the mind (the soul as we now know) that does the thinking, but if the mind has been excluded from the process, obviously only the brain is left. Note that this assumption becomes a fact and the main reason for associating this rationality with biological evolution.
2. Churchland has denied the existence of the soul, as reflected by the desire to limit the scope of metaphysics. With the evidence of the soul's reality, this entire statement, which reflects neuroscientific thought, is completely false. The soul's retention of past memories and its use of them in thinking and demanding physical activity proves that thinking and rationality remain part of the soul, not the activity of the previous dead brain.
3. Note the guidance by the logical lead-in to the argument that the more we understand about the evolutionary development of the brain, the more reason to discard the metaphysical. But understanding more about the physical changes does nothing to reduce the ignorance about the non-physical. Stressing the understanding of the brain seeks the acceptance of physical reality as the main issue. The simplicity of this acknowledgement is stressed to gain an acceptable mood for what follows immediately. "We reason and think with our brains" implants the idea that it is the brain, not a mind that thinks, which paves the way for concluding that the ability to reason and think is a

biological evolution. This is the deception that is necessary to promulgate the absence of a non-physical entity.

4. The fact that our brains are as they are sheds no light on the fact that our cognitive faculties are as they are. The same can be said with the soul's reality in the picture. If you noted, the word "they" is italicized in the quotation, as it appears in the book. The italics imply that there are two entities being addressed, the brain and the cognitive ability. Otherwise, a "they" without italics would have been sufficient to address the two aspects of the brain, its physicality and capability. I wondered why this word, referring to the cognitive faculties was emphasized. Reference to *them* seems to imply a separate something from the brain, which I could interpret as subject to a different development than biological evolution. Was this an unintended philosophic slip? If the non-physical is also considered, according to the psychoneural translational hypothesis (PTH), the brain and the mind can interact because they are complementary aspects of the same transcendental reality. So Churchland has it in reverse because the biological evolution was in the brain's ability to understand the language of the mind and to translate it into the language of the body. With the soul's reality established, this hypothesis has more credibility than materialistic guessing.
5. The final statement about the evolutionary pressures on our cognitive capacities is correct, although not as intended in the quote. The evolutionary history of the soul, through reincarnation, does have an effect on the cognitive capacities as the evidence of experiences has shown in the new-body reactions versus the old. If reincarnation is a repeating process, the soul may have an evolutionary history. Was it already part of hominid Lucy (3.2 million years ago) or as far back as Ardi (4.4million) when fossil records indicate parenting and social adaptation?

The above nit-picky skepticism is intentional to show how the denial of the metaphysical and the use of a preconceived solution can obscure reality in the quest for conceptual acceptance. But all is not lost. In fairness to Churchland's actual perception of the big picture of the mind-brain problem and dualism's possibility, I want to give credit where credit is due by quoting one of the conclusions she makes at the end of the Introduction to her book regarding the topic covering the nature of the self, consciousness, free will, and knowledge.

> Ultimately, its soundness will be settled by what actually happens as the mind/brain sciences continue to make progress. Conceivably, it will turn out that thinking, feeling, and so on are in fact carried out by non-physical soul stuff. At this stage of science, however, the Cartesian outcome looks improbable.[23]

This concluding concession that there may be soul stuff reveals the lingering uneasiness in the neuroscientific mind that there must be something missing in the materialistic approach. Evidently this uneasiness has not been sufficient to analyze the effects of soul stuff on the mind-brain problem.

Brain Damage

Brain damage resulting in functional deficiencies like amnesia is used as an argument against dualism as is the use of drugs and alcohol. The argument states that if a soul, or mind, uses the brain for nothing more than sensory inputs for volitional responses, then it should follow that emotion, rationality, volition, and consciousness should be invulnerable to the effects of brain damage or use of drugs or alcohol. But the opposite is true. Both of the resultant brain conditions affect the use of the non-physical properties in behavioral response. Therefore in the materialistic concept it is obvious that emotion, rationality and consciousness are activities of the brain since they are affected. Assigning them to a soul or mind makes little sense.

In a physically oriented analysis that supports the materialistic mandate for negating dualism this is a non-question-begging solution. But it does have a significant flaw. In the assumption above that the mind has nothing more than the ability to use the brain for sensory experiences it is clear that the argument is restricted to the materialistic approach. By generalizing the argument with a broader but consistent assumption, it can also be said that the mind can use the brain for initiating the motor commands necessary for physical activity *under normal conditions*. However, when certain elements of the brain that are involved in this train of commands become damaged and incapacitated or affected by drugs, this train is broken and no activity can result from the mind's commands. The information for the activity is being generated and transmitted by the mind but it stops at the damaged brain element and ceases to be effective. It is the physical brain that is responsible for the impaired or lost activity without affecting the signal initiating function of the mind. The mind, still existent, and fully operational, is blocked by the damaged brain from completing its desired activity. It is like a damaged TV set. The signals continue to be transmitted by the sending station and are available at the TV but the receiving set is incapable of producing the picture. This argument is more logical than the materialistic one that limited the mind's functions to deprive it of a deconfirming response. If a mind is to be the subject of an argument, all its properties must be included.

Folk Psychology (FP)

Folk psychology is often used as an argument against dualism due to dualism's reliance on commonsense rather than scientific facts and theories. The use of FP as an argument is often effective because its subject matter is events and beliefs, some that have been proven false in the past. References to this faulty past in analogies and comparisons may be illogical, therefore, unrelated and irrelevant. Unless those inconsistencies are noted, however, the arguments are effective.

The definition of folk psychology is given by Dr. Paul Churchland on page 3 of his book, *On the Contrary, (*1998*)*.

> "Folk psychology" denotes the prescientific, commonsense conceptual framework that all normally socialized humans deploy in order to comprehend, predict, explain, and manipulate the behavior of humans and the higher animals.

The neuroscientific explanations of human behavior are also given in a conceptual framework to do exactly what the above definition imposes on folk psychology. The conceptual explanations of the brain's ability to control human behavior are not supported by factual data and are only optimistic predictions. Manipulating physical data and computational analyses to explain human behavior are precisely within the scope of the above quotation. The obvious implication is that the nature of neuroscientific research is also akin to folk psychology, but apparently not from the scientists' viewpoint. A further clarification of folk psychology, following the previous quotation, presents a similar paradox of the inclusive conceptual nature of neuroscientific research.

> The term "folk psychology" is also intended to portray a parallel with what might be called "folk physics," "folk chemistry," "folk biology," and so forth. The term involves the deliberate implication that there is something theory-like about our commonsense understanding in all of these domains. The implication is that the relevant framework is speculative, systematic, and corrigible, that it embodies generalized information, and that it permits explanation and prediction in the fashion of any theoretical framework.

This clarification also includes neuroscientific explanations in this framework because of the use of generalized (admitted lack of adequate factual support) information with predictions and explanations given as theory rather than proven solutions. My conclusion, based on Dr. Churchland's own definitions and my knowledge of the soul's reality, is that the speculative research about the brain's causal control of human behavior is akin to folk psychology and false.

4.6 THE EXISTENCE OF GOD

The six Meditations of Descartes dealt as much, if not more, with the existence of God than the existence of the soul. This topic was not covered in Chapter 3 because it is not directly relevant to the dualism debate: except for its association with the spiritual nature of the soul. The certainty of the soul's eternal afterlife introduces the questions about that eternal existence in the spiritual domain. What is the spiritual environment and more importantly on an individual basis: Is there a God to whom each person must account for the behavior in the concluded earthly life? As free will has been a demonstrated property of the soul, the responsibility for earthly behavior is the only baggage each individual can take along. The reincarnation evidence has created an uncomfortable burden for the average person, especially for those who have convinced themselves otherwise.

So does God exist to make this situation a reality? That is matter for another book. But there is a similarity in the requirement for physical evidence to prove the existence of an immaterial entity, this time for God. I present such evidence, some of which reinforces the credibility of reincarnation on a spiritual level. It is the kind of evidence that Descartes could have used in his arguments for the existence of God.

There is a difference between the arguments for the existence of God between Descartes and his objectors and the modern philosophical question of the reality of a Supreme Being. In Descartes' time, the philosophers believed that God existed and were searching for reasons to verify the belief. In the modern world, the reverse is true; God's existence is denied and the arguments against the denial are minimized. In the earlier times, God was a Being to whom man turned for help, mercy and understanding. In the modern world, except for the religiously inclined, God is an imposition on man's freedom and is to be argued down to relieve man of the terrifying thought that there is to be an accounting to that God in some afterlife. The negativity shows in the argumentative logic of the examples that follow. These examples also illustrate what must be overcome by any new evidence.

Arguments for and against the existence of God are similar to the arguments in the dualism debate. These arguments are even similar in that they propose concepts of God that cannot be verified or denied through philosophic rhetoric. The same type of evidence that proves the soul according to the descriptions in this book is not possible in God's case. It is too much to expect God to provide the first person evidence of His existence.

If God exists, what evidence is there to confirm that existence?

The first cause argument

Every effect must have a cause. Since the universe exists, there must be a cause for its existence. That cause must be the Uncaused Cause, God. Since it is not known, but possible, that there could have been another cause, why say that God was the Creator? Why must it be a supernatural cause rather than a natural cause like the Big Bang? This is presented as an argument of the philosopher David Hume (who died in 1776), by a neurophilosopher (whose name I will kindly refrain from mentioning, since the argument is so flawed and an example of thoughtless reaction). The Big Bang is a relatively recent scientific conclusion and could not have been used by Hume. It was proposed by George Gamow, who was born in 1904. Furthermore, calling the Big Bang a natural cause is scientifically and logically incorrect because it was the start of all natural causes, therefore, there was nothing like it to cause itself. However, the skepticism is shallow because the obvious response to the negative argument is that God caused the Big Bang. But note the rush to any argument that might appeal to a scientifically oriented audience.

The second part of the cause argument is that the series of causes cannot be infinite: they must stop somewhere to be the first cause. If the beginning of the series is assumed to be the First Cause God, why cannot there be a cause prior to that to create God or a first cause of all creation before reaching back to God? Since neither of the two is a certainty, the existence of God is not proven; neither is the non-existence of God. The existence of God, to be proven, must resort to a different argument than the first cause argument.

The benevolent Creator

As Creator and Ruler of the Universe, God is omnipotent, omniscient, and omnibenevolent. To control the Universe, He must be all-knowing and all-powerful. He has provided all of man's needs and cares for him with infinite mercy and compassion. But the skeptic argues: if all this is true how do you explain, evil in the world, pain, illness, sorrow, and natural catastrophes that beset man? In His benevolence could not God alleviate these conditions since His creative work is perfect, incapable of inflicting pain or causing evil? Obviously He is powerless to stop evil, like the wars that kill and destroy. Perhaps He doesn't know of all the evil, whereby He is not omniscient. If He can't stop the evil,

He is not omnipotent. If He does not care to stop the evil that is so disastrous to mankind, then He is not the benevolent, loving Being. Therefore, He does not exist as the all inclusive perfect Being that is the claim of the theists.

This argument makes God responsible for all of man's activities disregarding the fact that in His benevolence, God gave man free will. Without free will, man would be condemned to the commands and predestined conditions which He would impose but could not reverse. With free will, man has a choice in what he must do to avoid any adverse conditions that affect his desires, or how to live in a society that is not always friendly because others also have free will. Man has the ability to prevent illness by adhering to the necessary conditions for good health. The free will at top governmental positions creates the environment for wars.

If God interferes in any of the above activities, He is reneging on his most bountiful gift of free will. What a magnanimous gesture by a Creator to allow the creations to disobey any conditions that He could have imposed on them. Although He mandated the Ten Commandments as His order for obedience to His will and peace in the world, the creatures have the ability, and do, to act against them. Has anyone ever considered this as an act of love by the Creator toward His creations by depriving Himself of the divine right to force obedience to His commands?

In creating the universe, the Creator had to establish the natural laws for universal order to prevent chaos. Those laws from the Big Bang to the present have not changed. Their interactions vary, causing dynamic changes but they always adhere to the initial conditions. These dynamic actions, whether hurricanes, volcanic eruptions, tsunamis, droughts, floods or meteoric impacts are all governed by those basic natural laws. There is no divine programming for their occurrences. There is no need for such programming; they just happen. To blame God for an inability to stop them is merely a typical human reaction of finger pointing to place blame somewhere in an irreversible situation. It is irreversible because to reverse it each time for the convenience of man would be undoing the natural laws that have existed since the Big Bang. The same is true in cases of illness resulting from poorly regulated and contaminated food, or human-caused accidents, or behavioral causes like smoking resulting in cancer or other debilitating conditions. These are the dynamic changes brought about by the consistent, unchangeable natural laws, or human behavior, not the lack of divine benevolence or power to reverse those laws. Once set in motion those laws must prevail to prevent chaos and to assure a balance of earthly, as well as universal, energies and conditions.

The arguments against God's benevolence also argue against other criticisms, which illustrate the lack of consistency in the arguments. If God is to intervene to stop the natural disasters and the sorrows caused by illnesses, He must rescind the natural laws that He created which cause these disorders as predictable effects. But such action is the definition of a miracle. The conflicting argument is that miracles are not possible. So what is the net result of the arguments since the arguments negate each other? Both are meaningless. The critics can't have it both ways.

Physical Evidence

As was questioned for the evidence of a soul in the chapter on evidence, what physical evidence can be produced to convince the scientific community that not only the soul exists but also the Creator of all souls? A non-physical God cannot be seen, touched, heard, or felt (as by a wind or bright light). In spite of this lack of physicality, there are scientists who concede that God's existence should not be denied on this one criterion. There is other physical evidence that many believe points to a divine reality. Doesn't all of creation point to a Creator as the cause for the physical universe? But that argument has already been countered by the possibility that the universe came into existence by chance. With that argument out of the way, regardless of its dubious rationale, what other recourse is there for finding physical evidence? Just as I was able to find the evidence for the soul, I present two cases of physical evidence that may run into the same rhetorical obstacles by those who have preconceived denials regardless of any rational credibility. On their own merits, however, I believe these two cases have relative substance that should be evaluated objectively.

The most powerful factor of the soul evidence presented earlier is the spontaneous, first person demonstration of reality. There is no question, due to the repeatability and the detail of the evidence, with observed verification, that the evidence was authentic and the reporting properly documented. It was scientific evidence reported in accordance with scientific guidelines. To expect a similar demonstrated performance by a divine Being seems out of the question. If it did occur, what effect would it have not only on the scientific community but on the entire world? What would be expected of God as credible evidence-a display of supernatural acts? The answers to both questions are found in universally accepted evidential testimony. It is recorded in the most read volume of all published works, the Bible. The divine Person did appear in physical form and mingled with human beings. This was *incarnation*, a first personification unlike reincarnation which is a repeating personification. Nevertheless, it was the appearance of a spirit in human form. This incarnated Person claimed to be divine and proved His claims through control over the natural laws that can only be altered by the Creator of those laws. In spite of this demonstrated display of divine power, He was rejected and to this day neither the scientific community nor the secular communities pay much attention to the reality of His divine performance. That physical evidence of the existence of God was, and is, Jesus Christ. The spontaneous, first person performance, demanding acceptance as a specific Person, as the soul has demonstrated in reincarnation, is historic, documented evidence of the visit to earth of the prophesied Messiah. Aside from the religious implications and viewed specifically as physical evidence of an existence, there should be little doubt about Christ's reality and divinity.

If the Bible is history (except for creation) based on archaeological evidence, as historian Ian Wilson has written in his book by that title, then it should be given evidential credibility. In the Gospels of the New Testament, historical credibility should be greater because in addition to verified period accuracy, they are also biographic accounts of the teachings and activities of Christ by actual witnesses, like John the Apostle. It is in John's Gospel that Christ's claims to divinity are very specific and undeviating: "I and the Father are one." (10:30); "For I have come down from heaven, not to do my own will, but the will of him who sent me."(6:38); "I know him because I am from him, and he has sent me." (7:29); and "If I do not perform the works of my Father, do not believe me. But if I do perform them, and you are not willing to believe me, believe the works, that you may know and believe that the Father is in me and I in the Father." (10:37-38). The existence of one Being verifies the existence of the other Being since the divine power exhibited by Christ is that ascribed to the Supreme Being.

There were also incidents during which Christ made His claims of divinity with respect to the specific issues that were raised, as in the following two examples:

In John 9:25-26 as Christ talks to the Samaritan woman, she says, "I know that Messias is coming....and when he comes he will tell us all things." Jesus says to her, "I who speak with thee am he."

In John 9:48-58, Christ describes His divinity in a manner that relates to the Mosaic period. When He says that anyone who keeps His word will never die, the Jews react by saying that Abraham and the prophets were dead and accuse Christ of claiming to be greater than Abraham. To which Christ replies, "...before Abraham came to be, I am." The term *I AM* was the name God gave to Moses to describe who sent him to lead the Israelites from Egypt.

By insisting on these associations with God, the Father, and demonstrating them through the healing of all types of illnesses and deformities, and even bringing the dead to life, Christ tried to convince the Israelites of His identity. This demonstration convinced some but not all. The acceptance of Jesus today as physical evidence of who He was/is should be judged by the same acceptance standards accorded other historical figures whose existences are based on witnessed and recorded accounts of their lives. It is by some but not all.

If biblical references were accepted in the Cartesian era, why didn't Descartes use them in his proof of the existence of God? The reason was his objective of rejecting any information of the outside world in an attempt to reach answers to all his doubts strictly through his ability to think, to rationalize the principal problems of existence.

The other case of physical evidence of God's existence is not comprehended due to the confused rationale of related arguments. As I connected the existing evidence of the soul to dualism, which no one else has done in published form, so I have also found a rationality gap in the continuing disputes between evolution and creation, which involve God not directly but as a subconscious fear. The misapplication of the various concepts to the physical facts has produced a confused array of assertions and contradictions. In the process, the importance of one pervading factor has been overlooked, or conveniently disregarded: the relevance of DNA to the creation of life. The function of DNA in continuing life and altering its conditions is no longer questioned. However, the question of how it was created is pertinent to the debate but is not addressed. (The details of this difference are discussed later in order to address Intelligent Design as part of the confusion.)

The introduction of Intelligent Design (ID) into an otherwise endless debate about evolution and creation has obviated any possibility of reaching a solution. As Descartes formulated in his rules for analysis (see Chapter 2) it is necessary to not only keep in mind the objective of the analysis, but also to proceed from the beginning and progress in an organized manner to a solution. In the confused debate, which includes educational and legal participation, it is no longer clear what the fundamental objective is.

The first fault with the ongoing debate is the expansion of the theory of evolution to include creation. Evolution is a process that begins after creation, not before or as the actual process. Evolution attempts to show how changes appear in living material, not in how any matter was created: it can give no solution to the creation problem. This misconstrued association has been further exacerbated by the inclusion of the Creationism theory, which is a fundamentalist acceptance of creation as a historical 6 day fact according to the Bible. This theory changed the original thinking about creation as the inception of life in primordial organisms to an idea of creation of life about ten thousand years ago in the more recent forms of living matter, ignoring all prehistoric forms such as the dinosaurs. Consequently, Creationism defied Darwin's natural selection in favor of creation in fully developed species. Intelligent Design, as a result of being associated with such creation, took on a meaning that was never intended for it. Religious groups who advocate both ideas are responsible for linking the two, which is not part of ID theory. Even the debaters are not aware of what they are arguing about, based on a license of philosophic freedom for expounding any theory. To top off the confusion, the introduction of the scientific certainty, DNA, merely sends the arguments in all directions instead of aiding in reaching some sort of agreement, at least, in portions of the debate.

According to Rule III, we should start with what is credible not with what anyone else says. Therefore, beginning with definitions, only ID has not been defined above. In common understanding, design is necessary in the development of any device or any plan. The degree of intelligence demanded by the new design is defined by the intended functions of the end product. An obvious corollary is that for every ID there must be an intelligent designer, which is a disturbing contribution to the creation problem.

The reason for the increased controversy by the addition of ID to creation is immediately apparent: it means a different type of design for each creation. In the Creationist version, ID conflicts with evolution because it eliminates natural selection by the designed creation of *developed* forms. But ID does not apply to Creationism. Therefore, the argument bogs down by evolutionists claiming the introduction of ID is a religious attempt to inject God as creator to replace evolution theory. Since evolution has been battled through the courts and in the education system with the complete support of science, the battle is fierce as to who can control the worldview of how life came into being. The frustration of this fighting has reached the point that in the year 2007, there was an attempt at a truce to cease the controversy in order to pursue a more important question for human survival: the global warming and environmental protection issues. For the debaters in the creation argument, it may be a welcome respite, but beneath the surface it will never go away because the educational and legal problems remain.

Due to archaeological evidence, the Creationist theory lacks scientific credibility. Therefore, for our needs in seeking evidence, it is the other definition of creation and the involvement of ID that is more pertinent. The involvement of ID in the creation of life in earth's primordial slurry has an entirely different meaning because it is involved at the cellular level. Dr. Michael Behe, Professor of Biochemistry at Lehigh University, gives the reasons why it is only at this stage that life's beginning makes scientific sense. In his book, *Darwin's Black Box, The Biochemical Challenge to Evolution*, he details the complexity of the biochemical processes that determine life's development. It is a continuing process based on DNA and not natural selection, although there are relationships. Since Darwin did not have the information about DNA, he did the best analysis he could by observation. As a microbiologist, Dr. Behe explains the complexity of cellular structure and the functional operation of living forms based on this complexity. Nowhere in all of published literature, according to Dr. Behe, is there an explanation of how evolution was responsible for the development of life at the cellular level. There has never been a meeting, a book, or a paper on any details about the evolution of the complex biochemical systems. Since there is absolutely no such explanation, evolution (with respect to creation) is merely an unsupported theory which is being used as a scientific crutch in the absence of a valid explanation.

Without attempting to solve any part of the debate, I will draw out the pertinent facts that lead to my evaluation of the evidence for God's existence and then supply answers to some of the anticipated and inevitable skeptic's question-begging attacks.

It is a scientific fact that the helical spiral of DNA is present in all living matter. When it was first being explained on television, I remember the statement by the lecturer that the DNA's in his nose and in the banana he was holding were the same. The difference was in the genetic information derived by the combination of the four nucleotides labeled by the letters A, C, G, and U. The presence of the same basic DNA in all living matter allows the tracing of the genealogy of species and the presence of substances to various times in history. Thus, it is a fact that apes and chimpanzees have most of the same genes as the human being.

That there is only one DNA, an established fact, is an essential element in my rationale for the physical evidence of God's existence. It is physical evidence. Since there is no other DNA, it has been the pathway for analyzing the genealogical development of any species including man. The difference between man and animal is a story of its own, not directly pertinent to the search for evidence of God's existence. Along the path to genealogical discovery, this common DNA is present in all living matter through the billions of years to a common ancestor. But that ancestor is not of any genus or species; it is the *one original living cell.* There had to be only one because there are no other descendant DNA's. I will answer two skeptical questions now. Why only one? Because if there had been more than one, the obvious, almost trite, answer is that the others would still be replicated in today's environment. After a billion years, there is but one DNA. But what if there had been others that combined with the "one cell" to create the present one. If so, (a) that would have been the only change and that would still lead to the original one, as well as to the similar complex creations of the contributors; (b) if there had been a change at some point, the original cell descendants and the modified ones would now constitute two DNA's which has not happened. What if the first cell had developed by chance rather than as a created one? If so, others could also have originated by chance, at any time in the billion year period, resulting in a multitude of different DNA's. But they haven't. Since the DNA is part of every cell, there could not have been an original cell without DNA because there would be living matter of some sort without DNA, which is not a fact (except for viruses). Since every cell has DNA that can be traced back to the original cell, that original cell contained the DNA as a completed design, which had to have a designer. Such complexity, which is still not fully understood, *could not happen by chance*. There had to be a Supreme Designer, an existing Being we call God. This is the physical evidence of God's existence in addition to the other physical argument.

Evolution began with the first living cell because life is the division of cells. The second cell may have been slightly different from the first but that would have been the first evidence of evolution. Evolution, therefore, is no part of the creation process; it is an inevitable consequence of the first created cell. Even Charles Darwin admitted that the Creator created the first living cell. After completing the *Origin of Species*, in the final statement of the Conclusion, Darwin wrote:

> There is grandeur in this view of life with its several powers, having been originally breathed by the Creator into a few forms or into one; and that, whilst this planet has gone cycling on according to the fixed law of gravity, from so simple a beginning endless forms most beautiful and most wonderful have been, and are being evolved.[24]

After all his work and reflective writing, Darwin made three significant points in this final statement that is conveniently ignored by opponents of creation: (1) There was a Creator, (2) the Creator breathed life into the first form, and (3) there was one form, or a few. In this statement, he acknowledged that the creation of a single form preceded his conclusions about natural selection of developed forms. His use of the expression "breathed life into a form" sounds much like the Biblical account of how God created man: "Then the LORD God formed man out of the dust of the ground and breathed into his nostrils the breath of life and man became a living being." (Gen 2:7) Had Darwin known about DNA, would he have concluded that the appearance of the same DNA in all his observed specimens pointed to a theory other than natural selection?

All the above facts point to the creation rather than chance origination of the first living cell. With the same DNA as is now confirmed, because there are no others, it is obvious that the complexity of that cell was the same as it is now, and which we still do not fully understand. This complexity, according to Dr. Michael Behe, is positive proof that intelligent design was involved in the creation of that first cell.

> The result of these cumulative efforts to investigate the cell-to investigate life at the molecular level-is a loud, clear, piercing cry of "design!" The result is so unambiguous and so significant that it must be ranked as one of the greatest achievements in the history of science.[25]

> Why does the scientific community not greedily embrace its startling discovery? Why is the observation of design handled with intellectual gloves? The dilemma is that while one side of the elephant is labeled intelligent design, the other side might be labeled God.[26]

An interesting confirmatory event about the rationality and universal appeal of ID to metaphysical and religious issues is the important role it played in the life of Anthony Flew, professor of philosophy and the most renowned atheist in the English speaking world of the late twentieth century. After fifty years of debating the positive side of atheism, professor Flew reversed his view of the existence of God in 2004 to the great consternation of his also recognized fellow atheists. In his dedicated "pursuit of truth, no matter where it leads", it was the realization that the evolving scientific discoveries pointed to intelligent design of life and the universe and inevitably to the Intelligent Designer, God. However, Professor Flew must still be convinced of the soul's reality and a life after earthly life. In his pursuit of these truths, perhaps this book will help since they are related to a divine existence.

My arguments for the existence of God, with physical implications, are intended to be similar to those for the existence of the soul; attempting to show that the existence of the non-physical can be argued in a physical manner. Exposing the evidence is the objective here, not its generation.

The *intentional* appearance on earth by the *Son* (because He was sent by the Father, whose existence is doubted) is the physical evidence of that doubted existence. Christ's supernatural power over natural laws verified his claims of divinity. Human refusal to accept Christ's divinity or God's existence does not alter or negate the evidence; it only demonstrates a refusal to evaluate it objectively.

The complexity of the living cell is a physical complexity, not requiring any assumptions or future verification. DNA is a physical reality, accepted by the entire scientific and non-scientific communities. Its use as a trace of identity and genealogy supports the argument of continuity to the first living cell. Finally the complexity of the first living cell absolutely defies the possibility of chance in its creation. This assemblage of facts confirms my logical deduction that the creation of life could only have been by an Intelligent Designer, which constitutes positive evidence of God's existence.

4.7 RESOLUTION

The soul does exist. The ultimate evidence that science could demand for its existence would be the actual appearance of the soul in a physical form for demonstration and verification. That such appearance did occur repeatedly should now be a demand for science to accept the soul's reality and apply it to the research that so far has suffered by its exclusion. The appearance of the soul through reincarnation should not be an arguable issue due to doubt about reincarnation because both realities were confirmed and demonstrated in the same act of voluntary and spontaneous revelation. The spontaneous demand for recognition as a previous person with lucid memories of the past life is conclusive proof of an independent entity, a person. The appearances of birthmarks or defects related to incidents in the past life are corroborative evidence. The meticulous investigations by scientific researchers of the numerous comparable incidents, with methodical scientific reporting and disclosure, fulfill the requirements for the conduct of scientific research. Reincarnation, therefore is a scientific reality and its demonstration confirms the passage of a spirit, call it a soul, with all its non-physical properties from a dead body to a newborn body by passing through an immaterial medium.

The nature of the evidence leads to conclusions about the human intellectual and behavioral characteristics that are functions of the soul and its mind, conclusions that are in direct conflict with the claims of neuroscientists and cognitive scientists. Logical arguments are used to link the reincarnation evidence to a definition of a non-physical mind that is independent of the brain. It interfaces with the brain in initiating physical activity. Memory, rationality, thinking, and volition are mental activities independent of the brain but interfaced with it for information sharing. The brain also has independent functional responsibilities for continuing life support activities which require no conscious contributions by the mind. The result of these interactions is a holistic and integrated system operation of mental/physical efforts in human activity and behavior.

There are other than scientific issues revealed through reincarnation. The evidence of a reincarnating soul confirms that there is a continuing life after earthly death, a disturbing fact, perhaps, to the secular community that has been relieved of this concern through materialistic claims. This leaves open the question of what that means to each human being in terms of respect for life, the meaning of life on earth, moral responsibility, and the belief in God.

The differences between science and metaphysical philosophy were pointed out to emphasize the extent of the separation of the two domains. Although there is interdependence between the two disciplines, the basic functions of the two are incompatible. This incompatibility highlights the inevitable struggle for supremacy between the two in resolving the interface problem created by the inclusion of a non-physical soul and mind into the mind-brain and body-soul relationships. Will the scientists say what is required for the integration or will the philosophers be able to define what is available? The new philosophy must be confirmed by both disciplines before any meaningful research projects can be identified.

The probability of conflict within the scientific disciplines can be anticipated when discord about the current approaches already exists. The evidence of the soul will, however, strengthen the arguments of the critics of physicalism. They may be able to tie together the loose ends of their analyses which now cause them to doubt their own work.

The three examples of current arguments were included to show that some arguments will continue indefinitely as long as the wrong assumptions prevail. Evolution is not an argument about creation, and with the evidence of the soul, cannot address the development of the soul nor the point in time when the soul became part of Homo sapiens. Evolution can't even date the origin of man.

Brain damage as a reason for denying the existence of a separate mind is only valid if that mind is excluded as a basic assumption. Including a mind in the problem leads to different conclusions about the effects of brain damage on mental performance.

The argument about folk psychology should continue since it is used frequently as a convenient crutch in any argument. The definitions of folk psychology revealed that the unsupported conceptual explanations of physical causality are as "folksy" as any commonsense solutions.

Some of the evidence for the existence of God as presented in this chapter as being physical evidence could have been used by Descartes in his arguments. The incarnation of Jesus Christ as a human being is comparable to the reincarnation that produced the evidence of the soul. The evidence of a non-physical entity in a physical form is extraordinary evidence. As it proves the existence of the soul for a human being, a similar demonstration of duality in Jesus Christ is a logical comparison of physical evidence since it too was an incarnation by an existing Being. The rejection of Christ's divinity is similar to science's denial of the existence of the soul. The confirmation of the latter implies a similar conclusion about the former, especially since it was prophesied for thousands of years.

The following 4 chapters present the arguments why the current research of the mind-brain problem is misdirected; they are a combination of a backward glance at the worldview of the current materialistic approach with a forward look at the effects of the soul's participation. This seeming reversal of order in presenting the arguments after the soul's reality has been presented, is intended to instill a feeling for the direction in which the debate was continuing on the assumption of a soul's non-existence. Although there was no evidence for or against the denial of the soul, the scientific view of physical causality appealed to the scientific and secular communities and allowed the free development of a conceptual structure. Not only were the neuroscientists and neurophilosophers responsible for this misdirection, the theologians and traditional philosophers were as responsible through their inability to prove otherwise. Only in this order of presentation can a proper evaluation be reached about the misleading nature of the ongoing research with its implied certainty of eventual verification. In this retrospect one can perceive the magnitude of misconception society has been led to believe when the information that could change the direction of the research was available.

5

Conflicts and Contradictions

There are disagreements about
the Elephant's descriptions.

The most troubling aspect in the research for this book was the uncovering of so much misleading information in the developing approaches for solving the brain-mind problem. The philosophic optimism of the neuroscientists about the positive future vindication of their claims lacks scientific supporting evidence in too many conceptual approaches, *which is even acknowledged by the researchers.* In addition, there is no systematic plan of subsystem integration. In spite of these deficiencies, the research continues on the basis of a false fundamental assumption, the denial of the existence of a non-physical soul (and mind). The result is a mandated philosophy of materialistic certainty. *With the scientific verification of the soul's and mind's realities all those conceptual claims are negated.* When the major basic assumption for the conceptual research on physical causality is false, how can any of the related physical research be credible? It galls me to realize that the minds of so many intelligent, and even brilliant, scientists have been misused and so many resources expended in meaningless attempts to prove concepts which are but substitutes for not understanding the all-inclusive problem. It is not my lack of neuroscientific expertise that causes me to make such statements in spite of the purported advances in cognitive research. Instead, it is the knowledge and understanding of the pertinent evidence, available to the scientific community but ignored by it, that allows me to dispute the claims and envision how that evidence could have redirected the research so that the results by now would have been much more meaningful.

Since I am challenging the scientific community to rethink their optimistic conceptual conclusions, I feel that it is my responsibility to show examples of the inadequacy of what is being proposed to support physical causality. Its inadequacy is shown not by my criteria but by dissenting and contradictory logic of other scientists and academic experts who are not bound by the need to conform to any dominant philosophy. Such adversarial contact is the format for this chapter. By this process, I present the individual views of the authors and contribute my critiques based on my professional background. This chapter is my attempt to convey to the reader the implications of a misdirected effort. It is also an opportunity for the reader, now informed about the evidence of the soul, to detect, in addition to my comments, where this evidence would have made a difference if it had been included in the basic assumptions. These comparisons are exposures that conceptualization without verification is worthless.

A complex subject, as we have here, has associated problems of comprehension if the language and terminology are confusing, semantically vague, or esoterically unclear. And simplification of technical names and terms is not the only problem. It is frustrating to try to interpret the meaning of invented terms, which, though defined, do not always convey the reason for their invention when conventional usage would suffice as in the substitution of representation for idea, or mental state for mind. It is a convenient ruse to make a theoretical inconsistency appear credible when using terms whose meanings can be manipulated. This is especially true in philosophic discourse where semantic differences create arguments that often obfuscate the basic issue being discussed.

In a comparison of theories and arguments (as will be the pattern in these conflicts) clarity is essential or the opposing views will not relate. If the intended meaning of a conceptual disclosure or an opposing argument is not evident due to esoteric doublespeak, the resultant conclusions may be misconstrued and can detract from the main line of thinking. As an example, in the following quotation, a conceptual approach is broached. For the reader, uninitiated in the pertinent technical background that led to the manipulation and eventual choice of the quoted words, it may be a struggle to integrate the terms into a meaningful understanding of the intended thought.

> You can probably see what is coming. If the external environment is represented in the brain with high-dimensional coding vectors; and if the brain's "intended" bodily behavior is represented in its motor nerves with high-dimensional coding vectors; then what intelligence requires is some appropriate or well-tuned *transformation* of sensory vectors into motor vectors.[27]

Obviously, the meaning and intent are perfectly clear (to that author). But let's look at it as an example for the average reader, uninitiated in the terminology of cognitive science, who is also new to this concept and is just trying to appreciate the magnitude of this ingenious contribution to science. What could be the confusion here? First of all, beginning with the word *if* immediately introduces the possibility that what follows is not fact but presumption and that there possibly is an *if not* (which will probably not be given). With enough assumptions, any theory can be rationalized. Even assuming that coding vectors have been, or will be, well defined, the use of the word "intended" in quotes raises suspicion that the word is used to imply a special connotation that leaves the meaning of the entire sentence unclear (again with an if). The italicized transformation also raises the question why the italics are not also used with "well-tuned" or "appropriate" since all three provide a wide latitude of illusory interpretation?

The bottom line, simply stated, is that the brain needs some mechanism to transform sensor signals into motor signals when the brain decides on the intended bodily behavior. Obviously vector control is being pushed as a worthy concept. But that exposes the fact that intelligence, a non-physical property is needed as the interface between physical sensor signals and physical motor signals for bodily behavior. This is hardly a materialistic solution since "appropriate" (undefined for complete coverage) transformation is needed to help a non-physical intelligence to complete the physical loop in a concept that is supposed to be only physical. All arguments against these statements are already answered by the inclusive words like appropriate and well-tuned. Otherwise, there would be doubt that a reasonably well thought-out theory is being presented. This interface mechanism, requiring such elusive descriptive terminology, lacks the supporting information about how this integration occurs.

If this is a materialistic solution (note my use of "if"), why does it need a non-physical intelligence in the loop? Can intelligence be physical? What is being described here is a non-physical mind that perceives the sensory information produced by the brain, interprets it, makes a volitional decision for bodily behavior and orders the brain to generate the motor signals for body activity. All the vector control nonsense is an attempt to explain the mind in physical terms. Since we know the mind exists, the operation becomes clear although the interface is yet to be defined in any case.

The above "picky" analysis of one description merely demonstrates that the uninitiated readers who are trying to follow a technical description can lose contact with the author by the ambiguities of esoteric and apparently veiled meanings. In presenting the statements of either side in the conflicts that follow, my hope is that I have not been confused by such hidden meanings and guilty of erroneous conclusions. It is through my battles of trying to understand all the complex theories, that I have become aware of the possible misinterpretations, especially when invented terms are used which cannot be found in any dictionary. For this reason, I inject direct author quotations to assure that the exact terminology is theirs, not mine.

To benefit from the many conflicts between theoretical and substantiated data, it is necessary to understand the basic material involved in these conflicts. Some knowledge of the brain and the nervous system is mandatory, therefore, preceding the examples, definitions and summarized descriptions are given. Otherwise, constant explanation of sensory activity that is meaningful to the story would derail the train of logical thought. This material is included to aid those readers who are not too familiar with the physical and functional aspects of the human brain and nervous system. If these summaries are inadequate for any reader, there are encyclopedias, hundreds of books, and the Internet that can fill the need for more information. For those readers knowledgeable in these subjects, the material may be bypassed.

Of some importance is a top-level definition of the brain, its components, and areas of information processing. An explanation of the neuron and its connections is mandatory due to the constant references to these components. A definition of the mind is especially important because there are so many variations and for different theories they have different meanings. In fact, if the reality of a mind as a separate entity is denied, the reader can justly wonder how a human being can function without it. Searching for answers to such questions is the path of discovery that makes the journey necessary. Such a journey, however, has its risks in hopeless conjecture when non-physical, intangible entities are involved. The soul is another such entity that must be addressed in dualism. A soul must be defined because for many philosophical scientists it does not exist. For others it does.

Brain

The human brain is the principal organ of the nervous system. Its 100 billion gray and white nerve cells perform the functions of transforming the outer world information produced by the sensing receptors (eyes, ears, taste, smell, and touch) into useful signals for activation of physical activity and mental awareness.

The brain is divided into distinct areas that respond to signals from their respective sensors and generate the commands for bodily response. The largest portion of the brain is the cerebrum, which occupies the entire upper portion of the skull and is wrapped around the brain stem which is attached to the spinal cord. To visualize this, make a fist with the thumb under the index finger. The top of your hand, the fingers and the thumb represent the cortex. At the tip of your thumb is the upper end of the brain stem. This brain stem was the early part of the brain in its evolutionary progression while the rest of the brain grew with the evolutionary need for greater body response to the environment. The cerebrum is divided into two equal hemispheres, each being further divided into four lobes. All information processing occurs here in collaboration with the sensor functions. The sensor signal passes along a set of dedicated nerves to its designated location. The brain function at that location transforms the signal into an output command that will be integrated with any other required output signals in the center of the brain, the thalamus. All sight, taste, smell, and hearing information is processed in the proximity of the brain requiring only short transmission lines.

Any sensor information from the body-touch or pain-is conveyed to the thalamus for integration with these other brain signals for the generation of the necessary responsive motor commands through the spinal cord. There is an immediate feedback from the muscle to the brain to update the required motor action. This directing and feedback operation is extremely fast, as you experience in the continuous movements you make in any action responding to a given sensory input.

The amygdala lies at the base of the brain stem and is actually part of the spinal cord. Its function is to control the involuntary activities such as respiration, digestion, and circulation. The connections between the brain and the spinal cord cross in the medulla and the result is that the nerve centers on one side of the brain control the sensations and movements on the opposite side of the body.

Nervous System

Man's nervous system is a very complex and extensive network of special tissue that coordinates and controls the actions and reactions of the body in its adjustment to the environment. It consists of the brain, the spinal cord and the peripheral nerves. An analogy of the functions of these three parts is a business office communication system. The brain is the switchboard, the spinal cord the main trunk, and the nerves the various plug lines that receive and send messages.

The nerve is a slender cord made up of nerve fibers. It is surrounded by loose connective tissue called epineurium. Nerves are the communication paths between the central nervous system, the brain with its billions of cells, and all parts of the body. Neural action is initiated by generation of sensory information, which is channeled to the dedicated areas of the cerebral cortex. After transformation, it is integrated for delivery to the various body organs and body parts. Sensory (afferent) nerves carry sensory impulses from the sense organs to the central nervous system and the motor (efferent) nerves lead to muscles and other effectors. Most nerves of the body are mixed nerves containing both types. The generation and updating of information occurs in milliseconds, even at extreme distances like toe to brain.

There are special nerves, autonomic, which regulate the smooth and visceral tissues of the body (digestive organs, heart, lungs, kidney, etc.). These are like programmed instructions in any electronic device, requiring no conscious control except in taking corrective action during functional disorders. When the body needs swift reaction in emergencies, sympathetic nerves tend to tense and constrict involuntary muscles and blood vessels and step up the activity of the glands. In the reaction to a fight or flight situation, the stomach tightens, blood pressure rises and adrenaline pours into the system. The parasympathetic system works with the opposite commands tending to slow and relax the same organs. Between the two types, the organs work smoothly, yet adapt quickly to change.

The basic structural unit of the nerve system is the nerve cell, or neuron, composed of a nucleus and a cell body from which branch out threadlike structures. These extensions, dendrons or dendrites, divide and subdivide into smaller extensions and pick up nerve impulses for transmission to the cell. The nerve cell also has a principal connector, axon or nerve fiber, which may or may not subdivide. It transmits the impulses to other cells, organs or tissue. It is protected by a myelin sheath and by a thin outer cellular layer called the neurilemma. The end of the axon branches out in a bulb which is the transmitting presynaptic connector. The connection to a receiving postsynaptic dendrite or axon is across a gap one millionth of a centimeter wide. This gap is called the synapse. It is across this gap that information is passed in the form of electrical signals generated by the potential between positively charged sodium and potassium atoms, or ions, and negatively charged sodium atoms. In this close encounter lies a world of potential to hand off the signals that science claims find expression in thoughts, emotions and sensory perceptions (Remember, this is the best science can do without the evidence of a mind.)

Information is transmitted in the nervous system by nerve impulses. An impulse is the progressive transfer of a condition of excitation along a nerve fiber, initiated by a stimulus acting upon a sensory organ, by a cell within the nervous system, or experimentally by direct stimulation of the nerve fiber. There is measurable chemical change and energy consumption in the generation of an impulse, accompanied by changes in electrical potential associated with membrane depolarization (like exchange of sodium and potassium ions across the membrane). The passage of the impulse across the synapse is accompanied by the action of acetylcholine, a crystaline compound, while at terminal nerve fibers either acetylcholine or sympathin, which is closely related to adrenalin, accompanies the transfer.

As an impulse travels from nerve cell to nerve cell it is received by the dendrites and is sent along by the axon. A number of axons bound together with connective tissue form a single nerve and a bundle of nerves bound together form a nerve trunk. A collection of nerve cells along the peripheral nervous system is called a ganglion. Its function is to act as an intermediate coordinator of impulses for transfer to the higher order central system. Motor nerves do not contain ganglia since they are only transmitting commands from the central system.

In especially sensitive areas, the reaction is immediate because the impulse does not have to travel to the brain for a responsive command. In these sensitive areas, the sensory nerve is interlaced with a motor nerve in a mini nerve center, a reflex arc. Thus when a finger touches a hot object, the command for the instantaneous response is generated locally, much faster than a request for a response from the brain. This is known as an involuntary reflex action. Many voluntary actions become reflex actions through continued association of a particular stimulus with the same result. These are conditioned reflexes, just as with Pavlov's dog. Acquiring conditioned reflexes is the basis for habit, learning, and education.

Mind

A mind is difficult to define specifically for two reasons. The first is that it is not a physical entity and cannot be described in physical terms. The second reason is that it has different meanings for the different branches of science, different philosophies and for theology. In some definitions it is a separate entity from the brain, in others it is not. In both, however, it is not a physical entity because it cannot be located as part of the brain. The problem, therefore, is in attempting to analyze and describe how it functions or what effect it has on the brain. Due to this difficulty, most scientific analyses wish it away to simplify their analyses. It is easy to sympathize with them considering the problems of analyzing an unknown quantity and determining its specific effects on the physical brain and the resultant human behavior. But because it does exist as a separate entity, to eliminate it from any analyses is to do an incomplete study.

For some philosophers, mind is a separate entity and as such is considered to be the soul. For others it is a separate entity but only as part of the body with no spiritual implication. Still others assume the mind to be part of the brain, not the body. There are those who deny the existence of a mind basing their claims on the belief that human behavior can be adequately described without the use of such a vague and indefinable unit.

In spite of all the above conflicting versions of mind involvement in human consciousness, the high probability of a mind association with the brain, as will be referred to constantly in these conceptual battles, demands an attempt at a basic definition. Therefore, as used here, the mind is that non-physical source of activity that generates all willful thought. Thought initiation, though, is by either of two prompts: by sensory information from the environment or meditative (reflective) desire. Such triggers produce desires, volition, language, meditation and free will, all of which, from an experiential aspect, are non-physical "things". This definition will be contested in the various concepts of dualism or monism, but it will also be proposed as the argument against such concepts because the mind does exist as a property of the soul.

Soul

The existence of a soul as a mind or a "being" is more contentious than the existence of a mind. To the scientist who cannot prove the existence of a mind but must accept its apparent effect, denial of the existence of a soul comes much easier. It is the philosophers and theologians who are the proponents of a soul as the companion to the body in dualism. What is the soul? Again, defined basically, it is a spirit that unites with the body for the lifetime of a human being and continues its existence in a hereafter when the body ceases its earthly life.

Heretofore, it has been easy for the scientists, and anyone else, to deny the existence of the soul based on the inability to see, hear, or touch a soul. Of course, they say, even if the soul does exist these actions are not possible since a spirit has neither mass nor any other physical properties which human beings possess. How could the presence of a soul in a body, outside of a body, or in the hereafter be detected? It can only be done by personal contact with another soul outside of one's own. (According to the evidence presented in Chapter 3, there is no doubt about the soul's existence. As discussed here, that evidence has not been recognized by cognitive science and is, therefore, not used to disrupt the accepted lines of reasoning.) Reports of unusual personal experiences like NDE and ESP cannot withstand the inevitable ridicule of the skeptics and are therefore rarely attempted. But there are reports of such contacts and other evidence, which are beginning to create a need for admitting the probability of the soul's existence.

Self

What is the definition of self? Generally it is a subjective term with various meanings. It is seldom used technically but when it is used it refers to a person with incorporeal properties and obviously consciousness. As applied to the mind-brain or soul-body issues in philosophical, psychiatric, and psychological explanations it is either materialistic or dualistic with associated meanings. The principle difference is in the interpretation of subjectivity; it is either inherent mind awareness or a conceptual brain-generated capability for imitating the functions of a mind. When volition, which is intrinsic in subjectivity, is claimed to be generated through physical brain manipulation, subjectivity loses its basic meaning.

For the materialist, self is a brain dependent awareness in which the brain has a causal organization by which its patterns interact and are interpreted by other brain patterns to initiate and execute an individual's behavioral activity. Brains presumably construct a self-concept to coordinate activities for self preservation.

For the dualist, self is a mental image of a total integrated person. The soul works with the body just as the mind works with the brain. Each has specific properties that implement the capabilities of the other. Definition of the interactions is not important because only the resulting behavior is of any consequence. With the introduction of definable properties of the soul and the mind, this convenient explanation of unity will no longer be adequate. The effects of the non-physical on the physical, although still classed as one self, will have to be defined for future research. That calls for the definition of such a self with two separable entities.

Which function performed by self defines who the self is? It is the controlling factor that determines self. In the monist it is the brain; in the dualist it is the complete unit because sub units are not prioritized. In the definition of a new reincarnated self, the controlling factor is easy to distinguish. It is the soul because it survives the first body, thereby controlling itself and subsequently controlling an immature new body. As the experiences of the past life are still fresh in memory, the guidance and control are obvious. The reincarnate provides the direction for defining the subsequent self. Through reincarnation the soul demonstrates that it has faculties which can function independently but rely on the physical properties of the body to create a self. There is a need, therefore, to identify these properties and their respective causal powers, not only with respect to each other but also upon the external world.

Although there are two separated entities, how can self be explained as a singular entity? The self, according to the two cases above, is defined by the controlling element. In the new self, the soul as the residual but changing personality is the one essential entity; the body is only the temporary "vehicle" for the current earthly trip. This may sound unattractive to one who is physically oriented, but as reincarnation is a reality, the continuing personality (actually the eternal person that may develop into another personality) is the only meaningful entity.

Since this is the first book that reveals the reality of the soul with respect to Cartesian philosophy, it is possible to define self on a factual basis. In the new philosophy, the definition of self should be a top-down controlled hierarchy with the soul at the apex of a conceptual pyramid which contains all the elements of the union, both physical and spiritual. The causal powers act downward and upward due to the intertwining of physical and nonphysical elements. It is one integrated unit but with complex interfacing. This is the self during life on earth. However, since there is an inevitable separation of the two entities at death, all the interfaces are discontinued and the remaining self is a person (with a lingering personality), which then includes the soul, mind, memory, free will, and the ego, all as one. This must be the true self of the human being, the self that can move on to another life.

This concludes the introductory material to the intended purpose of the chapter, the examples of conflicts within the debate.

To set the stage for the three examples of conflicting views, I quote the following explanation by Dr. Paul Churchland, a leading neuroscientific philosopher, of the basic approach to how the brain "represents" reality, from his book, *On the Contrary.* This basic explanation current at the time of the book's publishing (1998) may have changed since then but such a firm statement should remain unchanged since it was a fundamental approach in materialistic neurocomputational research. Note that this is a computerized concept of what a computing brain should be able to do, although in a later example of conflicting positions, an opposing argument shows that the brain does not function like a computer. Furthermore, the reality of a separate mind also argues against this basic position. However, my point in presenting this position is to submit an authentic framework for the three examples of conflict that follow.

The activation vector in the following statements is the pattern of activation across the sensory receptors, the eyes ears, etc. The vector quality derives from the variation of signal strengths at the receptors.

> ...the brain's basic mode of occurent representation is the activation vector across a proprietary population of neurons – retinal neurons, olfactory neurons, auditory neurons and so forth. Such activation vectors have a virtue beyond their combinatorially explosive powers of representation. They are ideally suited to participate in a powerful mode of *computation*, namely, vector to vector *transformation.* An activation pattern across one neural population (e.g., at the retina) can be transformed into a distinct activation pattern (e.g., at the visual cortex) by way of the axonal fibers projecting

> from the first population to the second, and by way of the millions of carefully tuned synaptic connections that those fibers make with the neurons at the second or target population.
>
> That second population of neurons can project to a third, and those to a fourth, and so on. In this way, a sensory activation pattern can undergo many principled transformations before it finally finds itself, profoundly transformed by the many intervening synaptic encounters, reincarnated as a vector of activation in a population of *motor* neurons, neurons whose immediate effect is to direct the symphony of muscles that produce coherent bodily behavior appropriate to the original input vector at the sensory periphery.[28]

If this is the basic approach on how the brain uses physical sensory data to produce responsive behavior, it reveals that there is no coherent understanding of either the requirements or the solutions for human behavior. The escape from this lack of understanding through computerization is evidence of slavery to the reliance on computational algorithms rather than rational use of pragmatic experience.

This basic approach sheds no light on the initiation or control of human behavior because it is only a series of steps without any factual support of the implementation of the steps. It is merely an exercise in computer programming in an attempt to create an illusion of knowledgeable ingenuity. But it relies on assumed capabilities that are unsupported by scientific fact. The transformations are assumed to be possible and the profound landing of the sensory activation pattern in a population of neurons that "know" how to transform into motor neurons that dictate the proper responsive behavior is absurd.

This approach is not representative of human rational behavior and only proves that there is no valid physical approach to a mental state solution. It is a computational procedure with inputs of conceptual allocation of unexplained transformational abilities. Who is the computer operator? It cannot be the brain because it is the computer.

The operator is the programmer who uses the rational mental mind in concocting such nonsense.

The assumptions that meaningful transformations anticipating an intended behavioral response are made by unintelligent neurons "by way of axonal fibers" demands verifiable logic since neural signals are simply on-and-off impulses. There are no vector magnitudes involved. (Look for the validity of the vector coding concept in the following examples of conflicting statements.)

There is a basic fallacy with this approach. Although a computer-like brain could handle the millions of sensory signals generated at the retina, there is no provision for the selectivity of specific portions of the total input. The eye may see at the same time one or more people in an environment that could include a street intersection with moving cars, and both red and green traffic lights. If all the visual sensory data are processed through the various transformations, how is the specific information selected for the intended response? To what information from the total mass of visual data does the activation vector respond? Is it to the red or green traffic light or is it to both concurrently? How does it avoid responding to the number of people? As set up, it responds to all the data, since there is no "self" to select the required response. What happens when there are no sensor data? If I close my eyes and cup my hands over my ears, what sensor information initiates the transformational cycle since there are no sensor data. But I still continue to think and plan my behavioral activity without any vector activation. How can a brain, as a computer, intend a response without sensor data?

This is a misleading chain of activities as coded transformations through finely tuned (computer trial and error derivation) synaptic connections, etc. The naive assumption that all this can be accomplished because a computer analysis predicts it is a wishful concept that has no physical supporting facts. Yet it is the *basic approach* to how the brain causes behavioral activity. The resultant behavior can only be robotic, deterministic, and chaotic because the response is to all the data. There is no inclusion of the selectivity in the voluminous sensory input as to what part of it is to be acted upon for an "appropriate" behavioral response. That requires volition which has no part in this basic approach.

Such is the framework upon which extensive neuroscientific research is being conducted. And the scientific community is contributing to it by not demanding scientific verification of the basic concepts.

The format for the remainder of the chapter is organized to track the contributions of the authors and to isolate my comments from them. Several concepts are used as illustrations for the arguments between professionals in the debate. I have used representative subjects to illustrate the discrepancies in what is basically a scientific belief (same as folk psychology) in the validity of physical causality. The order for the conflicting opinions is a conceptual claim followed by a response and my comments. The summarized claims and responses are based on material presented by the authors. Within the summaries are supporting book quotations to prevent misinterpretation of the author's intent. The originators of the concepts and the dissenters are identified by a parenthetical code, e.g. (5-1-S) chapter 5, concept 1, Smith. This will allow subsequent references to the specific sections. Even though I have done the summarizing as I interpreted the conceptual thinking of the authors, I have tried to avoid distorting their views. I reserve my comments to an evaluating section following each encounter beginning with (C) and ending with (CE). I will analyze the conflicting action, like a TV commentator in the control booth of a sports contest, reviewing the completed action.

The validity of the arguments will be evaluated by you, the reader. Since the presentations are based on the assumption that there is no soul involved, you might note your reaction, as I do in my comments, to how the argument might have gone had the soul's involvement been considered.

The authors of the books from which the comparisons are made are:

Dr. Paul M. Churchland, Professor of Philosophy at University of California, San Diego and a member of its several research programs in Cognitive Science, Science Studies and the Institute for Neural Computation.

Professor Joseph Levine, Professor of Philosophy at Ohio State University.

Professor Ian Glynn, Professor of Physiology and Former Head of the Physiological Laboratory, Trinity College, Cambridge, England.

Professor John R. Searle, Mills Professor of the Philosophy of Mind and Language, University of California, Berkeley.

The three issues in contention are:

1. What is the extent of the brain's ability to determine and control human activity and behavior?

This issue is about the dominance of the brain in translating sensory stimuli into motor commands for human activity and for generating the skills and knowledge required for social interactions. The brain functioning is not observed by any non-physical "self" or mind; there is only the physical brain. The opposing view is that traditional belief and personal experience in rational causality, which cannot be attributed to any physical phenomena, verify the existence of a non-physical entity in the human being. However, there is an assumption about the nature of the non-physical entity: the mind is devoid of any spirituality like a soul.

2. How does the brain interpret the sensory information provided by the environment?

This is an attempt to define how the brain uses a vector coding method to interpret the meaning of sensory signals for use in developing motor signals for human behavior and activity. The rebuttal exposes the limitations of the proposed concept by imposing biological reality on the theoretical model.

3. Is the brain comparable to a digital computer?

The third issue attempts to show that the brain operation is comparable to a high order computer. The complex nature of the cognitive operation demands a Parallel Distributed Processing (PDP) type computer capability. The opposition counters with the various reasons why the computer comparison is unsound.

Issue 1

The first presentation is a concept described in *The Engine of Reason, the Seat of the Soul* by Dr. Paul M. Churchland.

(5-1-Ch) The traditional mystery of how a human mind-brain combination functions has been demystified to a certain extent by neural network research in showing that the biological brain is solely responsible for human activity. If this implies that physical causality is responsible for human activity, the mind-brain problem becomes strictly a brain problem. This work, based on animal studies and conceptual computer models, has increased the understanding of what causes humans to behave as they do. Experimentation with animal brains has shown that animals respond physically to sensory information in a similar manner. The net result is that the biological brain has been shown to be the only entity responsible for the initiation of physical activity. There is no other non-physical "self" watching, controlling, or motivating such activity.

How is this accomplished? It is through the interaction of the brain's 100 billion neurons in 100 trillion synaptic connections that the sensor inputs from the external environment are processed, interpreted and integrated to produce motor signals for body activity. This sensory information, picked up by the five senses, is passed through nerve fibers to the various cortical brain areas where it is perceived, interpreted, and processed into outputs useable for integration. It is then integrated by the cerebellum and transmitted to the various body areas. Through this process, the information of the outside world is translated into perception of the social and physical spaces within which, and how, the body is to navigate. This includes recognizing one's own position with respect to others in these spaces and how to move within these spaces without harm. Entailed in this awareness are the human emotions, desires, and necessity for planning the future.

To accomplish this awareness and to structure a nervous system to respond to the imposition of the worldly environment the brain must "learn to generate behavioral outputs".[29] It must learn the interactive social culture, within which it must live; the ability to move about and survive; the relative importance of positions, events, structures and processes; and the moral relativism of its activities. It does this by configuring a set of prototype responses in the synaptic connections that can react to various situations. By changing the "weights" of the synaptic connections, different and ever changing environmental demands can be responded to. What is a synaptic weight change? In the computer network simulation (which is the basis for this concept) it is the setting, through error analysis, of the values "of the synaptic connections so that collectively they will perform the desired transformation".[30]

The awareness of belonging to a social space begins in infancy. As the brain matures after infancy, it "carefully" tunes the strengths of these connections into a "lasting" configuration with more useful values, which then dictate how the brain is to react to the worldly environment. These adjustments are guided by genetic influence and by encountered experiences, mostly in early life. The configurations are stored in the synaptic connections through the learning process. The major changes to the configurations occur in early life, diminishing in the rate of change with age.

To adapt to the ongoing moment-to-moment changes required by ongoing activity, "These more fleeting facts get represented by a fleeting configuration of activation levels in the brain's many neurons, such as those in the retina and visual cortex."[31] The connections remain unchanged but the internal activity within the neurons accounts for the required changes to produce motor changes.

Standard circumstances, like chairs to be moved, dangers to be avoided, and telephones to be answered, require standard but plastic modes of apprehension and behavioral response. To be able to recognize and respond to these circumstances, the brain must learn about the causal effect of this world input. This knowledge must be part of the trillions of synaptic connections. "This is where intelligence begins in the brain's capacity for executing principled sensorimotor transforms, in its capacity for doing the right thing in its perceived circumstance. Here is where skills reside, where know-how is embodied, where smarts are buried."[32]

Thus the brain responds to the external environment by utilizing this knowledge to produce motor commands to respond to that environment.

The contradiction of 5-1-Ch is based on *The Purple Haze* by Joseph Levine.

(5-1-Le) Why is there a mind-body problem? It is because the conceptual storehouse of proposed solutions is ill equipped to supply the concept that closes the gap between the materialistic obsession with the physical brain and the unsubstantiated denial of any dualistic natures. The concept 5-1-Ch above, that only describes the physical part of the dual nature and conveniently imposes the functions normally associated with a mental complement on the physical interactions of synaptic connections with proper weights, is a deliberate escape from treating the entire problem. There is a serious gap in how this transfer could occur. Without a detailed explanation of how a physical reaction becomes a mental experience with volitional capability, such a concept is merely a naïve attempt at denying the existence of a mind (an example of why the definition of a mind is important). How strange that a mind with the volition to generate an original concept can propose that it itself does not exist to initiate that concept! What kind of environmental physical sensory signals are capable of initiating and formulating brain functions that lead to conceptual innovation and behavioral intuitions? Could it be the sight of a shelf of books, the examination of a dissected brain, or the smell of a chemical laboratory?

> What is it about a physical state, such as a sequence of neural firings in the brain that could give rise to a representational feature, such as my thinking about the red diskette case on my computer table? That is, what could make something in my brain be about the diskette case? Furthermore, what is it about an event in my brain that could give rise to my having an experience of red? The relations between the two sorts of phenomena seem baffling. It therefore seems plausible to adopt the hypothesis that the reason we can't understand how mere matter and energy can support these features is that they can't.[33]

The attempt at physical causality (physical motivation for body activity) completely ignores the very mental features that define a human being: the ability to override a reflexive reaction, to conceive future activity in the absence of an environmental input, and the conscious experience, the awareness, of any external contribution. These three features need elaboration to show why a strictly physical system must be explained in how a sensory signal input is transformed into a meaningful, non-automatic, deliberate activity. Unless such a transformation is explained, there is no validity to a seemingly intelligent concept of only physical totality.

Overriding a reflexive reaction is a "first person" activity that cannot (normally) be verified as a third person observation: but it is a reality. If a physical neural system (like in 5-1-Ch above) causes a reflexive action due to the "stored environmental configuration" without an override capability, it becomes a simple Pavlovian response. A human being, however, has that override ability, even if this decision to ignore a potentially detrimental situation proves to be harmful. How often has an unkind remark demanded a response in kind, but for the sake of avoiding a confrontation, that natural urge is overridden? This is a deliberate decision by a non-material entity. This is human behavior that is not the result of a myriad of synaptic connections in "anticipating any future activity".

Conceiving a plan is an originating operation, not reactions to a set of neural signals from a synaptic configuration. It is drawing on information that may not be stored in memory and has to be generated as a developing process. It must include the situation that requires the plan, the different courses of action, and then the deliberation on which of the proposals promises the least risk. All of these actions require rational thought which in no way can be attributed to a system requiring external environmental signals to get the process going. Finally there is the decision to put the plan into action, another deliberate act. The resultant plan may need confirmation because it may not be realizable due to non-existent facilities. So how could a previously configured synaptic system be able to do any planning on unavailable information?

The third feature, the awareness of conscious experience, is the ability to absorb the environment mentally, evaluate how much of it is of importance for consideration and to act only on that portion. With our eyes open most of the time, all types of visible signals are entering the brain. It is impossible to act on all the information. There must be a deliberate awareness of what is necessary for the immediate activity. A system that would react to random, unfiltered sensory information for transformation into motor signals would create chaos. There must be a discriminatory act to evaluate all the information and decide which is of primary importance. As I work at the computer, I am flooded with information on the screen and within my angle of vision I occasionally glance at my diet drink and decide to take a sip. I look at the screen and select that which I want to pursue in the search mode. My radio plays its music uninterrupted and it provides an environmental background that does not disturb me nor interrupt my thoughts unless I deliberately stop what I'm doing to consciously listen. My back aches because I slouch. I correct my position. All of these acts are the results of awareness, a conscious experience, and a decision to act, none of which are triggered by the volume of information that is pouring into my consciousness.

The above three examples describe human mentality in terms of rationality, intentionality, and consciousness. With respect to rationality and intentionality, the use of formal logic and computer science can in some way and to some degree explain the physical processes that lead to the transformation of sensory signals into motor signals. This does not imply that rationality has been fully explained by the use of these formal processes because rationality includes analyzing and synthesizing ideas not dependent on sensory inputs. With respect to consciousness, there is no known definition. There is knowledge of what it seems to be, how it is responsible for human activity, how it is necessary for life to exist, and so on. But no one can adequately define its fundamental totality.

> .....the explanatory gap between physical properties and qualitative properties is a symptom of the subjectivity of consciousness.[34]

There are two sides to the problem of subjectivity. On the one hand, qualia (qualitative features of experience) have a dual nature, two states—conscious experience and non-conscious mental state—that serve as the relevant sources of phenomenal access. But there are also conscious thoughts that are intimately connected and include qualia in their representations in a cognitively special way. How can materialism resolve this duality? The only basic relation it can contribute is the causal relation, which provides information transfer. This causal relation can account for subjectivity in either of two strategies: as a conceptual relation or as a distinct functional role with the same physical state (explained in great detail in *The Purple Haze*). However, subjectivity is an unexplained property, which implies that when qualia are instantiated, they are necessarily objects for the subject in whose experience they appear.

Whether mentality (the mind) is of a natural order or not remains to be resolved. The mind is a problem partly due to its many definitions. If its function is ascribed to other physical parts of the human body like the brain or a configuration of synaptic connections, the answer to its physicality is obvious. Otherwise, it is a problem.

Is the generation of the initiating signal for motor activity a physical or a mental process? What is the link between these two processes that can justify consideration of the mind as a natural entity? Can all the non-basic properties—those that are considered purely mental rather than physical—be realized as basic physical properties? They can if the mind, representing the non-basic, can be "naturalized" through the relationship of the two properties in the "causal interaction". The rationale for this approach is as follows:

(a)Sensor information can cause mental activation which can result in physical activity. I see a friend approaching me on the street; I decide to stop him because I remember I have some information to relate to him.

(b)Mental activity independent of sensor information can cause physical activity, which can result in further mental activity. After considering the alternatives, I decide to watch TV rather than read a book. I then move my body to the TV set.

Thus, the causal interactions are continually in action and it is possible to consider that they are essentially natural (therefore physical) phenomena. By implication, the mind, the generator of the mental phenomena can be considered (not by everyone) a natural entity. This interaction of the two properties appears to naturalize the mind if no separate mental entity can be proven to exist.

Although the neural system of synaptic connections (important in deriving motor activity) is explained scientifically, consciousness is still a dividing issue. It is that awareness of the "self" which thinks, creates concepts, makes decisions, overrides intuition, and meditates within and with the help of a physical environment that makes consciousness indefinable.

There are no positive answers to some of these fundamental questions. Hopefully, continuing research will supply information to supplant the many concepts, scientific and philosophical, which currently only provide bases for argument. Knowing what we are and how we function would dispel some of the outlandish concepts that attempt to explain our behavior.

(C) The description of the 5-1-Ch concept begins with the assumption that the mind does not exist as a separate entity. The physical brain does all the "thinking" and controlling for the human being. The convenient justification for excluding a mind from the analysis is that the existence of a mind cannot be physically proven. This is hardly a valid reason because it is the same reason for arguing the possibility that the mind does exist. The truism holds here that absence of evidence is not evidence of absence. If the mind does exist as a separate entity, then the denial of it makes the entire concept worthless because the effects of such a mind have not been considered. In presenting this concept, Dr. Churchland often unconsciously slips into "Word language" that, like Body language, discloses thinking contrary to the intended message. For example, in referring to "your thoughts" he subconsciously admits of a mental state outside of the description of his computerized brain, admitting that you the reader are in reality not subject to his proposed concept. Or his own, for that matter, since he could not claim conceptual originality for his concept if it were automatically conveyed to his writing hand by signals from the external environment which might be a shelf full of books or the words in a study report. Could it be a major slip of "word language" that in spite of the denial of any external entity in this purely physical concept that the title of the book includes the description of the content as the "Seat of the Soul"?

This concept, 5-1-Ch challenges rational understanding. Jumping from the physical to the mental by implying that the analytical and anticipatory activity automatically stems from physical elements by adjustment of the synaptic connections is a wild guess without scientific proof. There appears to be a missing link, and a vitally important one, in this chain of signal processing from physical manipulation to mental awareness and that is the exclusion of volition. Anticipation requires knowledge of future, even unknown, perhaps non-existent, possibilities that cannot be achieved by any adjustment of the synaptic connections. All types of random future conditions must be anticipated from which a selection can be made. Anticipation involves unrestrained imagination and rational selection. Recognition also entails an option of acceptance or rejection, hardly an automatic response based on synaptic connections.

How can physical synaptic connections using physical signals put together knowledge, sufficient to spell out required know-how activity? Even if they could, where is the volitional capability that decides whether to act on any perceived sensory or synaptic information? Experience testifies to the inclusion of an override capability. This convenient and unexplained insertion of synaptic knowledge-ability through the gradual buildup of credibility by the use of terms such as, carefully tuned, fleeting configurations, plastic modes of apprehension, are unacceptable substitutes for scientific verification, the type of verification imposed on non-physical entities.

I believe that the use of the word "circumstance" rather than "activity" is an attempt to circumvent such volitional requirements. The circumstance of having to answer a telephone that is ringing merely entails the conditions relative to the event (as defined in any dictionary). The actual activity of answering the telephone, or any other activity, demands volitional continuation of the act as in the conversation during a phone call. The selective responses to the demands from the other end of the telephone line can in no way be determined by fleeting configuration changes of the standard configuration of trillions of physical synaptic connections. (The activity can be simulated on a computer if the volition requirement is excluded.) The responses require a dynamic awareness of the past and an anticipation of what an ill-defined future might involve. This play attempt of how the brain postures itself to respond to the worldly demands through the use of terms such as: "careful tuning", which can be done in a computer simulation but has to be explained in biological terms, lacks the scientific verification demanded of any coherent theory and is an insult to human intelligence. Figure 5.1 shows the required mental interactions that are involved in the activities attributed to purely physical synaptic connections

Physical	Activity	Mental	Override
Audible signal of phone ring	Phone ring	I should answer	I don't care to
Continuing signal	Continued ring	I will answer	
Ring stops	Pick up phone	Who can it be?	
Nothing	Response	Hello	
Audible signal-message	Recognition and selection of memory content	I know who it is	Continue or terminate
Continued audible speech	Conversation	Dynamic interplay of past, current and future information	Change and update own information
Continued audible signal	Conversation	Continue talking	Time to end call
Sensory signal ends	Phone back on cradle	Continued thinking about call	Change subject or activity

Figure 5.1 Physical versus Mental Interaction in a Simple Activity

The illustration is a simple one for a specific reason: to show how easy it is to isolate the mental from the physical activities, those that defy the concept of necessary synaptic configuration updating. The sensory signals in the left column contribute only to the physical activity of ringing and the removal of the phone from its cradle. They could be responsible for the Pavlovian response to stop the ringing of the phone by answering it. But there is the override that negates the automatic reactive urge. The decision could be made not to answer the phone. Once the phone is picked up, all activity ceases to be a synaptic causality because a priori knowledge (the implied synaptically included smarts) is insufficient for responding to and generating unique, first-time information. Volition is required to do this with a continuous override (decision) capability throughout the conversation. The conceptual permanently tuned synaptic configuration is too simple an answer for this activity (implied circumstance).

The evidence of the non-physical mind obviously changes all the above.

The rebuttal to 5-1-Ch is concise in saying that the simulation of physicality is meaningless because it is unreal. Physical neurons cannot produce an awareness of the mental properties of desire, volition planning, and accumulation of knowledge. This awareness is subjective and not due to stimuli from the five senses. Subjectivity, like consciousness, is undefined but is experienced and can react in different ways to the same stimuli at different times. The three features of rationality, intentionality and consciousness cannot be programmed into a conceptual computer simulation with a claim that the simulation represents human behavior. Anything can be simulated on a computer but the results cannot be randomly generalized to adhere to a mandated materialistic doctrine.

Subjectivity and its effects on human behavior are covered extensively in *The Purple Haze*, not as a brain dependent property but as an independent, undefined quality that must be taken into consideration in the mind-brain problem. It is interesting, with the knowledge of the soul's reality, that subjectivity can take on a completely different meaning. It is part of the personality that endures from one life to another and may help to explain the dual problem of qualia as independent phenomena and at the same time as contributions to rational thought. (CE)

Issue 2
The second issue is about the process of transforming sensor signals into motor signals. It is a continuation of the concept in 5-1-Ch.

(5-2-Ch) The sensory experience begins in the sensory cortex of the brain. Each cortical area must "make sense" of the perceived information and transform it into neural command signals.

> ..if the brain's *intended* bodily behavior is represented in its motor nerves...then what intelligence requires is some well-tuned transformation of sensory vectors into motor vectors.[35]

For this to happen, the information from the sensory receptors must be evaluated as to intensity across the sub-elements (sensor cells) of the specific sense. As an example, in the sense of taste, there are four types of cells on the tongue, each dedicated to a component of taste: sweet, sour, salty and bitter. To convey coherent sensory information from each receptor for proper identification of the stimulating effect, the activation in the receptor must be of a type that can be used by the cortical area. This is accomplished by vector coding. What is vector coding? It is the identification of the proportional strengths of the four signal components as vectors. Vectors represent magnitude (level of excitation) with a direction (toward the cortical area). This is shown graphically in Figure 5.2. Suppose, as an example, that this is a representation of the reaction to a bite of a peach.

> As the juice hits the receptors on the tongue, it affects their levels of excitation—their activation levels—but it does not have the same effect on each of the four types. Cells of Type A, for example, respond strongly, almost maximally, to the presentation of a peach. Type B hardly respond at all. Type C respond robustly, although not so much as A. And type D cells react politely but without much enthusiasm.[36]

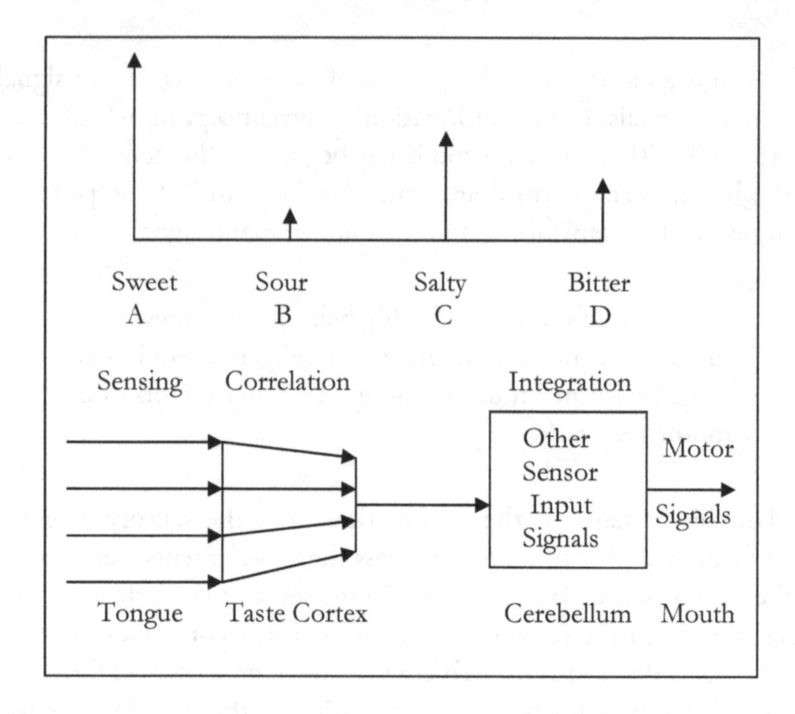

Figure 5.2 Translation of Sensory Signals to Motor Commands

The importance of this vector coding is that it represents a "collective pattern", not the reaction level of any single receptor type or a mixing of the four but a signature of a peach to the brain.

After correlation of the values, the information is sent to the cerebellum for integration with other sensory inputs. The resultant motor signal is needed to either chew or swallow the food, or if unpalatable, to spit it out. And how is this accomplished?

> By vector coding, of course. What has proved to be so spectacularly useful at the sensory input end of the system proves to be just as useful at the motor-output end. In order to control a large population of distinct muscles, the brain uses a vector-coding system that sends to them, all at once, a pattern of activation levels. Each element in that activation pattern arrives at the

> appropriate muscle fiber within the large population of muscles and dictates its tension according to the magnitude of that one arriving element.[37]

Similar vector coding is modeled for the other senses in which the complexity increases because the brain's perception requires more integration of the various signals before generating an appropriate response. As an example, for sight with its requirements for color, movement, depth and shading, the path from the receptors to the vision cortex is interrupted by a preliminary sorting of the signals to coordinate the signal characteristics for each of the requirements.

> In the sense of smell, the ability to accurately discriminate among the multitude of aromas resides again in the combinatorics of vector coding. Humans possess at least six distinct types of olfactory receptors and a particular odor is coded as a pattern of activation levels across all six types.[38]

The descriptive coding is similar to that for the taste illustration given above. There are estimates that the human being can discriminate at least ten thousand distinct aromas.

Vector control, therefore, lends itself to the simulation of brain activity. This system of vector control is a convenient method of representing the combined effects of multi cell contributions to sense identification.

How does this neural network represent time in the updating of motor signals in response to continuing environmental events? Although there are many undefined brain features responsible for this activity in the model network, there is a good representation as follows:

As the brain sends instructions along the nerve complex to all parts of the body it requires a response from the activated muscles. The transformational sequence originates in the brain and develops along the neural path to the intended muscles. However, this sending of signals in one direction does not account for any feedback, which the brain needs to update the motor signal for continued activity. Such feedback is possible through a myriad of recurrent network loops, which are located in the brain and along the nerve fibers.

Activation by motor signals always moves in a forward direction along "ascending" pathways. But the reactive information must be fed backward to update the continuing stream of motor signals. This happens along a feedback pathway that runs in the opposite (descending or recurrent) direction. How can a feedback network perform this function?

> A feed-forward system is a pipeline, a pipeline of information. The farther along in the pipeline one samples the flow of information, the farther back in the past must that sampled information have first entered the pipeline, and the further back in the past must lie the events that the sampled information depicts. Since a recurrent axon originates in a cell farther along in the pipeline, a descending or recurrent pathway therefore makes information about the network's past activity available for current processing, specifically at the layer of neurons where the descending pathway touches down for a landing.[39]

But how does the brain present the specific sequences of events in time?

> Just as a temporal sequence of pixel-patterns on a TV screen represents some temporal sequence of events in the world, so does a temporal sequence of activation vectors in the visual cortex, for example, represent an unfolding sequence of events in the world.[40]

Of course, any or all of this recurrent operation is not possible unless the synaptic weights have been properly configured. But it is a logical explanation of the time related brain response to updated sensory stimuli.

The contradiction of vector control is based on information in *An Anatomy of Thought* by Professor Ian Glynn

(5-2-Gl) The brain is the most complex object we are aware of in the universe. Whether it is assisted by any other simple or complex entity does not affect the physical operation of the brain. It must physically process the sensory information of the outside environment for the use of the physical body in responding to that environment and its effects. The sensory input must be converted by the brain into motor commands for responses by the various parts of the body. This physical activity must somehow combine with a mental process that can direct the type of response but not the nature of the physical action. The sight of an approaching tornado is processed as a dangerous picture of the environment through the optical system, but before any responding action can be initiated, there must be a decision on what action to take. Until that decision is made, there can be no generation of motor commands to respond to the environmental threat. The description of the physical process is given here with the realization that there is this intervening step, but by ignoring any interruptions, the continuity of the complete physical process can be better comprehended.

How does the brain get the information from the receptors? It's not like getting a message through a telephone line wherein the message is independent of the carrier. Each message across a telephone line is different. In a nerve, the impulse is the message, created by the stimulus at the receptor.

The nerve is a long thin tube filled with a salt solution containing mostly potassium chloride with a little sodium chloride. This tube is bathed in a salt solution containing mostly sodium chloride with some potassium chloride. The wall of the tube, the membrane, is permeable to both sodium and potassium ions. What happens when the nerve receives a stimulus from a sensor cell or somewhere along the nerve route? There is a change in the permeability of the membrane, causing the sodium ions to enter the tube from the outside. This unbalance changes a resting nerve into an active nerve creating a local electric current. This local current then activates the adjacent resting nerve, which in turn will do the same and the repeating activation moves the impulse to its destination. To increase the speed of transmittal, this step by step movement is enhanced by grouping many steps within a myelinated sheath which allows the local electric circuit to proceed in longer steps by circumventing the individual steps with these much longer steps. The current thus passes along the myelinated course rather than through the small inner tube. This describes the activation and movement of the impulse but does not explain the encoding of the message that it carries to supply useable information for the brain.

The message activated at a receptor is always of the same nature. It is an all-or-nothing message. This was determined by testing a nerve having only eight or nine nerve fibers. An increase in the strength of the stimulus produced a stronger reaction in the tested muscle. However, the increase occurred in a stepwise order, never exceeding the eight or nine steps. This indicated that each step increase was due to the identical message being produced by an increase in the number of fibers that were activated. This all-or-nothing nerve reaction to a stimulus is the nature of any impulse that travels from the point of activation to the nerve end. Furthermore, unless there is a gap of a few thousandths of a second between shocks, there is no second impulse, further proving that if the reaction increases (the message changes), it can only do so as a result of a series of all-or-nothing impulses rather than in a continuous stream.

> However complex the information that is to be sent along the nerve, it can be sent only as streams of impulses whose size cannot be controlled at the point of stimulation. Information relevant to the most subtle sensations or to the most complex actions can be transmitted to and from the brain in no other way.[41]

So how is the complexity of the message interpreted by the incoming message from the receptors?

> Because of the all-or-nothing behaviour of the nerve fibres, the size of the impulses carries no useful information, so the only information available from the impulses arriving along a given fibre must come from their timing and from identification of the fibre by which they arrive. ....In fact only two

> straightforward codes appear to be used: a *labeled-line code*, used mainly to convey information about the site and nature of the stimulus, and a simple *frequency code*, used mainly to convey information about its intensity and duration.[42]

Labeled-line coding is possible because of the variety of sensors throughout the body (eye, ear, nose, tongue, skin). For example, there are skin cells that respond to heat, others to cold, and still others to pressure. Thus, the brain recognizes where the message is coming from and the type of message it is.

Frequency coding is effective because a stronger stimulus generates more frequent impulses while a maintained stimulus produces a nearly constant frequency. Since there are many cells involved in any message, the number of cells stimulated define the size of the affected area. The persistence of the stimulation determines whether it is a stimulation of short duration or long. The impulses are not continuous as in a liquid stream but as a series of short pulses, the frequency varying with respect to the duration. In the case of pressure on the skin, the initial stimulus generates a high frequency message. However continued pressure without an increase reduces the frequency to a steady state condition providing no additional information.

(C) There is a stark contradiction between the conceptual description 5-2-Ch and the opposing account. Vector control is not the actual representation because the individual cells cannot generate the relative magnitudes of the vectors. As the vector approach is used in the conceptual analysis, by definition, the qualities of a vector, magnitude (value) and direction, must be used. But as the rebuttal indicates, each nerve carrying sensory information is an all-or-nothing carrier to the cortex with no specific value. It is merely a labeled line with the frequency defining the strength of the sensory input. Since the same vector control concept is used with the other senses, it is obvious that the concept <u>fails to describe how the brain works.</u>

Furthermore, in the concept of recurrent networks, an all-or-nothing input cannot update the level of response by "vector activation" because there are no vectors. Obviously the network simulation is intended to support a neuroscientific approach but fails by its inability to come even close to explaining the complex operation of the brain (or mind?).

(The vector magnitudes are also used in the conceptual analysis as locating values in a three dimensional representation of the varying interpretive capabilities of the brain. This is useful in understanding how an activity can be visualized, but it does not explain the actual transformation of the sensory signals into motor signal commands.)

The net effect of this comparative evaluation is that the response based on actual experimental data is more compelling scientifically than a conceptual unproven computer simulation. If vector control is the accepted explanation of sensor stimuli identification and processing for all sensors, then neuroscience has failed to explain the process. (CE)

Issue 3

The third issue, a comparison of the brain to a computer, is a continuation of the concept in 5-1-Ch.

(5-3-Ch) The neural system can be simulated to show the functional operation of the integrated brain-body. An "artificial network" is used to study the functioning of the brain and its causal neural control of the human body. Through the synaptic connections the brain reacts to the incoming sensory information, becomes emotionally involved in this information and uses it to induce activity and comprehend both the present necessity and the future needs.

> Like the assembled pixels on a TV screen, the overall pattern of neuronal activation levels at any given instant constitutes the brain's portrait of the local situation here and now. And like the TV screen, the temporal sequence of these ever-changing patterns constitutes the brain's ongoing portrait of an ever-changing world.[43]

> But who....can be watching this pixilated show? The answer is straightforward: no one. There is no distinct "self" in there, beyond the brain as a whole. On the other hand, almost every part of the brain is being 'watched' by some other part of the brain, often by several other parts at once.[44]

The brain is the most complex organism in the world. Despite its complexity, its reaction to the sensory inputs of the external environment is instantaneous. In this response, neuroscientific analysis has shown that there are trillions of "acts" performed each second of the brain's waking day. This continuous neural operation in terms of reliability based on the numerical probabilities is unmatched in the material world. And yet, these operations are similar to the processes followed in the analyses of other scientific problems. Could it be that the brain operates on the same principles that the scientists use in their computer simulations? The answer is yes. Not all computers are up to the required tasks, but some computers can do the job. This concept claims that the brain does use the computational procedures of a computer.

Not any computer can do the necessary computations. A serial computer cannot perform the required reaction speed of the brain's functions. Not only the number of "acts" but also the required times for serial operations preclude their adaptability to the brain's tasks. But with Parallel Distributive Processing (PDP), representative neural network models can be developed. This style of computing is what is needed in transforming one neural pattern into another by passing it through a large configuration of synaptic connections. This is a standing requirement in the entire animal kingdom.

The information processing in an artificial network supports the analogy that the neural network is such a computer and that it operates in a similar manner. Therefore, the brain functions can be analyzed through computer simulations. Such simulations tend to verify how the brain functions and that mental states do exist. These results have been verified through human response tests in the laboratory. Based on such tests, the theory proceeds to define the capabilities of such a simulated human network and by inference actual human activity. These cognitive capacities include pattern recognition, information recall, recognition of sensor qualities, activation of body motion, and initiation of social and moral behavior. These capabilities, if initiated by the physical neural network with the trillions of synaptic connections, argue against any dualism of the human being inasmuch as they are the capabilities ascribed to an external entity controlling such human motivations.

There are two reasons why PDP is a requirement. The first is the need for speed in the transformation of the sensory signals into motor commands.

Without the almost instantaneous reaction to the environment a living creature would be unable to protect itself or to hunt for food efficiently. These and other routine functions require the processing of millions of individual computations simultaneously. No serial type computation could ever produce the necessary reactions. An example of the magnitude of the task is in the processing of the visual signal received by a human being at the retina. To transform the many individual nerve stimuli from the physical light signals into a perceived picture with stereoscopic effects and in color requires a multi stage process. At each stage of the process, there are millions of individual and combined transformations from which the subsequent stages must develop coherent signals. Although it would take a desktop computer running at 12 MHz about 15 minutes to perform 100 billion computations, a single stage of the human visual system does the same in 10 ms.

> As the original pixilated pattern across the many retinal neurons gets passed inward from one specialized neural population to the next, and to the next and the next, the original pattern is progressively transformed at each stage by the intervening configuration of synaptic connections. This is where the bulk of the brain's computation takes place. This is where past learning shows itself, where character and insight come in, and where intelligence is ultimately grounded.[45]

The second reason that PDP is required for the neural network is the reliability of the parallel performance. As the individual nerve signals combine at succeeding stages of a PDP type operation, the effect of the failure of single nerves is minimized since the multitude of other cells are still contributing to the total transformation. If the entire transformation were to be a serial operation, the net result by a failure at any stage would be a complete failure. With PDP, the transformation continues to completion in spite of the failure of a minute portion of the nerve signals.

> A PDP computer can suffer the malfunction, inactivation, or outright death of large numbers of its

> synaptic connections, and yet suffer only a marginal degradation in its performance.[46]

Because the brain is like a PDP computer, there are many cognitive properties that can be displayed by this style of computing. Some of these cognitive properties of living creatures are: seeing analogies, recalling relevant information, selecting solutions, recognizing subtle human qualities, moving one's body, and navigating one's social self. Such biologically salient capacities are the characteristic signature of a functioning PDP system.

Although the PDP system is representative of how the brain must work to be as effective as it is, there is an unknown factor that is independent of the PDP system but does affect the simulation of the brain as a computer. This worrisome feature is the inability to understand how to configure the preferred synaptic connections required for pattern recognition, sound interpretation, language formation and interpretation, facial recognition, etc., which are done instantly. The best available approach often used by neuro-modelers is called "synaptic adjustment by the successive back propagation of errors, or just propagation for short."[47] In this technique, random values can be used initially to detect the degree of error. Then these errors are reduced incrementally until the desired synaptic weights are defined. The adjustments to the synaptic configuration and synaptic connections must be made in actuality through experience. But the simulation is not representative of the neural reality. Thus it is biologically representative in that biologically, the adjustments are made. "Sadly, it is biologically unrealistic in just about every other respect."[48]

The opposing view is based on information in *The Rediscovery of the Mind* by John R. Searle.

(5-3-Se) The answer to the mind-body problem is:

> Mental phenomena are caused by the neurophysiological processes in the brain and are themselves features of the brain. To distinguish this view from the many others in the field, I call it "biological naturalism." Mental events and processes are as much part of our biological

natural history as digestion, mitosis, meiosis, or enzyme secretion.[49]

How does this concept differ from the concept that the brain is a unit analogous to a computer that takes sensory data and develops into motor activity? It differs mainly in the denial that a computerized model explains the actual functioning of the brain. Instead, the brain is a naturally functioning entity. But it is a complete unit; it does not depend on any immaterial being or substance to perform its function.

The cognitive scientists claim that the brain is a computer. Their claims are based on the apparent similarities of the brain operation to that of a digital computer. These similarities encourage them to explain their theories about the physical causality of the brain through computer simulations. Because they have these theories about rational behavior, and modern technology has supplied the electronic components to test these theories and devise supportive computer simulations, the neuroscientists are quite certain that these theories will be confirmed. But there are gaps in their analyses and computer capabilities that argue against their certainty.

If the brain is to be thought of as a computer, it must have characteristics that are clearly similar to one. Let's begin by establishing a definition that applies to our needs. A commercial computer is hardware with a syntax of symbolic representation that is used to perform defined computational operations. This syntax is intrinsic to the physical system, which stands ready to be used by an operator or to be used in the execution of the commands of additional software. In either case, the results generated by the computer are interpreted by the operator, who is not intrinsic to the computer. The computer does nothing until it is commanded by the operator to use its symbology to derive results. In its basic operating role, its computing can be considered unconscious as compared to its conscious performance with additional software or instructions from an operator. The influence of software or operator's instructions is the injection of semantics, the meaning of what is to be processed by the computer. Ultimately it is an observer outside the computer who must interpret the results of this meaningful operation.

According to such a definition, the brain also has two states, unconscious and conscious, with some differences in the interpretation of these terms. In its unconscious state, it, unlike the commercial computer, is active. Its activity is the control of those vital body functions which are continuous, even during sleep and in a coma. This state cannot be invaded by an operator directly except by activities which are independent of the basic functions of pumping blood, digesting food, and breathing. An operator can change the performance by incidental activities like, running, eating, and relaxing but these are not conscious activities whose sole purpose is to control the rates of body functions.

It is in the conscious mode that the dissimilarity is greatest between the computer and the brain. Whereas the commercial computer is inactive until commanded to operate, the brain is continuously active until commanded to stop, and even then the shut-down is only partial. Closing the eyelids or cupping the hands over the ears can stop the input of some sensory stimuli (but not all, as the sense of touch is retained in the cupping of the ears). Those actions do not interrupt the ongoing life support functions. Furthermore, when the brain is "awake" the information pouring in is not restricted by any program conditions. It must accept all of this information and transmit it to the interpreting cortical areas where perception of the outer world occurs. This is a link with the outer world that does not occur intrinsically in a commercial computer. But the brain must deliver its output in accordance with the sensations of the outside without any commands of the operator. But who is the operator of this alleged computer? There must be an operator because the processed sensory information is acted upon after manipulation by the brain into perceived modes. Such an operator receives the output of the integrated sensor information without requesting it, unlike in the commercial computer from which the operator must ask for the output. In the commercial one, the operator gives the command to generate the output. But in the brain, the output is generated without operator command because it is produced without the operator and is generated solely for the operator's use. Except in cases where the response is automatic, as in dizziness, or in reaction to touching a hot stove, the operator (?) then uses these perceptions to generate the responsive motor actions. Note that the brain action is again with respect to an undefined operator. The cognitive claim of physical causality is based on this lack of a defined participating operator.

Further indication of the presence of an undefined operator is its ability to discriminate about which part of the total input it wants to concentrate on. The eyes can select a specific area of the retinal picture, like specific words on a computer screen full of words. All the sound signals are being processed but can be disregarded by the operator while selecting the replacement words to clarify a statement. And when a distraction interrupts the concentration, this undefined operator must resolve the situation.

The brain follows a pattern of sensory interpretation. But how does it do this transformation of a physical signal into a mental experience of sensation? This is a dynamic transfer of ever changing inputs, not with computer type 1's and 0's but with meaningful transformations. Which brain parts or operations are responsible for these transfers so that they can be designed into the algorithms that are supposed to represent the brain's functioning? As in the concept 5-1-Ch above, these transformations are conveniently lumped in the synaptic configuration of the brain without any operator involvement.

With these differences in mind, let's see why the brain is or is not a computer. There are three aspects to consider: hardware, computation, and commands.

Hardware

The commercial computer's main elements are the microprocessor(s), electronic circuits, data storage, screen and keyboard. The syntax of the commercial computer system is designed by a programmer. It is implemented, tested and refined and is intrinsic to the computer because it defines that computer and causes it to operate per its instructions. It can be repaired. But if the brain is a computer, who can define the intrinsic operating syntax? And who can change or refine it? No one because no one understands the syntax for signal transformation.. This inability to explain the process is circumvented (as in 5-1-Ch above) by proposing that the transitional phenomena occur in the synaptic configurations.

The nervous system is comparable to the electronic circuits since there is information flowing along the nerves and across the synaptic gaps. The cortical areas and the cerebellum can be considered the microprocessors. But that's the end of the comparison. There is no known brain unit for data storage, the memory. There is no screen to show the output of any signal transformations although they are instantly available for application in motor signal generation. Mostly missing is the keyboard for the injection of controlling commands. (The brain receives all sensor stimuli like input from a keyboard but the controlling feature is missing.)

The brain, therefore, is not a digital computer from a hardware standpoint. It solves the very complex problem of transforming physical receptor information into perceptive sensations and awareness of the worldly environment and more efficiently than any computer could do.

Computation

What is the definition of computation? It is the operating procedure in a computer. "It is controlled by a program of instructions and each instruction specifies a condition and an action to be carried out if the condition is met."[50] Does the brain do this? Yes, in a general sense, because all its operations are controlled to respond in a predetermined manner. But its computational methods are so vastly different from that of a commercial computer that the two cannot be deemed alike.

Every computer has a program for performing its computing functions. Is the mind a computer program for the brain? (The mind defined here is the mental state as distinguished from the physical state of the brain.) In the physical computer, this program is unchanging with its logic based on 1's and 0's. It is a program in which the conditions, the syntax, have been finalized before being added to the computer. It will always respond in the same way to the same inputs. But the brain's program is different. Its response is dependent on the nature of the changing stimuli, and subject to a decision process which evaluates the input before producing the output. There are no 1's and 0's to control the processing of the data. The result may not always be the same for the same given input. For this computational reason, the brain is not like a digital computer.

The effect of syntax and semantics are basic in computer computation. Syntax, as mentioned above, is the symbolic representation of the computer functions and its operation. "The formal syntax of the program does not by itself guarantee the presence of mental contents".[51] But the semantics of the mind's thoughts involve mental effort. What symbology can adequately convey the mind's intent, the decision process, the precise meaning of the language of thought? This is the missing element of a brain simulation that is based strictly on a representation of transformation and control by a physical system. The difference "rests on the simple logical truth that syntax is not the same as, nor is it by itself sufficient for semantics".[52] The semantics of the ideas being included in the program are not easily symbolized.

The results of the computational process are observer related (the interpretations of the computed results). However, the computational features of the brain, if they can be called that, are not observer related in the same way. They are observer utilized through conscious, instantaneous interpretations. They are intrinsic to the brain because they have been incorporated along the evolutionary path of Homo sapiens. But here again, who is the observer of the brain's responses to the environmental inputs? The environmental inputs are constantly changing and more importantly are injected into the brain without any special instructions for acceptance or rejection. It accepts all the stimuli and processes them without any specific commands to do otherwise. The eyes see all the details within view, whether of any importance or not. The ears accept any sound without any discrimination as to which sounds to accept. There is no syntactical discrimination pattern. The commercial digital computer does not behave in this manner.

Can the operations of the brain be simulated? The answer is yes, but what does the simulation represent? With the brain, there are so many unknowns that a simulation can only represent the assumed conditions that a specific theory is attempting to prove. So is the brain a computer since it acts as one? The answer is NO.

Intentionality

What is intentionality and how does it fit into the computer comparison? "… intrinsic intentionality is a phenomena that humans and certain other animals have as part of their biological nature."[53] Hunger, thirst, fear, desire, etc. are these phenomena. This is John Searle's definition, which is based on the concept that mental states are physical states, which he terms biological naturalism (see his opening quotation above referring to digestion, mitosis, etc.). The intrinsic nature of physical processes like digestion can be reduced to simulated models. But the intrinsic nature of the mental intentional process cannot be simulated in a computer model.

With the commercial computer, the operator, the interpreter of the semantic results, is separate and outside the computer. The semantics of the software are unchanging; the intentionality is built in. But with the brain the intentionality due to the semantics is constantly changing with its override ability, which with the commercial computer is commanded by the outside observer. The brain can command the eyelids to close to stop the visual input, to cup the hands over the ears to silence the sound. The operator function is a controlling mental function within the brain! It can even change the syntax of brain operation. An example is the focusing of the eyes on a specific point. While driving a car on a bumpy road, the eyes can be focused on a point at a specific distance in front of the car even as the body bounces. It is also possible to keep the eyes focused on that same point while partially turning the head to speak to the passenger. The brain cannot be a slave to the constraints of a computer type operation.

What does this difference between a brain and a computer do for the thematic problem of dualism? Once again, it zeroes in on the question: who is the observer of the mental processes?

(C) The concept 5-3-Ch is another example of the neuroscientific procedure of extrapolation of factual physiological data to propose and promote impossible solutions. Using the brain's hundred million neurons and trillions of synaptic events to show the need for a PDP type computer with the implication that the brain operates in that manner is such an extrapolation. As John Searle points out, there is no justification for such an extrapolation because the basic operational differences show a disregard for fundamental facts.

A computer comparison must be consistent. Every operating computer has an operator. Neuroscientific insistence tries to convince you that the brain functions like a commercial computer, a very fast one, to cause cognition without an "observing self". It succeeds in showing that both the computer and the brain operate in precise and consistent procedures. But if the comparison is to be accurate and complete, there must be a watcher of the brain, just as there must be with a computer. It is the watcher that consciously determines the cognitive direction of human activity. This may not be the same with animals, but human experience is used in analyzing human activity in a first person determination. Therefore, the claims of neuroscientists in attempting to show that physical causality is responsible for human behavior through computer comparison fall short of the usual demanding requirements for scientific proof of conceptual theory.

The use of the TV analogy is as an aid in understanding the concept of the brain's operation, the interacting of the neural elements of the wiring network to produce an integrated picture like that on a TV screen. Such a TV picture is produced for the benefit of an outside, independent entity who can appreciate what is shown by either enjoying it, absorbing it into memory or taking action in response to it. This is the activity that the concept is trying to assign to the physical brain. But it is an outside activity. However, the TV related quote in 5-3-Ch is a straightforward denial of such external activity, which contradicts the reason for using the analogy. The picture, therefore, has no meaning. The internal "watching" by the separate brain elements cannot be watching if it is an integrating operation requiring participation rather than observation. The use of the word watching is a diversionary tactic intended to justify the use of the picture analogy and to conjure up a similarity which does not exist.

There is an example in 5-3-Ch of an assumption that is immediately turned into a conclusion. I repeat an abbreviated form of the statement here. "These capabilities, if initiated by the physical neural network........ argue against any dualism of the human being inasmuch as they are the capabilities ascribed to an external entity commanding such human motivations." Note that the assumption of physical initiation immediately negates the possibility of the opposite occurrence. Of course, if the action is a physical initiation, then it cannot be the opposite by definition. If the opposite assumption were made that the capabilities are initiated by an external entity, the physical neural network would not be responsible. The statement adds nothing to the argument but it does leave an impression of the certainty of the materialist approach. This is a good example of my reason for including Chapter 2 in the book.

I disagree with the Searle statement that mental processes are part of our biological natural history because in my view of intent (the root of the word intentionality) there must be the ability to discern the need for and to initiate an act. There is no intent in the digestive act; it is a natural process, as are the others. I prefer to define intentionality as a characteristic of a mental state that depends on awareness of current conditions to initiate activity. The basic difference between these two definitions is the intrinsic content of defining the ability to make decisions. This argues against the mental state being classed as biological naturalism. And it is this same intrinsic difference that differentiates a mind from a computer. Therefore, intentionality as I have defined it argues against the mind (generator of intentions) being an example of biological naturalism or the equivalent of a digital computer.

If PDP simulations are representative of brain operation, but biologically unrealistic, then what good are they and how can the jump be made to say that biologically the operation could be realistic in spite of the lack of scientific proof. The basic reason for assuming possibility is removed but the fantasy remains. This is an admission that the concept is a computational study of representative, not actual, operations. (CE)

Summary of Conceptual Comparative Analyses

I used the concepts of one neuroscientist mainly to show the consistency of neuroscientific promotion of physicalism. The practice of extrapolating physiological data is used too often to justify conceptual claims with convenient explanations that the scientific verification will eventually be supplied. This is typical in the need to substitute some explanation for problems that are not understood at a fundamental level.

The claims of cognitive science that the physical brain and neural network cause behavior patterns are just that: claims. The contradictions in the comparison between the concept and actual physical facts in 5-1-Ch with convenient ignoring of the effects of conscious experience illustrates that there are huge gaps between theory and fact. The use of a computer simulation to infer that the brain and neural network are understood is a ploy at attempting an explanation under cover of scientific methodology. It is a misleading leap from factual knowledge of physiological reality to conceptual guesswork in identifying such reality as the cause for total control of human social, moral, and intellectual behavior. This lacks scientific verification.

Cognitive science claims that there is no external entity involved in the total awareness and cognition in responding to the environment by the trillions of synaptic connections. This claim is made arbitrarily without any scientific support. However, since the external entity does exist, the control of human activity by this entity is also based on the availability of these same trillions of synapses for its use. How this is done is the problem to be solved in the future because due to its neglected application it is not understood now.

If the simulations of 5-3-Ch are biologically unrealistic, they do nothing to support the concept of physical causality as the initiator of moral and social behavior. They only add to the investigative work toward defining a concept of Artificial Intelligence. Its denial of mental intentionality and the existence of any spiritual entity are convenient exclusions of necessary elements in the analysis of the *total picture* of human behavior. This exclusion illustrates the incomplete approach to conscious behavior. Unless the total picture is analyzed on a "what if" or a "what if not" basis, the results of any behavioral analysis are meaningless.

The comparison of the brain to a computer is often used with various implications. The brain may be like a computer for regulating the life support functions of the living body where life sustaining "constants" contribute to the semantics of the computer. But the mind as a rational and volitional separate entity is a dynamic force with controlling properties that are a mystery unlike any computer.

The examples used in this chapter reveal the current misunderstanding of the behavioral nature of the human being. Even the dissenting arguments are physically oriented. The inclusion of the soul and the mind can only complicate future research but at least they will be based on fundamental reality.

The next three chapters continue the descriptions of the current conceptual versions of memory, free will, and consciousness as viewed by materialists. Those chapters, as this one, attempt to convey the prevailing attitudes about physical causality. Obviously, the effects of the soul and mind are absent, creating, as in this chapter, opportunities to reflect on how these effects may alter future research. There are no dissenting opinions in these chapters since the intent is to disclose, not argue. My comments will again be identified, as will be the names of the authors and the titles of their books.

6

Memory

Elephants have remarkable memories.

Memory is the key property of the soul's mind that proves its own existence as was detailed in Chapter 3. The non-physical memory negates all the neurophysical claims that memory is strictly a physical property of the physical brain. Therefore, memory with the soul's reality is an extremely significant reversal of the immense amount of conceptual work that has been done to promote physical causality. It is the memory of the past life that spans the gap between the old and the new lives and it is memory that is investigated for confirmation of reincarnation. Memory as a non-physical reality challenges the entire materialistic concept of human nature, affecting not only the scientists, but also philosophers, theologians, and most importantly the beliefs of the monistic secular world. This unsettling condition is a devastating consequence of science's unsupported and convenient denial of the soul's reality. The modern world, having succumbed to science's assumed role of interpreter of natural laws, accepts this appraisal of human nature but ignores the fact that science in its total occupation with physical matters knows absolutely nothing about the non-physical domain.

In spite of this knowledge of the proven location of memory, there remains the question of how the brain can control those automatic bodily functions, like digestion, heart operation, and breathing, since conscious volition is not involved. Unless it has some connection with a memory of basic functional rates and procedures, it cannot generate the motor commands for automatic operation. Perhaps it has a different type of memory, one not based on experiences, the need to learn, and volitional control, but one based on instructions of the DNA genetic code. This problem needs redefinition in the rethinking of dualism.

This chapter on memory is intended to serve two purposes. One is to present an overview of what memory's basic functions entail along with the various types and systems, like short and long term memories, that are used to explain these functions. The second purpose is to describe some of the conceptual approaches that are science's vain attempts to explain memory as a totally physical system. In writing this chapter, I (and you in reading it) have the advantage of knowing that the assumptions used in the concepts are invalid due to the soul's reality. Nevertheless, I present the conceptual thinking, sans a soul, to show the results the researchers produce when making the wrong assumptions. In reading the conceptual solutions and their expected effects, consider the dramatic difference between reality and conceptual exploration.

Why is memory a key element of the dualism controversy? Because its unexplained nature leaves doubt about the assumed claim that physical causality is responsible for the physical/mental integration in the neural network. Two overriding questions will eventually determine whether the nature and functioning of memory can be explained physically. For the scientist the question is: Can memory be fully described in terms of the physical brain's systems and subsystems and their interactions? The philosopher's question is: How can a physical memory perform non-physical functions and actions like storing and recalling experiences of learning, skills, and creative thinking?

Memory is either the brain's unitary data storage unit or a composite of several integrated functions. Is it the nervous system and several physical brain components that perform the activity? As an example, the brain's information processing subsystems can classify the incoming sensor information for proper storage. In some concepts this processing is based on content. In other concepts it is based on spatial identification. In still others, it is the recognition ability, like face identification, which determines how and where the information will be stored. These processing methods are assumed to be up to the tasks since in actuality they perform well.

There is no scientific approach that considers memory as part of a distinct non-physical entity, either with partial or total functional responsibilities. *Therefore, because a non-physical entity does exist, with an impact on conscious behavior, the entire scientific analysis of how the human system relying on memory functions is flawed by this exclusion.*

Since the materialistic descriptions of memory are only conceptual, what do they accomplish?

(1) They describe what appears to be necessary to achieve the desired results.

(2) They isolate the functional properties of the brain's subsystems needed for memory.

(3) They attempt to identify the subsystems associated with the conceptual activities.

(4) They propose the interactions between the physical subsystems that produce a logical path of information flow.

The information obtained from these activities may be conceptual, but it can be used in the ultimate definition of memory's physical aspects and an attempt at an integrated system. The current fractured approach, with no top-down plan, merely supports each concept's specific claims. And the uncertainty of the validity of these concepts is apparent from the many disclaimers about the available data to support them. In a 15 line paragraph of a conceptual description, the following words and terms appeared: could conceivably argue; could hold; presumes; however, there is no evidence to support such an assumption; it is clear; it is not at all clear; and could reflect. In spite of all these caveats the concepts have value in supplying information that will eventually find its way into the definition of the complete memory operation.

Current (2009) understanding of memory consists of theoretical conceptualization, analyses of experimentation with animals, and diagnoses of human brain disorders. These diverse attempts at explaining memory have resulted in explanations that are not subject to criticism because the store of irrefutable scientific evidence does not exist to validly contradict them. Interpretation of test results further dims the possibility of complete agreement. And the conceptualizations can incur only disagreements, not contradictions, because in the scientific world, the optimistic expectations of eventual verification are sufficient to satisfy current critics.

6.1 Background

The attempt to define memory is not new. The Greek philosophers tried to explain the difference between knowledge (stored information) and learning (accumulation of information), both requiring memory. By the seventeenth and eighteenth centuries, conscious and unconscious memories were the topics of mnemonic studies. Little was known about the brain and its functions so there was little motivation to delve into the memory's physical location, biological makeup, or functional performance. This lack of specific knowledge did not preclude the generation of ideas about the brain or its subsidiary function: memory.

The Greek philosopher Aristotle (384-322 B.C.) addressed memory in his physical treatise, *Memory and Reminiscence.* To him, the soul was a reality, a part of the human being. With this belief, he could explain the physical/non-physical relationships by the concept of "presentation". All objects that could be presented to memory were immediately and properly objects of memory, whereas those objects that only involved presentation were objects of memory incidentally. In other words, actual perception stamped the impression on memory but an occurrence during quick or very slow movement failed to leave impressions. If an intense pain lasted for a hundredth of a second, it would not be remembered. Also if it grew slowly from no pain to perceptible ache to full pain, the memory of when it started would not be remembered. The resultant existing pain, even if it eventually ceased, would leave a definite impression. No memory was involved in the mere fact of recurrence, such as the sound of rain or the chirping of birds.

Aristotle also made a distinction between remembering and recollecting. The former required no directive action but the latter was a mode of inference. A need to recollect inferred that an activity or specific information was stored and a deliberate search had to be conducted as a conscious effort to recall that information. Memory was aided by association of facts such as remembering people's presence and the locality details from an important meeting or discussion group.

Toward the end of the nineteenth century, William James, (1842-1910), an American psychologist, wrote his comprehensive treatise on psychology, *The Principles of Psychology*. In it he covered the many aspects of human thinking and behavior. Included in the many subjects such as the functions of the brain, consciousness, mind stuff, habit, attention, reasoning and attention, were two subjects that are pertinent to this book's interest in dualism: memory and the soul.

James reasoned that memory was only meaningful when time was involved. For a state of mind to remain in memory it must have endured for a length of time. He also postulated that not all thought is preserved in memory and that there are types of memory with different recall durations but he did not know why. He was already asking the questions which led to the research that is ongoing about what determines the type of memory, the process of encoding the sensory stimuli, and the differences in duration of memory storage. But the question that really intrigued him was how an immaterial thought could be related to and be the product of a completely physical brain.

> The ultimate of ultimate problems, of course, in the study of the relations of thought and brain, is to understand why and how such disparate things are connected at all. But before that problem is solved (if it is ever solved) there is a less ultimate problem which must first be settled. Before the connection of thought and brain can be explained, it must at least be stated in an elementary form and there are great difficulties about so stating it. To state it in elementary form one must reduce it to its lowest terms and know which mental fact and which cerebral fact are, so to speak, in immediate juxtaposition.[54]

To this day, this connection of thought to brain has not been explained. The scientific allusions on how this connection is possible have no valid scientific support. And if that connection cannot be explained, how can a thought be stored in a physical memory?

There was a theory at that time which attempted to explain this connection called the mind stuff theory. It stated that our mental states are compounds. Just as atoms aggregated to form planets, atoms of living material aggregated to form the living brain. However, this did not explain consciousness. To overcome this new wrinkle, the theory was implemented by giving each atom its own consciousness. Rather than having the brain cope with the multiplicity of conscious atoms, one central, superior cell was assumed to be the integrator of all the conscious atoms, developing the integrated thought bit for the brain. Thus only one point of contact was necessary. But there was no one central cell identified in the brain for assuming this total brain function. James had to reject the theory as unintelligible.

James's own theory in escaping from this psychic-atom concept was to consider the entire thought as the minimum to be dealt with on the mental side in relation to the entire brain process. But the entire brain process is not a physical fact at all: it is only the appearance to an onlooker as a multitude of physical facts. He found both approaches unacceptable. But if this were so, what explanation could he give for the thought-brain problem? His only recourse was to consider the possibility of a non-physical entity, the soul.

> I confess, therefore, that to posit a soul influenced in some mysterious way by the brain-states and responding to them by conscious affections of its own, seems to me the line of least logical resistance, so far as we have attained.[55]

James observed that although the idea of a soul did not explain everything, it was less objectionable than mind dust or the theory of psychic atoms. But in so doing, he also put aside the idea of a soul, principally because it had no scientific credibility.

Prior to 1950, memory, because it was undefined and not much was known about the brain, was believed to be a unitary function. It was in the last half of the twentieth century that the cognitive scientists began to accumulate data about learning and the effects of damage to those portions of the brain that are associated in some manner with memorizing. Questions began to arise about the mind's unitary function. Some of this information resulted from observing the effects of lesions in parts of the brain. Additional information was gathered by subjecting animals to memory type tests using electronic probes inserted in their brains (a method which could not be used on humans). Human tests using brain-scanning equipment to measure the levels of blood flow in the various parts of the brain during test procedures produced remarkable corroborating results.

In the 1960's and 1970's, memory became a short-term-long-term study and this peaked in the 1980's. (Some proponents of the unitary system still refuse to change their views.) Between 1985 and 1995 the nature and number of memory systems became the subjects of intense studies which led to various conceptual definitions. These are now being pursued through neuropsychological and neurobiological research. The studies of multiple memory systems have led to the distinction between systems and processes. Interactions are being proposed to integrate the evolving knowledge into an acceptable definition of the total memory function. Such integration is necessary for a full understanding of the system.

The principal unresolved issues are:

1. A unified explanation of memory.
2. An explanation of the translation of physical sensory data into storable non-physical information.
3. Conversely, an explanation of the transformation and encoding of generated ideas and thoughts into storable format.
4. The nature and number of memory system types, as opposed to a unitary multi-functional system.
5. The definition of the interactions between these systems and/or functions.
6. The rationale for interpreting the observations of memory-task performance and correlating it with theoretical conclusions.

There are still many unknowns that will make resolving these issues difficult.

6.2 Types of Memory Systems

This section is based on information from various sources in the extensive memory literature. Due to the paucity of factual information about memory as a physical part of the brain, the conceptual field about its location and functions is open to all cogent interpretations. Concepts about memory systems cannot be refuted due to the lack of deconfirming information. This section, therefore, is an informative one on the types and systems that are being proposed as logical explanations of memory.

If all the conceptual descriptions of memory must be physically oriented, it is possible to develop logical concepts showing how various brain components contribute functionally to produce memory. But the probabilities of the elemental contributions cannot be estimated. Thus there are proposed processes to achieve desired results, like where the storage is located, and there are theorized systems that produce specific results, like controlling the selectivity for long term or short term memory. The resulting mix of ideas calls for a classification system that will initially combine these processes and systems and eventually lead to an integrated definition of the total memorizing and learning process.

To begin to understand what is now perceived to be memory, it is helpful to define the accepted types of memory and the conceptual approaches to memory processes and systems. (Later in this chapter, two representative concepts will be described for an indication of the type of theorizing in progress.)

Memory *types* respond to the needs and expected results of the applications. Short term memory, for example, is needed as a working memory without a need for long term storage of the developing information. There are options for transfer to long term memory if future use is anticipated. It can also become long term through continuous use but is not an automatic enduring input.

Processes are the specific operations of brain elements that are required to achieve the desired results. Encoding of the receptor stimuli, activation of the inputs for storage, and retrieval are examples of such processes.

Systems, on the other hand, are combinations of physical brain elements or sub-elements that store information in a format that can be recalled at a later time. The system descriptions in turn define the functional sub-system requirements to assure system integrity.

As in any developing scientific endeavor, new processes and concepts come with new terms that have special meanings and connotations. These new terms, along with associated meanings of standard terms can introduce misunderstanding if not defined for their conceptual relationships. In fact, the general term *memory* must be defined within the proper context to limit what is being investigated.

Memory is a faculty by which information and experience are remembered for subsequent recall. This term, when used without modifiers, such as long or short, refers to the total function of information storage as distinguished from knowledge, learning and behavior, which depend on the stored information.

Intuitively, we know what memory is. We use it without thinking about how it operates or in which phase or mode we are using it. But in an effort to improve it or to learn how to use it effectively, understanding these modes can help to organize our efforts. For example, understanding short and long term memories can help in organizing how we store or recall information. (An excellent book for testing memory is *Your Memory, A User's Guide* by Alan Baddeley, which leads one through many practical exercises with grading of the results.)

Although the accepted definition of memory is as a physical system, memory is not a unique component of the brain or an organ like the heart. But because it exists and is involved in a physical system, it is assumed to be physical. A combination of brain sub-systems stores information physically so that subsequently it can be recalled for knowledge of the past and for planning the future. Studies of damage to any of these sub-systems have been the source of information about the effects of such damages on memory's effectiveness.

The broadened understanding of the nature of memory changes its functions in that memory is no longer considered distinct from information processing but rather as intertwined with this operation. This may be a vague acknowledgment of the need to transform thought stimuli into physical stimuli for acceptance by the data base; a complex and contradictory operation in a purely physical system. Conceptual explanations claim that there are more than one such system involved in the complex functioning that stores information and produces learning. Continuing studies have resulted in the need for classifying several types of memory to account for the differences between learning and storage.

Memory Types

This classification is used to define the different kinds and uses of memory: short term, long term, and working memories.

Short Term Memory

Short term memory is the retention of pertinent information for the time required to perform a task or in an ongoing situation. An example is your retention of the fact that you are now reading about short term memory as part of a description of the types of memory. Shortly, it will be forgotten. Remembering the names of participants in a group discussion is an example of short-term memory that can be improved through training. Such information may be discarded after the group meeting or it can be transferred to long term memory if the information is expected to be important for future recall. Preserving it as a written reminder is a good practice, providing you remember writing the reminder, because the information will be lost in short memory cleanup. Another example is remembering the right and left turns while driving to an unfamiliar destination. Upon departure, the effectiveness of short term memory is tested, especially if some time has elapsed between arrival and departure.

Working Memory

Working memory is a sub-classification of short term memory to emphasize the increased duration of memory required to perform a task to its completion. In a calculation like addition of a set of three digit numbers, the process involves remembering the carry-over digits from the right columns to the left columns. For those with lazy memories, these digits are marked in the left columns as part of those columns. Although the process of addition is stored in long term memory, with the increasing reliance on calculators and computers, this manual process may be destined for oblivion. However, in such an operation, when the calculation is completed, only the final result is remembered as long as it is needed, then forgotten. However, if there are some innovative ways of simplifying the manual process, like adding by groups of ten, this innovation will be continued into long term memory.

Another example of working memory is in the construction of a piece of furniture. The overall process is recalled from long term memory if this task has been done before. However, when building the piece of furniture, there are the various needs for short term working memory such as: Where are the cut pieces of wood? Where are the nails? Where did I leave the hammer?

Long Term Memory

In long term memory the duration is for hours, days, months or a lifetime. (This is what is usually meant by the single term *memory*.) The shortest duration that is still considered long term is about two minutes.

Long term memory differs from the other two memories because it is where knowledge and skills reside and provide the base for learning. If knowledge, skills and learning are one "package", is long term memory a unitary, dual or multiple system? This is a controversial issue. Perhaps it is because knowledge is subdivided into two categories: semantic and episodal. Learning depends on recall of either or both of these types of knowledge.

Semantic knowledge is the information about the world and the environment, not as experienced but as stored facts. Remembering songs and quotations belong in this category. Episodal information is that which has been experienced and remembered from specific events (episodes). The separation of knowledge into these categories raises the question whether separate systems are responsible for the accumulation and dispersal of these different types of information. There may be other categories such as the remembering of innovative concepts that are neither facts nor experienced history, but which can be stored for future recall to be used in reduction to practice.

Learning is accumulating knowledge for future activity. At times it is a conscious effort as the memory is filled with information that exists and may or may not be necessary for a specific activity. At other times, the accumulation is without a special effort: it just comes "naturally" by experience or unintended participation (also an experience). This distinction between the two types of learning poses questions about how the processing is done, especially when it is a mix of the two. Each of the recalls has a name and a limiting definition. Explicit memory refers to the intentional or conscious recollection of past experiences. By contrast, implicit memory refers to unintentional and non-conscious use of previously acquired information. Prior experiences, or tests, that do not require any intentional or conscious recollection of those experiences, produce changes in performance or behavior. The latter is like conditioned (habitual) activity.

Skills are the accumulated experiences of activities, which with recurring performances, reduce the need for stored pertinent information. Skills must be developed: they cannot be acquired through inherent desire without associated practice. The actual experience must be imbedded in the stored information.

Processes

The processes involved in the translation of sensor information into the required format for a functioning memory are: sensing, encoding, integrating, routing, storing, and retrieving. In the conceptualized versions of memory, the processes are assumed to exist because their functions are necessary. The descriptions of these processes, however, are not detailed because neither all the contributing elements nor the actual implementing details are known. Generalizing these processes is all that is descriptively safe. For example, the processing in short term memory was thought to be by visual rehearsal, which through item repetition was more likely to become long term due to the longer repetitive duration. But that concept was replaced by processing in short term memory. A variety of methods could accomplish this including: noting the visual characteristics of the printed word, through rehearsing it, attending to the sound when spoken, or by an elaborate system of coding to determine meaning. The latter would permit storing the information based on content and the integration of the simultaneous receipt of sensor inputs from more than one receptor.

Since perception involves more than one sense, there must be a process of correlation and integration of the receptor stimuli to present a combined effect for use and storage. When viewing a movie, this correlation involves storing the image for a brief period so that the gaps between frames in the film are eliminated to produce continuity. This is done through a retention period of about one tenth of a second during which time the image is retained as a brief memory. The same is true in hearing. The direction of the sound is determined by the difference in the arrival of the sound at each ear. Sensory memory, as part of the perception process, must correlate these time delayed receptions but the selection of the information for channeling to short term memory must be done by elements of the brain. That which is irrelevant is relegated to what William James called the "bottomless abyss of oblivion". However, from experience we know that somehow most of the sensory input is preserved although only part of it was of interest at the needed time. For storage, the discriminated information should be organized at this stage. The "unneeded" information sometimes helps us to recall information about an event, such as who was present, or the nature of the surrounding environment at the time, even though those items were not of specific importance when the scene was enacted.

If the information is coded on the basis of content, the task of coordinating the visual, the auditory, and any other sensor information prior to routing must be completed. After correlation, pattern separation must be coordinated with directional routing to the intended storage locations. The routing process may involve the same brain elements that normally route other non-memory nerve impulses.

Where and how is the storage process accomplished? Conceptually, this vital process utilizes the physical nerve complex with storage in the synaptic configurations. A concept of this operation is described later in this chapter.

Recall depends on the initial classification of information in a manner similar to that of a library. Normally it is a subconscious effortless action. However, as you have experienced, there are times when the memory needs some prompting. For names, have you used the alphabet process in searching through the classification? Or have you used the scene and time method of associating a person with a previous meeting experience? Classification by association with other factors will speed up recall by reference to those factors.

These descriptions of the memory processes are necessarily general. The details of the manipulation of information processing are not available. Each concept assumes that these processes are necessary and *in some manner* are implemented. Unless, or until, they are firmly defined, conceptualization of their potential functioning is a free game.

Systems

Memory systems differ from memory types and processes in that they are concepts of groupings of neural elements that interact to transform and retain sensor information for future recall. A system contains its own circuitry and parameters but can interact with other systems.

An example of the difference between types and systems is working memory. As a type it has a task to retain information for a short period in the performance of a specific activity. As a system, it is that part of the physical brain that performs that task by providing the temporary storage area for the information and subsequently deleting that information when no longer needed. It may also transfer that information to longer storage if it is intentionally desired for future use. (Note that intentionality, a purely non-physical and normally excluded concept, is used here to explain a physical system operation. This is a bothersome feature in any of the memory concepts.)

The uncertainty about which subsystems participate in the various memory tasks has led to system concepts that have specific requirements and perform specific tasks. Note here that the dependence on a subsystem generally defines the function of the system.

The following examples demonstrate why an integrated analysis of memory is needed.

Locale. The locale system uses the hippocampus to process and store spatial information like locations and shapes. This is information about the environment that is *always* stored in a map-like arrangement. It is easily acquired, and also easily eliminated, with the motivation being the desire to explore and learn about the environment.

Taxon. The taxon system does what the locale does not do. It is a longer acquisition system, slower and taking longer to eliminate. It is concerned with non-spatial learning and thus accepts more generalized information and does not depend on the hippocampus.

Declarative. Declarative memory is the conscious recollection of facts and events such as words, faces, stories and places. Because these facts are identifiable they can be the subjects of testing for amnesia.

Nondeclarative. Nondeclarative memory involves the non-conscious set of those types of memory that are not included in declarative memory because they are not discrete items. The diverse information is acquired during skill learning (motor, perceptual, and cognitive skills), habit formation, and emotional learning; knowledge expressed through performance rather than recollection. This nonconscious effect results in shaping behavior and motivation. Nondeclarative memory is retained in amnesiac patients.

Perception and Reflection. This is a conceptual system that separates the perceptual acquisition (P) of information (learning) from the utilization of that information by reflective generation (R) of goals. It coordinates multiple goals through interactive communication. The breakdown of the functional assignments within the two modes assumes that the cerebral architecture can accommodate the complex system requirements through adequate integrated mechanisms.

P1 is the learning process for finding and/or reaching for an object, which may be in a perceptually difficult environment. This process is involved in learning skills.

P2 is the process for identifying the object when it is acquired. It must be able to resolve spatial relations between the object and other nearby objects for proper evaluation.

R1 is the process that makes the initial effort at utilizing the object in setting goals for its applications.

R2 is the coordinating process when multiple goals have been identified. It has communicative ability to integrate the goals and set up agendas.

It is obvious that this conceptual system is a demanding one, but it encompasses the needs for a cohesive and progressive memory operation. The performance of such a conceptual approach can be analyzed through simulation and may lead to a better definition of the complete memory operation.

Although this perceptual/reflective description consists of processes, as a combined operation they are a system using several brain components to accomplish the task.

There are other conceptual systems, but the above examples are indicative of the conceptual activity in describing how memory works.

Subsystems

What is the difference between a system and a subsystem in the memory complex? To some cognitive theorists the terms are interchangeable, which leads to confusion. Normally, the terms are used in reference to a hierarchical arrangement to show the subservience of the subsystem to the system. In memory studies, however, the problem is that the differences between systems and subsystems are still fluid.

Subsystems are brain components with nerve circuitry that contribute to the achievement of a task or activity. Because the total memory function is not understood, only the results of its operation are identified. The results are obtained by observing human amnesiac effects and animal behavioral testing. (The correlation of human and animal participating subsystems is based on assumptions, not scientifically accepted facts.) These results point to a participation of the hippocampus as a subsystem, as verified by the effective results (memory analysis) produced by PET (Positron Emission Tomography) and MRI (Magnetic Resonance Imaging). These tests illustrate what is happening in the known parts of the brain as various tests and scenarios are suggested to the subjects. In addition, amnesic patients have been studied to relate the kind of lost information with the activity in the brain component responsible for the loss. Lesions on the hippocampus are an example of this type of loss.

The subsystems that have been identified through these tests and behavioral correlation are many. It is beyond the scope of this book to try to describe the extent of subsystem participation in the integration of the total system. Even the physiologists don't understand the relationships adequately. The best source would be a recently published book on memory which should cover the latest research efforts.

(C) The complexity and diversity of memory concepts preclude any critical comments because there is a dearth of verified information about memory. But there is one major criticism that must be responded to if the functioning of memory is to be coherently explained. The challenge is to explain the process of thought conversion into physical stimuli for acceptance by the physical memory system and integration of these stimuli with the other stored information derived from receptor stimuli. This is a serious gap that is ignored by researchers.

The defining of memory types and systems to explain memory functioning serves as an explanatory purpose; a "such as" approach. But this is all based on the initial assumption that it is all a physical system. With such an assumption, it is mandatory to provide an answer for any question about memory even though there is no evidence to prove the concept scientifically. Use of convenient concepts that are accepted by cognitive peers seems to be adequate scientific justification.

The problem is very clear. In an activity involving sensor inputs, even the processing of physical stimuli requires information processing, which is not always part of the memory concepts. It is presumed to exist, producing the required data for all memory activities to proceed. An outside observer is never assumed to be in control of the activity. There are casual references to the effects on an "organism", thereby admitting that there is an overriding influence on the physical activities. But these references are glossed over as unimportant. Only the physical mechanisms are addressed as solutions to parts of the total operation. That is why there are so many concepts. There is enough uncertainty for anyone to bite off a portion of the total pie and describe the taste in individual terms. What is lacking is the integrated picture of how any and all the information becomes part of memory. An example illustrates not only the underlying general process of the physical movement but also the lack of interpretation of nonphysical stimuli.

Consider what happens when you read a detective novel. Reading produces visual stimuli of the words on the page. This is neural processing of visual words for understanding and sending to short term memory or working memory. The words are interpreted through reference to long term memory (a skill). As the story progresses, the action is entertaining until something happens that you think is a clue to the solution of the case. This thought is not a direct product of the words since they have already been interpreted. The idea that a clue may have been interjected into the story is an original interpretation, an effect after having received the word stimuli. This thought must now be stored in long term memory for use in a later part of the book. This original thought, which may be one of many more, must be processed differently than the reading process and stored as a separate part of the story under "clues". This is a command by you to store the clue separately for a special recall when you evaluate all the clues before the solution is given in the book. Although there are two distinct operations, only the physical description is attempted. Obviously, there is a missing link in the analysis.

Although the systems are conceptualized with respect to the physical movement of the information, the concepts do not describe how the selection is made between any two choices. Obviously, the intellect supposedly does the routing. But it is the operator of the activity for which the information is being processed that makes the controlling decision about the future possible use of the information, initiating the decision to send the information to long term storage. Without consideration of this command function the movement of the information is strictly an assumed physical operation without bothering about any overriding features which would complicate the analysis. Yet the need for the routing command is obviously there. (CE)

6.3 How Does Memory Work in Theory?

(6-1-Gl) This description is based on information in *An Anatomy of Thought* by Ian Glynn, Professor and Former Head of the Physiological Laboratory, University of Cambridge, England.

The description of how memory functions has been influenced by physiological and biochemical findings. However, it is neurochemistry that promises to reveal more adequately what is happening within the human brain to store and recall facts and happenings and to learn. But the task is extremely difficult due to the complexity of the interactions of all the contributing elements, both physical and neurological. It demands collaboration of the psychologist and the neuroscientist to piece together the story of how memory works.

Memory is associated with learning since learning is the accumulation of information that is stored for immediate recall. Two problems must be answered before memory can be adequately described: How is memory stored and where is it stored?

Increasing the duration of the synaptic response after the receipt of new stimuli that change the synaptic strengths is believed to be how memory is developed. (The interactions between nerve cells were described in an earlier chapter.) The longer-lasting changes in synaptic responsiveness are enabled through the electrical transmission by glutamate receptors at the synapses. These chemical messengers, the neurotransmitters, allow one neuron to communicate with another through the release of calcium ions. Repeated stimulation reduces the release of calcium ions and slows the forward movement of the impulse.

Computer studies have shown that modification of synaptic strengths will produce recall effects like recognition of faces. However, these computer simulations are only representative and suggest possibilities, not actualities. These results suggest that this process of changing the synaptic strength is the basis for memory and learning.

The process begins with the receipt of sensory information that is to be stored in memory. Before learning can occur, this information must be interpreted (perceived) by the brain and correlated. Learning is not achieved by individual memory cells but by groups of cells that are engaged in the activity that leads to the learning. Correlation, therefore, is necessary since the sensor stimuli may enter simultaneously through several channels and must be integrated to produce useable output signals. The resulting stimuli are sent to the hippocampus, the amygdala, and the neocortical areas to be processed for storage in the synapses. According to a popular hypothesis,

> Memory storage occurs as patterns of synaptic strengths (and perhaps also of the firing thresholds of nerve cells) distributed across nerve networks. The neocortical zone and the hippocampal zone together form a single network.[56]

The hippocampus is the principal initial depository because it is believed that synaptic changes occur more readily in this zone than within the neocortical zone. Due to its limited hippocampal storage space, additional space is provided by adjoining areas of the hippocampal zone and other cortical areas. But the hippocampus is important in coordinating the distribution of the synaptic inputs. After repeated recall of the particular memory in the neocortical area, the need for the coordination diminishes, as does the need for the hippocampus.

The retention of the changes by the synaptic transfer is called potentiation and its duration is a function of the synaptic strengths. Retention in terms of seconds and minutes depends on different mechanisms than does retention for hours, days or weeks. Single brief bursts of repeated stimulation increase the synaptic strength but the effect diminishes in a short time, like minutes. However, several bursts in rapid succession produce a similar synaptic strength but the effect is much more lasting, like for hours or days. This potentiation, therefore, is involved in the determination of how long information is retained for future recall.

> Not all changes in synaptic strength are brought about by long term potentiation acting through NMDA receptors and learning can involve changes in the number of synapses, and even dendrites, as well as changes in synaptic strengths. But what is clear is that at the cellular and sub-cellular level machinery exists that is capable not only of simple logical operations but also of being modified by previous experience so that its behavior changes. It is this machinery that forms the basis of the ability of networks of nerve cells to learn and to remember.[57]
>
> (NMDA=N-methyl-D-aspartate)

Although this network capability of nerves is stated positively, it is much less understood than the functional happening at the synapse or within the nerve cells. It is that collaboration by the various disciplines and laboratories that is necessary to produce an integrated theory.

What is the experimental basis for this definition of how memory and learning are accomplished? The initial tests that disclosed this apparent learning process were performed on a simple small organism, a marine snail, *Aplysia californica.* The procedure was to test the snail's defensive mechanisms by stimulating its extended spout, which it uses to expel waste and sea water. The reaction to the first touch was fast. Repeated touches resulted in diminished responses indicating that the snail was becoming accustomed to the stimulus and acquiring a habit; the snail was becoming *habituated* to the stimulus. Further testing determined that the habituation was accompanied by a reduction in the amount of transmitter glutamate released by the terminals of the sensor cells at their synapses with the motor cells. These tests, although very simple, gave the first indications that information is stored by the nervous system for subsequent recall and that conditioning to a repeated stimulus allows a degradation of the response. The principal result of the tests was the information about the actions at the synapses under these test situations.

Other experimentation led to the disclosures that a mechanism known as long-term-potentiation (LTP) would enhance the duration of responsiveness, for days, weeks, and longer, possibly explaining long-term memory in learning. Learning depends on specific changes in synaptic effectiveness. This happens when the receptors on both sides of a synapse operate at the same level of activity: the firing depends on having exactly the right balance of ions in the receptor channel. When this balance between the pre-synaptic and post-synaptic is right, the nature of the synapse changes so that in the future a much weaker pre-synaptic stimulus will still cause the post-synaptic neuron to fire. This ability to respond to a weaker signal overcomes the weakening of the stimulus due to time.

The notion that consolidating LTP plays a key role in memory is attractive because if it does, then it should be possible to interfere with learning by interfering with the hypothetical mechanisms. This would allow making genetic changes to affect behavior.

(C) There are no rebuttals in what is really a Guessing Game because conceptual thinking is hedged with the admission much has yet to be learned about memory. Conclusions based on experimental and surgical/medical data are physical facts and interpretations of their meanings are valid based on the assumptions of the tests.

There is much room, however, for commenting about what is not addressed. These can be major comments because if they are valid, they may affect the basic assumptions and interpretations of even experimental data. The uncertainty about the actual operation of memory may make these comments important.

The current descriptions of memory are of a physical system. They involve parts of the brain, the nervous system with detailed descriptions of the synaptic performance in the storage of memory. The facts of how the system works physically are based on measurements of physical phenomena. The duration of stimuli retention can be measured. But this is retention of physical motor activity in a physical memory. It is a conditioning of body elements in how to respond to physical sensor induced stimuli. Merely saying that the brain "translates" physical receptor information into suitable stimuli avoids the need for explaining the process.

In the second quotation, the ability of networks of nerve cells to learn and remember is explained by attributing this ability to "cellular and sub-cellular machinery". This machinery, without any definition of how it works, is also a convenient way of evading the issue. But since learning is included, the question is: What is meant by learning? Is it just the accumulation of physical motor signals or is it the broad definition of mental learning? If the latter, it is just as convenient to say that a knowledgeable non-physical entity does this logical and mental operation. But the leap from physical reality to the retention of purely non-physical mental facts and thoughts, generated without any sensor stimuli and stored in a physical memory, is not justified without a complete explanation of how this can occur.

There must be a missing entity, the one that not only generates such thoughts but also stores them in its own separate memory system. This element of an obvious missing link is ignored in cognitive memory studies. It's understandable that if the analysis is beyond any scientist's ability, it will obviously be lost in all the doublespeak of mnemonic studies. But if it is that missing link that is responsible for the failure to understand the total picture of how memory operates, it must be addressed. Once the assumption is made that there is no observer/controller entity in which memory is a vital part, it becomes by default necessary, even mandatory, to explain memory as a strictly physical system and process.

These are the problems with the current memory research and this particular concept. The evidence of immaterial entities refutes the conceptual approaches and adds credence to what are considered paranormal oddities. In the case of Near Death Experience (NDE), every one reporting such an experience remembers activities and locations when out of the body, which is clinically dead (brain is not functioning). This is evidence that there is a memory function of a non-physical nature in a non-physical entity. The question to be answered is: How does the memory of the out-of-body experience get transferred to the physical brain after the return of the non-physical entity? How does the information get into the synapses when it is generated outside the body? It must happen for the person to be able to relate the experience with the use of what is claimed to be a physical memory. Or is that experience stored in the non-physical entity's memory that is separate from the physical memory? And would the memory storage process be entirely different from the physical one? Scientists do not accept these experiences as proof of the reality of a spiritual entity. But experience is reality although not reproducible as scientific rigor demands. There have been so many NDE's that they cannot be ignored. The non-physical laws of storage of mental information like thoughts, learning, skills, etc. in the immaterial mind are not known, but their effects are experienced. There are many fields of research to be initiated.

Although the tests on the marine snail were only rudimentary examples of the use of physical stimuli on a small, uncomplicated nervous system, they were interpreted as a revealing illustration of a memory and learning system. A simple physical touch test demonstrates the initiation of a defensive action which in subsequent repeated touches may lead to conditioning but it fails to give additional understanding about the type of recall necessary for intelligent problem-solving. The basic information must not be extrapolated to other more complex tasks.

When a recall is necessary, the physical process is described in conceptual detail. Absent completely is the progressive process between the need for the recall, the knowledge of the existence of the information, the initiation of the recall process, and the gathering of the related data from all the synapses. Who wants or needs the recall? It is not a random request; it is a directed call for the physical system (which is the storage facility) to produce the stored information. This problem is never discussed in any memory functional descriptions; it is assumed not to exist. The evidence of a separate mind will change this thinking. (CE)

6.4 Typical Memory Concepts

Two examples of conceptual theorizing are given below to illustrate the current attempts to understand what memory does and how it's done. The first is an example of multiple brain systems necessitated by the information handling capability of the hippocampal system. The second is that of a memory system with a general assignment of functional duties to physical brain components and areas.

Declarative and Nondeclarative Memories

(6-2-Sq) This summary is based on information in Chapter 7 in *Memory Systems* 1994 by Larry R. Squire, Professor of Psychology and Neurosciences at University of California School of Medicine at San Diego, California.

Declarative memory is the conscious storage and retrieval of distinct information such as words, stories, places, and faces. These are facts and events that are acquired with awareness and intent to remember. When a new word is encountered, the rush to the dictionary for the precise meaning is a deliberate, conscious effort with the intention of storing it for future use. If it is not used for a long time, a future need for the word or its definition requires a conscious search of memory. However, repeated subsequent use of the word relegates it to a nonconscious mode of automatic recall.

Nondeclarative memory is that automatic mode in which the stored information is used in behavioral or skilled applications. Continued use of a manual operation becomes a skill requiring no recall of the detailed steps to perform that operation. The acquisition of mental, non-physical, skills is also a nondeclarative function.

From the two descriptions above, it is obvious that there is an overlap between the two types of memory in that the dividing line is difficult to define either in the processing or the degree of consciousness required.

What are the differences between the two types of memory? Declarative memory is fast and available for conscious recall. Nondeclarative is slower and less flexible since it is a non-conscious accumulation of information based on performance rather than storage for recollection.

What are the characteristics of these two types of memory that determine how they function? The declarative memory is a biologically meaningful category that depends on a specific brain subsystem like the hippocampus. This is evident from tests with amnesic patients and animals. Nondeclarative memory includes several types of memory, thereby depending on various brain subsystems for functional compatibility. This indicates a need for more than one memory system. Although declarative memory relies on conscious storage and retrieval, nondeclarative depends on system performance rather than recollection.

The relationship of declarative to nondeclarative can be likened to familiarity without identity. There are times when a face is familiar but there is no recollection of identity. There is the subjective experience of familiarity (nondeclarative) but the need for identification must be met by recollection (declarative). The similarity is akin to remembering versus knowing.

Information is stored in memory for conscious recall or nonconscious application without a clear separation of the two activities. The search for a word is a conscious effort. However, when there is no specific need for finding the correct word, language is fluent without special effort but obviously with constant subconscious reliance on past acquired vocabulary.

Priming is a term used to describe the effect of using past experience to improve the ability to detect or process information. It is also credited with altering judgment as in distinguishing between variations in names, words, etc.

Theorizing about these memory functions is based on testing of animals and brain scanning techniques PET and ERP (Event-Related Potential). With this technology, scans of the brain can be viewed and the flow of blood in various parts observed as the different reactions to stimuli are experienced. The data, however, are subject to interpretation and correlation with the observations. Verification of information obtained in different tests cannot always be made due to the variances in the test conditions. Therefore, the data upon which the concepts depend are not always completely supportive.

In the routing and storage of information, the hippocampus accommodates descriptive information. This has been determined through testing and observation of amnesia patients. The nondescriptive information is not affected, implying that other brain components comprising other memory systems are involved. Some studies have also shown that priming depends on brain systems other than those supporting declarative memory.

This short description of a memory type concept is indicative of the attempts to explain the memory function of the brain. However, it is clear that different forms of memory need different parts of the brain to function properly. and consistently. In spite of such positive conceptualizing about multiple brain systems, there are still researchers who maintain that memory is a unitary system.

(C) The test data from animal testing and human testing of amnesic patients are convincing evidence that brain elements, like the hippocampus, when injured, do have a degrading effect on memory. This information is used by neurophilosophers to argue that such data are positive evidence that the memory system is a completely physical system. This is a reasonable conclusion and especially effective when it supports the materialist viewpoint that there is no non-physical entity. I will address this issue in a later chapter, but I wanted to call attention to the specific data described here about the two types of memory systems. (CE)

Working Memory

(6-3-Ba) This summary is based on information in Chapter 11 of Memory Systems 1944.

Working memory is a concept proposed by Alan Baddeley, Professor of Psychology at the University of York, England, that describes how short term memory is enhanced through additional processing to aid in comprehension, learning, and reasoning. Short term memory retains information for a very short period. But if information is to be retained sufficiently for use in, for example problem solving, there must be a system that maintains that information for the duration of the activity.

Working memory consists of three components: the phonological loop, the visuo-spatial sketch pad, and the central executive. This concept is the result of many tests, which addressed the functions necessary to produce the distinctions between short and long term memories in both visual and verbal information processing.

The *phonological loop* retains auditory information but only for about two seconds. To retain for a longer period, rehearsing, or repetition, is necessary. You have experienced this in working with numbers or phrases whereby subvocal repetition will keep the information from deteriorating. This loop seems to work in the occipital part of the brain as evidenced from the PET scans during controlled tests.

The *visuo-sketch pad* is the mechanism that retains a working image whether new or recalled from past experience. This implies some reliance on long term memory. In drawing a picture of an oak tree, does your mental image help you to picture the difference between an oak tree and a palm tree? That is the sketch pad component working through the central executive component. The sketch-pad seems to be located in the occipital lobes at the back of the brain and the parietal region at the center.

The *central executive* controls the verbal and visual subsystems and is the interface with long term memory. Coordination is required due to the differing delays in reception and processing of the verbal and visual channels. The location of the central executive component is not clear due to its need for coding and processing, which are functions of several brain components.

Working memory requires information from past experience, i.e. from episodes, which are stored in long-term memory. According to Professor Baddeley:

> I assume that this process of retrieval from episodic memory makes a representation of an earlier episode accessible to working memory, which allows the central executive component of working memory to reflect on its implications and choose an appropriate action.[58]

This assumption is in line with other assumptions that are necessary to present a credible explanation of how working memory could function. This model and other similar models strongly suggest that memory is not a unitary function due to the various interacting needs for encoded and processed data, which must involve other brain components.

Although in the quotation, episodic memory is an assisting factor, there seems to be a contradiction in another quotation from the same article. In this case episodic memory is not necessarily beneficial. Where is the line drawn between the useful and the detrimental? (I pursue this in the comment section below.)

> While a working memory system that coordinates information from a number of sources is likely to aid perceptual organization of the world, it would not necessarily benefit from experience. Hence, it would not form concepts such as would be necessary to recognize a cat as such, nor would it allow one to learn that cats tend to hiss rather than to bark. More important perhaps, it would not allow one to know whether cats were dangerous or, indeed, to recognize one's own cat or, of course to remember whether it had already been fed or not.[59]

> One additional quotation sheds light on the overall approach of physical causality.

> Before learning can take place, an organism must be able to perceive the world and take advantage of the fact that the information from the range of sensory channels is likely to be correlated.[60]

(C) The working memory system is a system with a stated objective of retaining memories for the duration of an activity when normal short term memory performance may cause temporary problems. The concept solves the problem by having separate mechanisms for auditory and image retention with a central executive function for coordinating information from those two mechanisms and acting as an interface with long term memory. As a system with subsystems it adheres to the definition of a system. Functionally, however, it steps out of the limits of its system responsibility by the words of the first quotation, which states that the central executive is allowed to reflect on the implications of the episodic memory and choose an appropriate action. The reflection and choice are not functions of memory but of a higher order subsystem. Retention and recall are functions of memory. By injecting the ability of reflection and selection, the properties of a higher order subsystem are added simply by statement. So in a total analysis of the concept, it is no different than the materialistic concept of a brain that relies on memory and makes a decision for appropriate action. With a soul in the loop, the mind is the central executive with the ability to accept information, reflect on the implications, and make decisions on how to act.

In my opinion, the concept reiterates the mandated materialistic approach in two respects. The first is that it is just another variation of the mind-brain problem through the use of invented terms that become the solution simply by implied functional capabilities. (Remember in Chapter 2 the advice to watch for invented terms that describe special conditions which then become the integrated solution?) The second reiteration is that the described system which is completely physical is intended for dualistic use. Although the central executive controls the subsystems, it is the only interface with long term memory, a physical element in materialistic theory. In the final quotation, however, an "organism" that must perceive the outside world is introduced and it must be able to utilize the correlations of the system in order to learn. This is an admission that the physical system is conceived for use by a higher order entity. This precise criticism exposes the subtle lack of concern about the vital involvement of perceptive users of the physical systems. Another example appears in the second quotation in the use of "one", a reference to the independent entity that learns, thinks, but is limited to the concepts formed by the memory system.

I continue to detect in the materialistic approach a determination to avoid using language which would reveal an underlying concern of admitting the need for an independent entity to control the mental-physical operation.

6.5 The Two Memory Solution

The scientific community admits that memory is not understood. Therefore, research of the memory function can only result in conceptual innovations and fragmented research. Many researchers are involved, each following a conceptual line that is not part of one integrated system. In a single book, *Memory Systems 1994*, there are eleven conceptual presentations, each on a different aspect of memory. But the researchers are all part of one community that has its own language and basic assumptions to which they can address their individual efforts. This allows conversation between the participants of the community, but to the outside observer, like I am, it is not easy to watch the game when I don't know the rules and must accept the absence of any penalties for transgressions. There can be no penalties when the players make their own rules (conceptual guidelines).

Understanding a concept is not difficult if it is objective and clearly stated. However, when the concept is based on questionable or unproven information, it is usually tolerated as a scientifically oriented objective. That's when the outside observer, not finding the expected clarity for understanding, becomes the skeptic, demanding more rigorous and straightforward explanations. He/she can challenge the assumptions, analyze the semantics of the terms and nomenclature that are used, and critically reject the unsubstantiated statements. In memory concepts, there is much questioning that is valid, not as destructive criticism but as a sincere attempt at exposing why a given description falls short of the necessary clarity.

However, as a unique and historic observer, I have an unusual but clear-cut advantage over the scientific community in knowing that experiential memory is part of an immaterial soul, not part of a physical brain. My observation is unique in that no one has made the connection between reincarnation and dualism by using the eternal quality of the memory. It is an *historic* observation because the evidence for dualism has been unsuccessfully sought for the last three hundred and fifty years. Mine is an unusual advantage because no cognitive scientist is aware, or willing to admit, that a soul's mind with a memory exists. The above statements support my contention that the theory of synaptic storage is invalid because *an immaterial body doesn't have synapses*. How it stores information is not known, but that *retention has been demonstrated*. Learning, acquiring knowledge and skills are functions of the mind with the information stored in an immaterial memory that is disassociated from a physical body because *it survives the death of the physical body*.

With this demonstrated evidence of the location of experiential memory, it is pointless for me to try to negate the concepts described above. The comments given after each concept serve mainly to point out inconsistencies when physical causality demands that a physical description be substituted for the missing link in human behavior. The causality approach is misguided and needs rethinking.

The inconsistencies and contradictions in the current research, however, need not suggest that the current research is *completely in error*. Although experiential memory is not part of the physical brain, there is still a need for the retention of physical information and an interface with the experiential memory. Storage of sensor data and physical information needed for recall by the physical body's functional needs must still be translated and processed physically. And there must be a capacity for transforming the skills information stored in the mind's memory into physical activity as was described in the evidence in Chapter 3 when a child exhibited the sewing skills of a past life without any instructions in the new life. Such an interface must exist in each human being from birth because there was no training to develop that skill.

Another vital need for a physical memory, which *may be stored* in the conceptually described synaptic configurations, is for the storage of life support information and controlling motor commands. The required heart rate, the limits within which blood pressure must remain, the digestive discrimination between acceptable and faulty food, the lasting effects of vaccinations, etc. are all physical, controlling functions that must have a physical data base stored for physical use.

For the above two reasons, *I propose the possibility that there are two memory systems* which are confusing the researchers into trying to explain both with one explanation. It is logical that for two such diverse entities as a body and a soul, each should have properties that function properly in its own domain. But since both are united in one "configuration" an interface between the domains is an absolute necessity. The need for the definition of the interface is what is stated in the opening statement of the Introduction to this book.

In the following two chapters, similar presentations are given for the same reason as in this chapter: to provide an insight into the direction current research is taking. This information is interesting, but as is becoming more obvious with each chapter, the research is misleading.

7

Free Will

Belief in the Elephant is a choice.

In Chapter 6, the conceptual solutions to the memory problem could be contained within the materialistic framework of the neuroscientists. There was physical material that could be used as the location of memory—the neural configuration of synapses within which the data storage problem could be conceptually resolved. Describing the conceptual approaches to free will presents an entirely different situation. There is no part of the brain that is assigned the free will function. It is an integrated function which can only be traced through images of brain scans that show activity in different parts of the brain during decision type tests. Analyzing brain scan images does not result in adequately identifying the actual physical elements involved in the brain or the functional activities that result from the decision process of free will. So how do the thousands of pages of esoterically descriptive material explain the scientific knowledge of free will? Through philosophizing, which like any other philosophical explanation is only a belief in what the truth, or reality, might be. The physical version of free will is not, or may never be, explainable.

What is free will? Its meaning and acceptance as reality depend on who is addressing the issue. The range of beliefs extends from that of an illusion to that of reality. The philosopher will question what it is and whether it is an entity in itself or just a part of the mind, whether it is part of consciousness, and to what extent it controls human behavior.

The neuroscientist has no definition for it since in his cognitive analyses it does not exist as it has no physical qualities. Brain-created mental states determine human behavior, eliminating the need for the decision process. Lately, however, experimental tests of motivation are cause for rethinking of free will, or volition (a more acceptable scientific term), since the neural activity is not caused by external physical stimuli but by mental processes. For the theologian free will does exist and is the basis for his claim and teaching that moral behavior is determined by it. For the average human it is merely the ability to make a choice.

In the dualism debate, free will is one of the fundamental issues because it deals with desires, wishes, and volition, which determine human behavior. Can a physical desire, followed by a decision to pursue that desire, be *initiated* by physical stimuli? What if the desire is initiated without any physical stimuli but by a purely mental motivation stemming from a perceived need? The two situations cannot be explained by the same rationale. For the philosopher, the situation is so open-ended that it must be explained rationally as both possibilities. Not so for the neuroscientist, for in science everything must be explained in physical terms. Although the physical is intermingled with the non-physical, the scientific explanations must sound logical. For the theologian, the task is much easier because the consequences of believing that free will is a God-given gift to man, the starting point is that free will is a fact. The theologian's objective is to explain the consequences of man's having free will. This involves the moral responsibilities for the volitional acts performed by man.

Since the theological argument for free will as a given does not come under the either/or type of argument for the dualism debate, I want to inject a thought that has occurred to me with respect to free will that does not affect the debate. I have never heard or read any statement by anyone, not even a theologian about the magnitude of God's love and trust in giving man free will. By giving free will to man, God has limited His own power and control over human beings. As Creator of all the natural laws, He also gave man the ability to control some of those laws if he so chooses. By imposing the Ten Commandments on man as the moral law for his behavior, God still allows man to disobey those laws by choice. But by depriving Himself of full control of man's actions, did God also limit His own ability to know the outcome of man's decisions? If so, this dual act of relinquishing power for the love of His creations should be recognized for what it is (in human terms): an act of unselfish giving. A father, giving his son and daughter the freedom to do as they wish while still under his responsible care is similar. In order to retain some control, the father makes them responsible and accountable for their actions. Is it possible that such a price is what we pay for this freedom? Such a conclusion has no meaning for the atheist and the neuroscientist even though they also have been given that gift.

Let's return to the dualism problem. Like memory, the conceptual explanations of free will are given in physical terms based on the parts of the brain that are involved in those activities related to making a choice. Four actions associated with voluntary activity have been identified experimentally through observation of the blood flow in parts of the brain. These are: 1) awareness that something must be done, or that there is a need to be met; 2) the decision to act or not to act based on available information; 3) performance of the willed act through physical movement, verbal response or behavioral reaction; and 4) assessment of the success or failure of the willed act – satisfaction or dismay. Obviously, the only physical contribution to these four steps is in the possible initial presentation of a stimulus to require step 1. The movement of any part of the body and the verbal response, although physical, are not contributions but resulting actions of the mental process of a free will. In normal situations, other than testing to study reactions, the entire procedure may take anywhere from an instantaneous decision to one spread over a long period of time, as in the accumulation of adequate information to make the decision.

The importance of establishing that free will is a reality is that it is the basis of the Western cultural, legal, social, religious, and governing systems. The intent, *a free will decision*, determines the guilt of an action in the legal system. Morality is the willful expression of good or evil as *determined* by each individual. Even relative morality is based on the individual's *assessment* of the good or evil of any act. Social value is determined by a *choice* of what is best for the "common good". Governments are responsible for their *decisions* to go to war or to *refrain* from doing so.

So how can the idea that free will is an illusion be entertained when it is practiced by everyone every day, even in the inconsequential daily activities? It is even a deliberate, decisive act of free will to subscribe to such an idea. But since there is no physical evidence to contradict such a claim, or any other similar claim, it can be made by anyone. If the basis for the claim is that the subatomic activities of the brain do not support the existence of a free will, then a similar statement can be made that the physical anatomic activities of the human body preclude the existence of the undefined activity called life.

The following presentation is interesting because it is an example of a direct response by one researcher to another researcher's conceptual explanation. The timeline which is referred to in the Libet free will concept is the one that was illustrated in Chapter 3. It is reproduced here for the reader's convenience.

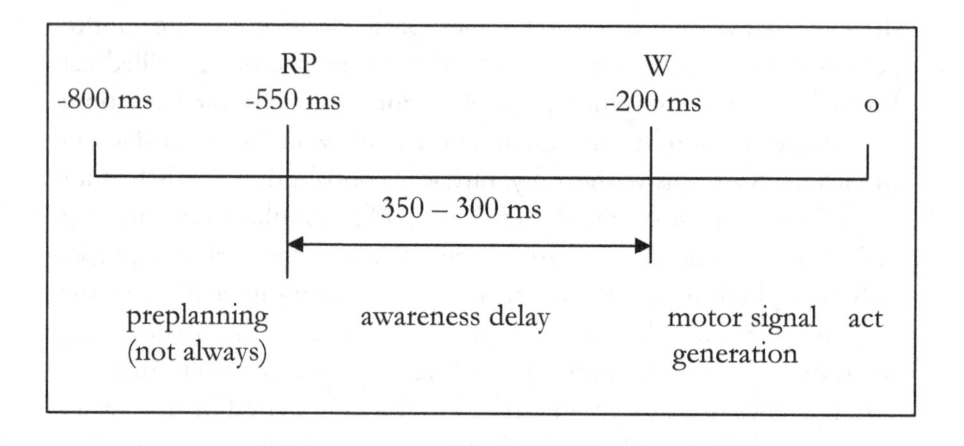

Figure 7.1 Timing for Voluntary Action (Same as Figure 3.1)

Do We Have Free Will?

This description is based on an article by Benjamin Libet from the Journal of Consciousness Studies, 6, No. 8-9, 1999, pp 47-57, and included in *The Volitional Brain* with permission for the quotation of brief passages in criticism and discussion. Benjamin Libet is Professor Emeritus of Physiology at the University of California, San Francisco and a member of the Center for Neuroscience at the University of California, Davis.

(7-1-L) There are several concepts that attempt to prove different aspects of, or lack of, free will. A controversial one is that of Dr. Libet, which we have encountered in an earlier chapter. It is the result of experimentation which analyzed the timing of the processes involved in voluntary acts. Dr. Libet proposes that the initiation of the voluntary act seems to originate in the brain unconsciously, rather than as a consciously willed act. This means that the initiation of the decision to act occurs *before the person consciously knows he wants to act!* The initiating process begins 550 milliseconds (ms) before the actual act. This is the readiness potential (RP) for the activity. The conscious will does appear 200 ms before the act, at which time awareness for the final decision becomes real. Free will then has this remaining time of 200 ms to either allow or to veto the action (actually only 150 ms are available due to the time required for the generation of the motor signal). It is in this short period that the veto, involving the decision to act, identifies the presence of free will because this decision process is a conscious act of "control". The control activity cannot be part of the preceding unconscious action. It is possible that the unconscious initiatives for voluntary actions are "bubbling up" in the brain preparatory to the impending decision to act or not to act.

The experimental work of actually timing conscious and unconscious volitional events does not firmly establish the reality of free will. These results do not address whether determinism by natural law's control of brain functions is responsible for conscious acts or whether conscious activity is independent of natural law determinism. If the former is true, then free will is an illusion with the brain having no causal powers. Determinism has worked well for the observations of the physical world. This had led many scientists and philosophers to assume that it precludes any subjectively explained solutions to free will and volitional control. But there has been no evidence or any experimental effort to substantiate that approach. However, if the latter must be explained, it can only be done in non-physical terms with conscious subjectivity outside the physical brain.

There is an unexplained gap between the concepts of physical phenomena and subjective phenomena. This gap could be bridged by a "conscious mental field" (not amenable to physical studies, although Libet proposed tests that could be attempted), which could influence nerve cell activities in the brain to promote the unconsciously initiated action or prevent its activation. Such a test has yet to be performed.

> If it should turn out to confirm the prediction of that field theory there would be a radical transformation in our views of mind-brain interaction.[61]

(C) Since you read the evidence of the soul's existence in Chapter 3, you can react to the final Libet quotation above: that the conscious mental field already exists. That transformation is what the challenge of this book is calling for. This subconscious hope by the scientists and philosophers for a resolution to the mind-brain controversy should be the inspiration for someone to recognize that the effort can be started immediately.

It is counterproductive to merely take a skeptical approach to any conceptual explanation because of the assumptions and interpretations involved. In a concept that is based on experimental data, however, there is reason for comments about the logical aspects of the concept. Furthermore, the knowledge of the soul's existence can also fill in some of the conceptual gaps.

The major comment about the interpretation of the experimental data is regarding the 350 ms after RP. The knowledge of the soul's existence changes the explanation for the initiation of the RP, and for the preparation of the imminent decision. To credit the initiation of the RP to an undetermined unconscious entity, which has the knowledge of the content of the forthcoming act tends to fall into the trap of determinism, in addition to being illogical. Determinism without volition cannot supply the information for the decision to veto or allow the act. A "bubbling up" effect sounds like a convenient unscientific explanation.

Volition and the Readiness Potential, a Response to (7-1-L). Based on an article from the Journal of Consciousness Studies, 6, No. 8-9, 1999, pp 59-76, and included in the *Volitional Brain* with permission for the quotation of brief passages in criticism and discussion. Article by Gilberto Gomes, CPRJ, Rio de Janeiro, Brazil.

(7-2-G) Gilberto Gomes responded to the controversial Libet concept in a lengthy article which touched on many key points that illustrate how perception of terms, interpretations, and assumptions can induce disagreement. An example of a basic disagreement was the initiation of a voluntary act. Libet based his concept of the unconscious initiation of a voluntary act on the 350 ms delay of conscious awareness before the participation of the free will to actuate the motor command. Gomes, in spite of agreeing that the time lag between initiation (RP) and awareness does exist, disagrees with the concept because, in his rationale, every voluntary act must begin with a conscious intention to perform the act. Consciousness must be involved throughout the period from initiation to motor activation, not being restricted to a censoring decision for completion of the act. Gomes agrees that the initiation may be unconscious but in a different sense than the Libet interpretation. If the agent is the conscious brain system initiating the activity with full intention, the effect is an unconscious one with respect to the subject. In Libet's concept, unconscious can mean causality by any external stimulus that initiates the brain activity. Both sides of the disagreement, however, seem to be based on relatively reasonable arguments.

Voluntary acts are a contradiction of the theory of determinism, in spite of the assumed conditions that are introduced to make them conform. As an example. quantum theory indeterminism, a probabilistic theory, concerns physical systems but it cannot affect purely non-physical mental activity, which is responsible for intentional initiation of voluntary acts. Relating voluntary activity to the naturalistic (materialistic) view of the world poses a difficulty that is easily answered by the dualistic theory. The voluntary acts initiated by the mind are spontaneous, not caused by external causality. However, the use of chemicals to alter brain activity (free action) poses a problem for the dualists. (I comment on this below.) What other factors affect free action, which should not be impaired in either the intention or decision of a voluntary act?

> A free action is an action that is not automatically determined by external events, but is determined by the subject himself, by his will. But what is the subject? What is his will? If we admit that his will is the functioning of some brain systems of his, there is no incompatibility between an action being free and its being causally determined. I make my actions; it is not something else that makes them. But what am I? If I am a functioning body, there is no incompatibility between my agency and its causal determination. My mind is free to choose whether to accomplish a voluntary action or not. This means that the final decision is not determined by anything external to my mind. But what is my mind? If it is the working of some brain systems, the incompatibility concerns only causal processes external to these systems, not the causal processes that determine the functioning of these systems themselves.[62]

Let us return to the RP for more information about what it does, or doesn't do. What preparation does the RP represent and what is the temporal relation between a voluntary decision (as opposed to the unconscious voluntary initiation) to perform the movement and the RP. If the subject initiates the voluntary action, must this conscious decision precede the RP? If, as in the Libet concept, the RP begins before the conscious decision to act, how can the brain start preparing the action before the mind decides to make it? How could the brain anticipate the mind's decision to make the movement?

Consider that the acts are determined by the self (the "I") or by the free will, identifying both as "the free agents". That creates two possibilities. First, free will is the activity of a brain system. The free agent is the brain system or some brain structures of a mind-brain relation according to materialistic theory. Since the RP always precedes the voluntary act, it is the expression of the workings of this free agent.

The second possibility is that the free agent is an immaterial mind, something that acts on the brain to make it perform the actions the free agent desires. This situation produces a temporal problem. Since the RP always precedes the voluntary act, it reflects the brain activity that prepares the performance of the act. But if the act is determined by a mind that is separate from the brain, then the mental decision must precede any preparation for the act, so that it may cause the act. *So the conscious decision must precede the RP*. However, since the RP occurs a half second or more before initiation of muscle contraction, there is a gap of this duration between this conscious decision to act and the motor act. But voluntary acts (motor activation) are caused by a conscious decision. *"And we are not conscious of such a long gap between our conscious decision to act and the act itself"*.[63] (Italics by Gomes) But in Libet's (and other) experiments, such conscious perception of the motor act only occurs immediately following the conscious decision (the veto point). In these experiments, the subjects performed the movements as soon as they decided to do so without being aware of such a gap. This is a problem for the dualist to explain. (I attempt this in the comment section that follows the Gomes theory.)

Libet does not discount free will. Although the experimental evidence showed that free will was not the causal factor in initiating the voluntary acts (they were initiated unconsciously), yet free will played a vital role at the 200 ms veto point. This is because free will is assumed to be a property of the brain free agent. But what about our conscious *intuition* that we initiate voluntary acts? Libet does not explain this intuition.

What if there is no apparent ability to respond to a stimulus and yet a responsive action occurs, as with a blindsight patient? There is evidence of a non-conscious perception of the stimuli presented in the blind field. The explanation is that the voluntary response is stimulated not by non-conscious perception itself but by conscious acceptance of the instructions for "guessing". It is only the content of this "guessing" that is influenced by the non-conscious perception.

In the conscious initiation of voluntary act, there must be intent to act. There is ambiguity in the timing of the intention because duration of the intention may be long or instantaneous, that is, planned action for the future or a "do it now" intention. The latter may also be further defined as either "do it now" or "irrevocably do it now" distinguishing a simple intention from an intention with a decision. However, either of the three is a precursor of any voluntary action. In the Libet experiments, there was also a distinction from the "do it now" in that some of the subjects had an urge to act but that was not followed by any movement.

Intention does not always become conscious as a separate event. To do so it must become conscious before performance has started. If it does not, consciousness of the intention to act now merges with the consciousness of the action itself. How do we experience this decision that immediately causes the action? I (Gomes) argue, we do not experience it as a separate event but as part of the experience of the action itself.

A voluntary act is a first person conviction of being the instigator of the act. Freedom to perform the act must be involved, that is, there must be free will. A third person observation of an act can draw the same conclusion, but does not guarantee the certainty of the fact. An example of an involuntary act, with no intent, is in surgery when an electrical impulse is triggered by the surgeon causing a reaction in some part of the body. The patient, if awake, is not aware of the organ activity or of contributing to make it happen.

(C) The description of the two free agents clarifies the difference between the Libet and Gomes versions of RP content but not initiation. Voluntary initiation is more logical but the gap problem needs explaining. What a difference a soul will make in the abilities of neuroscientists and philosophers like Libet and Gomes to apply a factual element to the constant questions and assumptions about what their concepts represent. In the explanation of free will, the soul is the dominant entity. How can this apply in the above conceptual disagreements? The following comments supply some of the possible solutions, but the complete picture, incorporating the conflicting aspects of the dualism debate, appears in Chapter 10. Questions raised above, such as: what is the subject, what is its will, what am I, what is my mind, will be addressed there.

With respect to the initiation of a voluntary act by either an unconscious or conscious input, the presence of a soul sheds light on both cases, uniting them through a new, but factually backed, concept. There is no question about a soul being involved; its presence must be included. For Libet, it may explain the inability to associate any consciousness or free will with the initiation of the RP. The experimentally measured 350 ms for the brain's preparation for the impending act is still a problem although the brain is a physical entity requiring time for the neuronal activity. However, at the time of awareness 200 ms prior to motor activation, there need not be any question of a possible veto within a 100 ms time span, because there is continuity in the soul activity of anticipating the act. In the Gomes concept, the presence of the soul would confirm the belief that it is a conscious intention of a voluntary act that initiates the RP if there be an RP interlude.

The soul's presence as the "other" entity besides the brain system can also explain the conflict. Since the soul is already integrated with the brain system, and is able to influence the operation of the brain (a stated assumption of a free agent and property of the immaterial mind, based on neural plasticity experiments) there is no need for an RP because there is only one conscious initiation by only one free will for any voluntary act. If the integrated immaterial mind can control the actions of the brain, there cannot be two conflicting operations in the same brain system at the same time. There is no gap; it is the same RP event. In addition, the presence of the soul, which is in continuous control throughout, assures that the motor act follows immediately after the decision, as proven experimentally. Thus, instead of being a serious difficulty for the dualist hypothesis, the soul's presence does the opposite; it confirms it.

The reference to the problem for dualism caused by the reduction in mental functional performance due to the use of drugs is based on the assumption that physical chemicals can damage the mind by damaging the physical components of the brain. The direct implication is misleading. Physical chemicals can only affect physical components of the brain. The immaterial mind is not controlled by natural (physical) laws. It relies on the infrastructure of the brain to direct the motor activation of the body's physical functions in speech, bodily movement, and so on. The mental demands cannot be fluid if the tools and mechanisms of the brain cannot produce the effects the mind is trying to produce. A driver of a car cannot force a stalled engine to start if the starter is inoperative.

In the case of blindsight response, how will the presence of the soul affect the explanation of the phenomenon? As mentioned in Chapter 3, reports of near-death experiences (NDE's) have shown that an out-of-body soul could see and hear the proceedings in the operating room in which its body was the patient. The observations by the soul were verified after the reunion of body and soul, although (I presume) the soul had neither eyes nor ears to observe and listen to what went on in its absence.

In blindsight, what is this "non-conscious perception" that supplies the contents of the "guessing" that Gomes refers to? This is a good example of a non-scientific, naïve explanation of a phenomenon being acceptable to the scientific community, while a more logical experiential explanation like NDE is disallowed.

I quoted a large part of the paragraph about free action because this is typical of how assumptions are transformed into arguments. This step by step procedure with leaps of unsupported justification and arbitrary assumptions becomes a flowing explanation. It begins with the free agent being the subject's will. A free action is "determined by the subject, by his will". The first flaw appears in the assumption, "If we admit that his will is the functioning of some brain systems of his." This assumption changes the whole argument. In the prior sentence, the will is making the action with the subject performing in an active mental role. By making a simple assumption, the will is relegated to a passive role from an active role, subject to the actions of brain systems. This major change is made through the use of an innocuous "if", a naive and convenient transformation for making the will a physical element. The net effect is the immediate jump to the conclusion that because of this assumption, there is no incompatibility between the free act and a causally determined one simply because both belong to the subject. But without the assumption, the argument is invalid because there are two different wills involved, subjective and deterministic. Just the opposite is true. The assumption is validating the argument.

The "subject" (third person) becomes "I" (first person) and an agent without any apparent reason. Why? The reason is given in the next sentence. By using "if" again, I become a functioning body whose "agency" becomes compatible with "its" causal determination. What does all this mean? Why is my functioning "body" being used instead of my functioning mind with will? And what does agency compatibility imply instead of body or brain compatibility? The net effect again is that compatibility is assured through an assumption of a nebulous relationship.

The very next sentence is a direct contradiction of the previous exchange of the will's dependency on brain systems. The statement, "My mind is free to choose whether to accomplish a voluntary action or not". But that freedom was given up to some brain states just a few sentences ago. (This is a distinct failure of short term memory.) The admission is made that the will is only free from external causality. It is still subject to brain system causality (not free), but now the incompatibility is between brain systems and external causality.

In this one quoted paragraph you have been led through an argument that is supposed to show that a non-physical entity, the mind, is compatible with a physical brain system while it is acting as a free agent. If the assumptions had not been made, the argument could not have been concluded with the same result. This type of arguing is the consistent method of cognitive science because the supporting facts are not available. The soul now becomes the only reality which the mind-brain argument can use in the search for compatibility. The above free act statement would merely say that the mind as part of the soul, being able to control the physical brain into activity, can pursue any free act it desires; this with no ifs or supposes.

Will the soul's reality initiate a change in the mental attitude of the skeptical scientific and philosophic communities to adopt an objective approach in the searching for the true explanation of human nature? The above examples illustrate that application of the soul's reality is a more logical argument for cognitive understanding.

In the above comparison of the incompatibilities between conscious and unconscious initiations, temporal gaps, and free agent activities with and without RP, I promised to attempt a resolution. I have the advantage of being able to use the knowledge of a soul's reality, including a mind, free will, rationality and a memory, all independent of the physical body and its brain, in the analysis. Therefore, the only free agent is the soul with its mind, the "I" as the initiator of a voluntary free act, a fact that neither Libet nor Gomes used but the latter was aware of the possibility.

To set the stage, at the expense of repetition, here are the two scenarios of the competing theories.

Libet - unconscious initiation of voluntary acts with a delay of approximately 350 ms before awareness of an ability (free will) to continue an action or discontinue it. This is in accord with the diagram shown in Figure 7-1. There is a recognized time before RP during which a priori planning may be performed although this is only an option.

Gomes – Voluntary acts are conscious acts that are followed immediately by motor commands if the decision is an "irrevocable do it now".

Problem – Is there an explanation of how voluntary and involuntary acts are initiated and completed that, in general, resolves the apparent incompatibility between the two theories? Furthermore, can it show the compatibility between voluntary and event-caused situations?

Analysis – An issue that is not addressed in either theory is the effect of the intent duration in the initiation of the voluntary act. There is time expended between the need, or desire, for the act and the actual initiation of the physical activity which requires the generation of motor commands. The importance of this expended time is that it may, based on its duration, fill the RP gap that seems to be a problem.

In the Libet case there is no intent in the initiation of a sensor stimulated voluntary act. Intent requires a conscious effort. Libet theorizes:

> Initiation of voluntary acts can arise unconsciously in the brain well before any awareness of any conscious intention to move.[64]

Consciousness is included in the Libet activity either as a choice between continuation and cessation of the act or in an act that results from a conscious thought.

In the Gomes theory, intent is present in the initiation of a voluntary act because it must be a conscious act. Intent prior to an activity consumes some time, except in a situation that is an unexpected action, like an accident. The duration of the intent before the actual initiation of the physical act may explain the expenditure of some of the "gap" time of the RP. In the gap-time, the mind can be alerting and informing the brain of what it will require the brain to do. This is the "bubbling" effect that Libet must use to update the brain's ability to react at the veto time. Gomes, therefore, has no problem with his demand for an immediate reaction to the mind's command for action. The continuity of the soul's presence as a separate controlling agent eliminates the switch from unconscious initiation to cognitive awareness at the Libet 200 ms point.

How much time does intent use up? Is the time beginning with the decision to initiate an activity and ending with the initiation of the act greater or less than a half second? If greater, then the gap has been used up. If it is less than a half second, at least a portion of the gap has been consumed. Consider an action like deciding to make a phone call with intent to "irrevocably do it now". Until the phone is lifted, the duration of the arm motion is intent time because the motion can still be arrested. Does reaching for the phone fill or exceed the gap time of a half second. Without arguing or statistically trying to prove it, I believe that the half second is exceeded. If so, what is the mind telling the brain (bubbling up) about the call? It is probably giving the phone number which will require the first physical action of dialing. When the phone is lifted off the hook, the mind is already consciously controlling the brain by commanding the pressing of the first digit of the phone number. This description explains parts of both theories. For Gomes, it explains the probability of an immediate response to the willed act. In Libet's scenario, it sheds some light on how the bubbling effect could be produced if a mind rather than the brain acted as the free agent. The continuous presence of the mind collects all the information for the upcoming decision to proceed or terminate the act. The decision does not have to be a switch act from non-conscious to conscious awareness.

In Chapter 3, Libet posed a problem in which his theory of RP activation had to resort to antedating information to achieve awareness of the act. As part of my problem statement above, this is the case of an event initiating the RP problem of awareness delay with an additional complication of a decisive action prior to any free will activity. The situation involves a child running in front of your car chasing a ball. You slam your foot on the brake pedal to bring the car to a screeching halt. Was that an act of free will since you intended to stop the car? Were you conscious of your action or was that an unconscious action you became aware of after you hit the brake?

> In spite of the presumed actual delay in the awareness of the boy and the ball for up to 500 ms, you are capable of slamming on the brake in about 150 ms or less after the boy appears...(reference to a figure here)...That action, therefore, must be performed unconsciously without awareness. Amazingly, your delayed awareness can be automatically and subjectively antedated or referred back in time, so you would report seeing the boy immediately.[65]

Figure 7.2 illustrates the situation as referenced in the quoted statement,

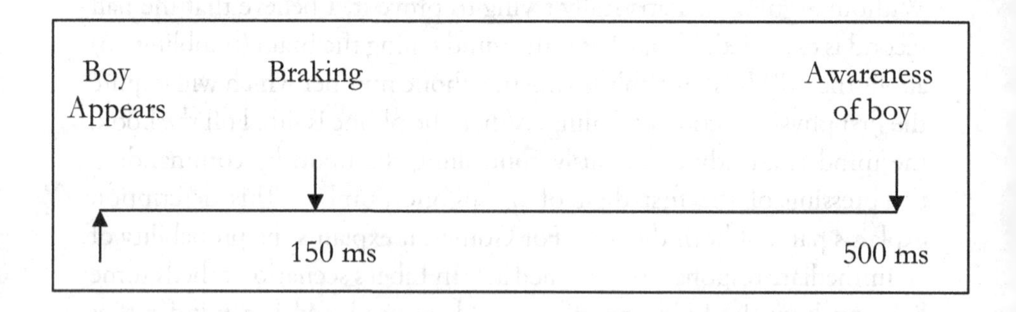

Figure 7.2 Sequence of Braking and Awareness

How can it be that by the time the act of braking is to be affirmed at the 500 ms point due to awareness, the braking had already been activated 350 ms earlier? The post awareness activation was even later by another 200ms!

The first visual stimulus of recognition 100 ms after perception of the boy's movement and 50 ms of neural response would account for the 150 ms needed for first reaction. How can the reaction begin without awareness? "That is, the brain starts the voluntary process unconsciously." Ibid., 93. The Libet assumption is that the brain is the free agent. But since we now know that the soul-mind is the free agent, it is possible to reassess the situation in a light that has never been done before: **using the properties of the soul to explain the braking situation.**

The automatic and subjective referral back in time is Libet's theory that since subjective timing and neural timing are not the same, there is a subjective referral of the timing for the sensory experience back to the time of the primary response. This produces an awareness of the experience as though it were in real time. The theory is explained in great detail in Libet's book *Mind Time*. According to the comments that follow, this dubious theory of time reversal may not be necessary for awareness because the immaterial soul can do in real time what antedating is proposed to do. But this calls for a redirection of the mind-brain philosophy. The problem is that subjectivity in the Libet theory is a substitute term for the soul. The soul as argued in previous chapters is the subjective "I". As the free agent it interacts with the brain throughout an activity. This should be the factual base upon which the reactive nature of conscious awareness should be evaluated. It should not be the juggling of brain functions in an attempt to justify a conceptual approach that cannot logically solve the mind-brain problem with respect to free will.

The Gomes theory of instant activation is closer to the answer due to the immediate activation of the motor signal when an "irrevocable intent to do" is imposed by an externally caused stimulus. This situation raises two questions. What is the nature of the intent and the second is how can a motor signal for the fast reaction be generated by the brain while the visual signal that is responsible for the neural response is still registering? This is difficult to explain in physical terms because it is in real time, not as an afterthought through time reversal.

With respect to the first question, there is no intent prior to the visual emergency stimulus of the boy running in front of the car. The Gomes approach of immediate response is still applicable but made difficult by the second question.

There may not be an answer to the second question because it may not be physically possible to slam on the brake in the first 150 ms. Even if it were possible as a stated condition in Libet's problem, the explanatory answers provide an insight into the functioning of the human being. The movement of the foot to the brake pedal may take more than 150 ms. However, if an explanation is to be attempted, it must be one that includes the soul as the conscious being with awareness of the situation. The physical brain only processes the sensor information. The soul as the only free agent directs the brain to generate the motor signals. Since a physical explanation doesn't seem possible, let us analyze the problem from the spiritual side.

In Chapter 3, I described the characteristics of a soul in a NDE. The soul, when separated from the body that was lying on the operating table, could see and hear what was being done and said by the medical personnel. Upon returning to the body, the recounting of what the soul saw and heard was verified by the surgical team. This verified that the soul without eyes or ears, at least the physical type, was still aware of all that was being done to the body. Translate this to the case of the driver upon perceiving the first sight of the boy moving in front of the car. If the mind (or soul) is in charge of the brain as the free agent, it perceives the boy's movement as do the physical eyes of the body. This perception, the conscious awareness of the situation triggers the chain of events. The free agent decision to act is immediate, the fastest possible reaction, since free will is one of the soul's properties. This non-physical perception and decision to act, not antedating of the response, is the reason for the immediate awareness of the required action. It happens in real time.

With a new philosophy of conscious awareness there is also a need to give credence to reports that have been ridiculed as paranormal illusions, like the NDE. These are first person experiences that are often verified by witnesses. Because these experiences occur only once they cannot comply with the scientific repeatability requirement for acceptance. But the innumerable reports of NDE, in a sense, are the repeatable evidence of its reality. These experiences provide an insight into the non-physical properties of the immaterial soul, and should be recognized as the only type of evidence that is, or will be, available for the study of the mind-brain and soul-body relationships. And I believe that they are applicable in the resolution of the braking example. (CE)

The above examples of the current explanations of free will show that the soul's contributions to the mind-brain problem can fill the unexplained gaps between conscious and non-conscious activities. In the next chapter, consciousness is presented in the same manner as memory and free will have been with examples of current concepts and commentary about them

8

Consciousness

Consciousness is part of the Elephant.

What is consciousness? A precise definition is impossible because consciousness is not a thing, a part of the brain, or an entity with distinct properties. It is awareness of the world, of other conscious beings, of one's own body, and even of one's own self. It is not a conflict in terms to say that it is awareness of being aware. There are four states of consciousness: awake, asleep, dreaming, and coma. But this contradicts the preceding statement because awareness is not evident in the sleep state. The contradiction lies in the interpretation of the word conscious. Is consciousness the awareness of the external world along with the interior self, or is it the condition of having life, which would include the sleeping state. As for dreaming, do dreams that are remembered after waking constitute awareness of personal activity within the sleep state? What is consciousness in a coma with life and perhaps recognition and awareness but a lack of movement and an inability to communicate? This nebulous quality of consciousness, without a physical attachment to anchor it, allows analysts and philosophers to adopt their own versions of its attributes and limitations.

In the earlier chapter on memory, it was possible to talk about a thing because the cognitive scientists ascribed physical properties to parts of the neural system which store the information. Although information is not physical and the transformation from non-physical to physical storage is not explained, memory is considered to be a physical storage area. The synapses are physical parts of the brain. In the chapter on free will, its definition was more difficult, but it was still tied to the neural system and it could be isolated functionally; it could do something. There the explanation was based on when and if it can be exercised (see Libet, Chapters 3 and 4). If it can be exercised, to complete the action or abort, free will must be an action by "something" in the brain, with certain parts of the brain responsible for the process. Of course, there are scientists and philosophers who claim free will does not exist. Their only problem, therefore, is not the explanation of free will but the defense of their position, that is, if they "choose" to do so. In this chapter, consciousness is neither a thing, nor an action; it is a condition. However, the condition, with its variables, may be easy to relate to, principally because we all experience it. This subjectivity opens the field for conceptual ingenuity without demanding a precise definition. (I succumb to this temptation in a later chapter.)

What is the function of consciousness? For memory, it is storage of information for future retrieval; for free will, it is the ability to make decisions. But consciousness doesn't actively do anything; it is passively present as a condition of life. That is how I see it but the proponents of physical causality explain it through conceptual innovations, such as those presented later in the chapter.

In the search for the true meaning of consciousness, there is disagreement about which discipline is best qualified or equipped to produce the most philosophically logical or scientifically verifiable version. Philosophers maintain that no matter how good the neurophysiologic data are, consciousness will remain unexplained because it is dependent on subjective evaluation. This is the domain of the psychologists, but to the neuroscientists it is rooted in neurobiology, "a coordinated pattern of neural activity serving various biological functions".[66] The physicists base their claims on theories of particle wave functions and quantum mechanics. The only way to understand consciousness, they say, is to research the subject at the microscopic level of atoms and at subatomic levels. But at that stage there are the indeterministic effects on those subatomic particles according to quantum theory. This, however, is an approach to understanding consciousness as a materialistic solution, because as a physical "thing" it can still be claimed capable of handling qualia. By treating the wave function (mind) and the particles (matter) together, consciousness and causality are intrinsic properties in the natural world. That immediately brings up the subjectivity argument as a completely non-materialist opposing view.

Due to such fragmented attempts at a meaningful definition, a holistic approach is needed but may never be undertaken unless there is better coordination and cooperation. The philosophers speak the language of logic; the physicists use the language of natural law; the neuroscientists ignore both and indulge in neural assumptive optimisms. It is time to redefine ground zero and adopt the Descartes rules for rigorous analysis.

Animals are conscious because they are alive, but how does that consciousness differ from that of a human? Man's rationality, language, and intelligence expand the limits of his awareness. To what extent do animals have the same capabilities? Awareness of their surroundings and social relations with man or other animals, and their ability to communicate indicate that it may be merely a matter of degree that their consciousness is similar to the human type. Experimental evidence confirms this apparent consciousness gathered from studies of brain deficiencies, which can be applied to human patients. The quantifiable levels and qualitative differences may be due to the evolutionary stages of the various species. Since plants also have life based on the developmental processes of the same DNA that governs the lives of humans and animals, do they also have a consciousness, perhaps of a totally different nature? They favor an orientation with respect to the sun that helps them absorb the sun's energy; they collaborate with bees, birds and insects to their mutual advantages; all of which imply a consciousness of the natural environment.

Like in any other neuroscientific explanation, there is no place here in consciousness for a non-physical entity, the soul. As stated in previous chapters, to describe the existing positions of soulless analyses without bias, the conflicts with a soul's presence, of which you are now aware, must be kept out. As before, remarks about the effects of the soul appear in the comments preceded by the parenthetical (C).

Consciousness Concepts

As in previous chapters, I want to show, by examples, the extent to which the scientists have tried to substitute for the presence of the soul in the human makeup. This is especially relevant here because life, soul and consciousness are so intimately combined that this relationship is the reason consciousness cannot be defined or explained. The physical aspects of life are understood through its reproductive continuation, but are there other aspects that are overlooked? For this reason, I continue to expose the ongoing concepts by experienced professionals in which questionable theories are justified on the assumptions of eventual verification. Since you now know that the initial assumption is false, it is evident that the optimistic future verifications have no chance of happening..

To accentuate the assumptive nature employed in these concepts I will identify those terms that, in my opinion, the conceptualizers use in closing the unexplainable features of their concepts or that stress biased implications. I will use their precise terms in the conceptual presentations and underline the words. An example is: by some sort of relaxation of activity. Wording such as that simply admits that the process has not been defined and may not be feasible; yet it is a key element of the solution. In my opinion, one such statement should negate the entire concept, but it doesn't to an optimistic and benevolent scientific peerage.

(8-1-T) Relational Consciousness

A summary based on information in the book, *Race for Consciousness* (1999), by John G. Taylor, Emeritus Professor and Director, Centre for Neural Networks, Kings College, University of London.

In the typical analyses of cognitive functions, there is a subjective element that eludes definition because the analyses are not of a holistic type. Portions of the total problem are studied separately, which is easier to do. Therefore, this subjective insubstantial condition eludes our understanding.

> There is something apparently insubstantial about the mind that gives strength to the dualism model of Descartes in which mind and body were separate substances.[67]

Somehow, this insubstantiation must be included in the holistic approach so that the qualia are represented in addition to the stimulated firings of the nerve cells. How can awareness be infused into the routine activity of "dumb" cells? The proposed concept to do this is through the relations between brain activities. It is through this relational activity of suitably connected neural networks that consciousness emerges. Nerve cells combine to enable the brain to support the mind.

For a relation of brain activities to exist, there must be more than one activity. A single activity is an independent occurrence of neuronal activity in a brain state devoid of any association with past activities. As such it is strictly a physical occurrence without non-physical characteristics. The relational feature is introduced by association with past experience. When two sets of activities are involved through reliance on past memory, a non-physical relationship is created if the relations between the activities rather than the activities themselves are emphasized. The reliance on the non-physical memory opens the possibility of the relation being representative of the unsubstantiated qualia in the brain activities, although not of a dualistic form.

The relations between neural firings are distinct from the firings themselves. How are these relations established? It is in the encoding of these relations by the strengths of synapses between the neurons that have been activated in related firings. Thus by activating neuron encoding for representation of an object by some sort of relaxation of activity in a network to a steady state by feedback, recognition is obtained from a previously experienced and categorized object. In addition, by activating representations of a broader range of objects than a given object, the encoded memories of various sorts, as synaptic strengths, allow expanded neural activity. This triggering of related activity gives a sense of insubstantiality to the total neural activity. It is clearly very important to justify this possibility.

From the above relational aspects of brain activity, it is obvious that memory plays a significant role in the associative features of relational consciousness. Memory in some form or other is quite extensive throughout the brain. This includes skills and values, enhancing the quality of the emerging consciousness.

The only way to solve the mind and consciousness problem is to develop a model in which this concept can be a good first approach.

There is a relatively easy part; the development of the neural network models that simulate the observed behavior of humans and animals. This can be done by using information processing modules in which the underlying implementation is not as important as the functions they perform. The difficult task is the one associated with the nature and origin of qualia and self-awareness; difficult because science cannot enter the inner workings of the mind as it relates to the physical brain and there are very few models in this area. A model representing this concept which can verify its feasibility is being developed. The thesis of the model being developed is: "The conscious content of a mental experience is determined by the evocation and intermingling of suitable past memories evoked (sometimes unconsciously) by the input giving rise to the experience."[68]

Relational Consciousness is conceptually feasible. However, phenomenal awareness with its property of being "intrinsic" to qualia poses a problem for relational consciousness. This property denies any relational characteristics for them (qualia), a direct contradiction to the relational mind approach.

The goal is to find the scientific solution to the crucial ingredient in the neural activity of the brain that sparks consciousness into life, the ingredient that adds value and guarantees the presence of consciousness.

(C) The evidence of the soul's reality is not the scientific solution that Professor Taylor seeks, but it does guarantee consciousness, for it is the soul's presence that guarantees life, not the brain's neural activity. Consciousness is not sparked into life; the opposite is more likely. Without life, there is no awareness. It is also life that guarantees brain activity. There is a conflict in basic interpretation when the goal is to justify physical brain activity as the generator of human activity, rather than an external control of brain activity. This concept still denies a dualistic form although it admits the existence of a mind.

In Taylor's words, consciousness emerges from brain activity. The entire approach has the sound of being based on a mandated solution, which has to be justified through a logical approach, one that can be verified through modeling. Like any of the other concepts that I have presented, this concept is merely an attempt at explaining away the human soul. To Professor Taylor's credit, he admits the existence of a very hard problem in explaining away the phenomenon of awareness. But having admitted the problem, his call for a holistic (global) approach to explaining the human brain, excludes the direct inclusion of the possible effects of a soul, the dualistic effect.

Since we have the evidence of the soul's existence, the real question, or problem, in consciousness is whether the soul is the consciousness that no one can define, because the soul's existence has been denied. A comparable problem is the relationship of consciousness to life itself. I have more on that in a later chapter.

There are two specific comments about the relational consciousness concept that, at least to me, make the concept weak on logical grounds. The first is that the relational aspect of brain activity, through which consciousness emerges, is only possible with two sets of related activities. The conclusion resulting from that limitation is that consciousness is not present when only one set of activities is involved. If meditation does not rely on any previous experiential content, does that mean that there is no emergent, or intrinsic, consciousness while meditating? It cannot be so, for consciousness is a continuing property of a living being. Presumably, the concept does not apply during sleep because there is no directed brain activity. This same reason could apply to meditation since that is a non-physical activity.

The second objection to the concept is that it depends on brain activity. By defining a relation as different from an activity, a major leap in acceptance is demanded. Brain activity is a physical operation; the neural signals, the routing and the synaptic transfers are all physical. The use of a non-physical concept of relation is an ambiguous reliance on a semantic crutch to provide justification for the concept. A relation that depends on synaptic encoding, as an intelligent operation, is reliance on what neuroscience claims happens in the synaptic weight configuring. There is no proven evidence of this conceptual claim, or of the synaptic storage of experiential information. We have the advantage of knowing that memory is not a physical "thing" as the evidence of the soul has shown, therefore, we can see the impropriety of the scientific approach. As I have proposed in Chapter 6, there may be two memories with different functions.

There is an aspect of the relational consciousness concept that, unknown to Professor Taylor, adds credence to the soul's connection, or identity with, consciousness. The need for experience in establishing consciousness in the relations between brain activities supports the explanation of the soul's possession of consciousness since the soul remembers the experiences of the past life. In a rethought analysis, the presence of such experience would contribute factual support for Professor Taylor's relational concept.

The statement that nerve cells combine to enable the brain to support the mind raises the issue of which of the two controls the other. The emerging consciousness from the relations of brain activities implies that the brain is the principal element. But if the brain supports the mind, the implication is reversed. The evidence of the soul acting in the reincarnated body leaves no doubt about which is the controlling entity.

The development of a model to define the brain faces a major obstacle because the effects of soul control of the brain can never be described, or modeled, in a physical modularity. If the attempt is to construct Artificial Intelligence, it can only be developed to a certain point, after which the philosophers will have to become involved, and even then its impossible outcome will be obvious. (CE)

(8-2-H) The Conscious State Concept

The summary is based on information in the book, *Consciousness*, (1999) by J. Allan Hobson, Professor of Psychiatry and Director of the Laboratory of Neurophysiology at the Harvard Medical School.

Consciousness is the result of higher order brain states. It does not reside in any single brain function, brain component, brain network, brain cell, or wave function. The higher order of the brain state is due to the integrated functioning of both the brain and the mind. Although the brain uses the biologically developed brain mechanisms, it is the subjective contribution of the mind that initiates the awareness of the world, other conscious beings, and of one's own self. The incorporation of the subjective element in the integrated mind-brain is not fully supported by the cognitive scientists. Their claims of complete physical causality prevent them from including subjective experience as valid performance evidence. On the other hand, philosophers seem to be saying that no matter how good the neurophysiological data, consciousness will remain unexplained because it is irreducibly subjective.

The experiential nature of consciousness is defined by its components. These are evident at either of two levels. At the first level are those that are triggered by external stimuli, which cause reactions like sensory responses, attention to selective inputs, and emotional responses. At the second level, the conscious response is internally generated as in information retrieval, reflection, learning, and reflective thought. The second level is more complex than the first. Consciousness, then, is a multifaceted array of components arguing against a unitary concept of consciousness.

This approach through synthesis as an explanation of consciousness is contrary to the scientific preference for the analytic approach. In the latter approach, information is being gathered about the parts, but it is the study of the total system that will lead to an understanding of the relationships of the parts. A meaningful step in this direction is the use of a graded complexity of consciousness. In the assumption that animals are also conscious living beings, the differences between humans and animals can be explained by the graded complexities that exist due to the different evolutionary histories of each and within species over the individual lifetimes.

In promoting the holistic synthetic route, the conscious state concept proposes that there are three major features of consciousness: its constant oscillation between unity and plurality, its extended modularity, and its graded quality within and across species. The concept attempts to explain all three features by focusing on the formal aspects of the three cardinal states of consciousness: waking, sleeping, and dreaming. By defining and measuring formal aspects of these three at both the phenomenological (mental) and physiological (cerebral) levels, the conscious state approach permits comparison across levels within each state as well as comparison of one state with the others. This comparison promotes the holistic approach by integrated explanations of the components involved within a state. Thus, if waking is characterized by accurate recent memory and dreaming is not, it should be possible to explain that difference. In fact, a brain chemical essential to memory is present in the brain during waking but is not secreted during dreaming. Therefore, memory is not a part of dreaming. In animals serotonin is required for learning. No serotonin, no learning. This is also true in humans. During waking, serotonin is released and perception and memory are present. During sleep, when serotonin is withheld, perception is possible but not memory. No serotonin, no memory.

Self-awareness may also be amenable to analysis since it is present in waking, but not in dreaming. There is an erroneous belief that self-awareness exists in dreaming. But in terms of the brain needing incoming informational stimuli to function, such self-awareness is not present. If self-awareness were present, the awareness that it is a dream environment should exist, but it doesn't. The imagery in dreaming is the result of the visual brain's input, which is sharper than creating images in the waking state. One factor that contributes to the disorientation with time and location in dreaming is the lack of external sensory signals on which the brain is dependent in waking.

In the conscious state model, coma is a state less conscious than deep sleep. A patient in a coma may be aware of the surroundings and the familiar people nearby but may not be able to move or to communicate with them. A man, who had been in a coma for six years after severe brain injury, was brought to a conscious state by the implantation of tiny electrodes in the thalamus. Although he could mumble a yes or no occasionally, he could not open his eyes or move his limbs before the addition of the "pacemaker". Within hours of the operation, he opened his eyes and was aware of the movements of people in the vicinity. His ability to speak slowly and to hold items in his hands improved with time, but it was slow progress. At the time of the report, he could only speak slowly and could not manipulate whatever he held in his hand. However, the fact that he had been conscious in the coma was verified through the reinstatement of the ability to process the external stimuli by electrical amplification. The reasons for the improvement could not be explained scientifically.

The conscious state model, AIM, is expressed as a three dimensional space in which it is possible to track the brain-mind as a succession of points. These points represent stages of waking, sleep and deep sleep, each of them occupying a region of the three dimensional space. The axes of the model are labeled Activation level (A), Information source (I), and Mode of processing (M). A represents the average level of neuronal activity in the various parts of the brain as obtained through scanning techniques. I represents the gating of information that determines whether the state is awake with a high information rate or asleep with a low information rate. M represents the intensity of the conscious experience by the amount of information, external or internal, that is being processed. The conscious state concept with the three dimensional model is an attempt to expand the investigation of consciousness beyond the limited approach now being taken by cognitive scientists. The model's representation of consciousness is possible because it is based on those physical characteristics of the integrated neural system. It uses the neurophysiological information derived by cognitive science as model inputs. An example is the brain elements that control the presence of consciousness.

> The reticular nucleus of the thalamus is as close to an on-off switch for consciousness as we have yet found. When the activation level of the brainstem falls, even a little, the thalamocortical circuits begin to oscillate, beating to an intrinsic rhythm that is a function of their own excitatory and inhibitory features. This kind of synchrony contributes to the global loss of consciousness that occurs in non-rapid eye movement (NREM) sleep.[69]

Consciousness has the capacity to create abstract models, the apex of which is human creativity. For in creativity, consciousness transcends the world by making its own worlds of science and art.

> But make no mistake; the path to creativity is a straight and direct one from the brain's intrinsic talent for representing stimuli in its neural discharge patterns and chemical codes. ......It is in this sense that I say that the brain-mind question and the problem of consciousness are already solved......Of course, we need to know more about how the world comes to be re-represented in primary conscious awareness, and how these re-

> representations come to be re-re-represented in the awareness of awareness in secondary consciousness.[70]

(C) The closing quotation is another example of the neuroscientific approach to the mind brain problem: we have the answer and it will become obvious when the mind- boggling unknowns are someday resolved. What kind of a solution is that when it has to be phrased in double-speak like the last five lines of the quotation? That is an outright contradiction of the claim of a solution.

I chose this concept for inclusion to show that not all neuroscientific efforts are directed toward specialty areas that promote the insistence on the dominance of physical causality. This system concept is an improvement over those concepts that deal with component issues because it acknowledges the need for a holistic approach.

With the evidence of soul reality, the homunculus, so despised by the scientific community, is back. There is an entity that is observing what is happening in the brain, and maybe controlling it. Without physical eyes or ears, it still needs a physical brain to control a physical body.

Throughout the description of the concept, Professor Hobson is continually concerned with the importance of the subjective nature of consciousness. But that is the contribution of the soul to the living being. The components that are involved in the concept, such as memory, and specifically the mind of the mind-brain concept, are not physical components as the evidence of the soul shows. The experiences, locations and learning that are retained by the memory are transported by the soul from the old to the new brain. Therefore, the old brain never possessed this information for it would have died with the old brain. This indicates that the holistic approach to the explanation of consciousness may well be strengthened by the inclusion of the soul with its subjective mental properties, if the ingrained opposition to the idea of a separate entity can be overcome. The concept is headed in the right direction and with reorientation would be amenable to rethinking dualism.

The concept claims that there is no memory in the dream state. Does this mean that the dream cannot retrieve information from the memory, or does it mean that there is no input to the "awake" memory about the dream events? In my dreams there are references to past experiences in general but in the actions only my self-awareness is real; all other associations are related to the situation I am experiencing in my dream. In fact, Professor Hobson admits the same interpretation by discussing his dreams in the book.

I experienced memory contribution to my dream that was very specific. I interrupted my writing of the Libet concept in Chapter 7 because it was late and time to go to bed. As has happened many times, my mind refused to rest while my brain fell asleep. In my dream, I continued my writing and it was identifiable because the name Libet was associated with it. That was a memory contribution although on waking I didn't remember the dream content for possible use in the book.

Not only do I remember the events in the dreams, but I am also astounded at what information, unrelated to past experiences, that appears in the dreams. There is the case of my listening to an operatic aria and marveling at the beauty of the soprano's rendition without recognizing the voice or the music. There is understanding, appreciation, emotion and control of the experience until the aria is completed. Where did this presentation come from and how was it implemented? Another instance was my reading a newspaper and understanding the mentally printed text about an unfamiliar subject, although I could only remember the act of reading upon awakening. These activities were not the result of information retrieval but of information generation and yet the format of the newspaper and the tonality of the musical aria, I believe are memories of external stimuli. So either self-awareness or memory must be part of the dream sequence. Unless I misunderstand the use of the word self-awareness, I disagree with the conceptual statement that there is no awareness in dreams because the brain is not receiving any external stimuli. I may not be aware that I am in a dream but my dreams are first person experiences. Isn't that self-awareness?

Since memory resides in the soul, and the soul may be the instigator of dreams, I prefer to think that it is soul memory for the input, soul-self for the awareness, and emotional reaction since the soul-self does not have a dream level like the body-self does. The soul continues to think while the body recuperates in unconscious *rest.*

The main objection to the concept is, as with the other concepts, that the presence of the soul, which could supply some answers, is ignored leaving any synthesis or analysis incomplete. (CE)

(8-3-S) A Neurophysiological Process
A summary based on information in the book, *The Rediscovery of the Mind*, (1992) by John R. Searle, Professor of the Philosophy of Mind and Language, University of California, Berkeley.

Consciousness is often confused with "conscience", self-consciousness, and cognition. It is neither of these. Neither is awareness the same as consciousness because awareness is associated more with cognition, an active function, whereas consciousness is a state that is experienced rather than rationalized. A vital element of consciousness is subjectivity, a subject that materialists refuse to grapple with. But consciousness will not be defined without the inclusion of its subjective nature. The precise definition of subjectivity that applies is the one that relates to the experience resulting from a specific stimulus, like that of pain, either physical or mental. The stimulus could be a thought caused by the observance of an activity by another being. Subjectivity as contrasted to objectivity of the truth of a debated belief is an example of a rationalizing subjectivity, not pertinent to the definition of experiential consciousness. Another pertinent quality of subjectivity is its inaccessibility to third person observance; it is an individual experience. This failure by the scientists and philosophers to take subjectivity into account in their assumptive meanderings is the principal reason why consciousness is so ill defined. A pertinent cause of this lack of subjective analysis is the worldview that all matter including biological systems are composed of physical particles, thus excluding from analysis any concepts that touch on non-physical, presumably mythical, conditions or entities. And yet what is posed as such imagination is life as it exists; consciousness with its subjectivity is experienced by every living human being.

Although the mental aspects of consciousness appear to be indescribable in physical terms, it is conceptually possible to view them as a result of brain processes. Since consciousness is caused by the behavior of the neurons, it, therefore, is caused by brain processes. Much has been learned about the neural system and the brain through studies in neurobiology.

> The brain causes certain "mental" phenomena, such as conscious mental states, and these conscious states are simply higher-level features of the brain. Consciousness is a higher-level or emergent property of the brain.... Consciousness is a mental and therefore physical property of the brain in the sense in which liquidity is a property of systems of molecules....The fact that a feature is mental does not imply that it is not physical; the fact that a feature is physical does not imply that it is not mental.[71]

Due to the use of traditional terminology, the above quotation could be misunderstood if interpreted in the usual neuroscientific usage. What is meant is that consciousness, as consciousness, as mental, as subjective, as qualitative is physical, and physical because mental.

(C) The inaccuracy of the closing sentence above is confirmed factually by the evidence of the soul's transfer from one body to another. If consciousness were part of the physical world, it would have remained in the previous dead body, or would have ceased to exist. Instead, it came across to the new body with all the subjective experiences of the previous life. If the necessity for considering subjectivity is an essential feature of consciousness, this retention of the subjective experiences in the reincarnation verifies the need for its inclusion in the concept but negates the physical aspects of the concept.

This concept, as well as all other current concepts, based on the denial of soul reality is false regardless of all the philosophical reasons that attempt to justify its presumptive aura of reality. It is an attempt at substituting another approach for the cognitive gap to explain the mind-brain relationship, and specifically consciousness. The inconsistencies of calling mental consciousness a physical reality are obvious, except to a philosopher. (CE)

(8-4-C) The Neural Substrate Concept

The summary is based on information in the book, *Brain-Wise*, (2002) by Patricia Smith Churchland, Chair of the Philosophy Department and UC President's Professor of Philosophy at the University of California, San Diego.

Consciousness is difficult to define in a single sentence because it has different meanings for different conditions. It is a set of states, which include sensory perception; states that are of a sensory nature but are not associated with the physical senses like remembering, knowing, and imagining; emotional states like fear and anger; and drive states like hunger and thirst. Other states are non-conscious in the sense that awareness is not discernable like in sleep, deep sleep and coma.

Attempts at understanding consciousness are of two types: direct and indirect. The former relies on existing experimental neurological data as the basis for formulating a conceptual version of consciousness. The latter strives toward an understanding of the integrated definition of the brain's subsystems with the goal of finding the effects of consciousness within this system.

The direct approach uses the notion that there is a physical marker that can identify a physical substrate with a distinctive signature. If the substrate can be identified as a "correlate of phenomenological awareness", it may then be possible to explain it in neurobiological terms. The substrate could be a pattern of activity by some structurally unique cell types in a layer of cortex in various areas of the brain, or it could be the synchronized firing of special cell sets and it could be distributed rather than in one location. "The lack of constraints is not a symbol of anything other worldly about this problem. It is merely a symptom that science has a lot of work to do."[72]

Such a substrate can be called a mechanism for consciousness. It may be very difficult to identify experimentally since the signature may not be obvious due to misinterpretation of test data, even though it may exist. "Or there may be other unforeseeable pitfalls to bedevil the approach." (Ibid.135)

In the indirect approach, there is a question whether memory, thought, meaning, perception, self-representation, etc. are somehow connected with being consciously aware. There is still much to be learned about attention, like the effects of different neurotransmitter systems on distinct aspects of attention, or dopamine suppressing conflicting information.

Another area that calls for intense research is in the contrast between conscious representation and nonconscious representation in the holistic economy of cognitive perception. The suggestion is that conscious representations are more accessible than nonconscious representations. Therefore, information distribution may be the explanation. Although we do not yet understand how to characterize the nature of information in nervous systems, "we can provisionally use this rough-hewn notion".

> Moreover, we do not yet fully understand how neurons code information, whatever information is. (Ibid. 170)

(C) In this final commentary, the intent in not so much on the specific items that are either contradictory or unsupported. The quotes are the main message, not only about this conceptual presentation but also about what has been obvious in the presentations of the conceptual explanations of other brain functions. The quotes are an indication of the consistent and persistent subconscious awareness by the neuroscientific community that there is something missing in their frantic search for an explanation of human nature and behavior. The final quotation is a frank admission of the basic shortcomings of the conceptual analytical efforts. The neural coding of information is used as a basic known fact to describe the many functional characteristics of the concepts. Yet, in this one statement is the apparent unconcern about the major negative effect of this lack of understanding on not only the results of past research, but also on the need for a rethinking of the basic assumptions in cognitive research. The consistent insistence against any Cartesian (other worldly) effect demonstrates the major concern that this missing element will turn up (pitfalls to bedevil). **And it has in this book.** The concessionary remarks about evaluating any new evidence will be put to a revealing test.

In the direct approach above, the search for a physical signature that would somehow bridge the gap between the phenomenological and neurobiological gap is an indication of that wishful thinking that it may be present but obviously remains undetected.

In the indirect approach, the system description, which the approach seeks, is obviously missing. The present approaches to its definition, which are based on faulty or non-existent assumptions, guarantee that it will never be defined without a redirection. (CE)

The definition of consciousness continues to depend on the interpretation by the user of the term. The above concepts illustrate how the interpretations can vary. There is consistency, however, in the attempt to make consciousness a physical property and rightly so because the brain must be conscious to function. This ceases when life leaves the body, which to all current analysts is the reason that consciousness had to be a physical property. But the real consciousness issue is that it remains unchanged in the entity that continues to exist, the "being" that retains memory, free will, rationality, and awareness, all features that are part of the consciousness issue. Since consciousness dies with the body but remains alive with the soul, are there two conscious conditions or is the soul the only possessor of consciousness that gives the conscious property to the body through its presence. Clearly the existence of the soul as part of the complete, complex being calls for a rethinking of not only consciousness but the entire duality issue.

This concludes Part II, the arguments and philosophies of the dualism debate. I found it difficult to describe current scientific approaches, knowing that they were incomplete and in many aspects completely "off base". The constant urge to disagree with these views, knowing that I had the advantage of the real evidence, caused me concern about criticizing the intelligence of the neuroscientific community. But then there was the consolation and technical support by the growing number of scientists, professors and professional experts that were also asking the same questions about current research. And they did not have the evidence of the soul to back their positions (at least they did not acknowledge having the information). Perhaps this book will provide material for their use in firming up their arguments.

In the next part, I will continue to challenge not only the scientific and philosophic communities but also the secular world to accept the reality of dualism. For the average person, including the skeptics, I include three analogies for understanding, not technical criticism. Analogies do not prove ideas; they explain them in understandable terms.

Having criticized what is now the dominant theory of physical causality, I fulfill my obligation to answer the question: "So what?" I propose a such-as synthetic approach to a new philosophy and holistic system approach to define the true nature of the human being. This first-ever approach will suffer the pangs of a new birth in the criticisms and disclosures of gaps but hopefully it will survive its birth problems and grow to full maturity.

Part III

The Deliberation and Verdict

In the thematic context of a judicial trial in which dualism is seeking its right to be recognized as being real, Parts I and II presented all the trial activity leading to jury deliberation or a simple judicial decision. Part III is that final set of activities that deal with understanding, rationalizing, and ruling on that requested recognition. A positive ruling, as in any judicial decision, may include conditions to be satisfied for the ruling to be justified, like defining some of the resulting effects. The three chapters of Part III perform that final set of activities. Since the soul's reality has been proven, this reality must no longer be questioned. It can be argued forever, like the arguments for the flatness of the earth and the falsity of man's landing on the moon. Eventually, however, everyone must accept the evidence of a soul objectively and rethink all previous viewpoints.

Chapter 9 paves the way for understanding the relationships and interactions of body and soul, the fundamental philosophic and pragmatic reasons why dualism can be understood as a holistic approach to human nature. Analogy can overcome the blocking tactics of intentional skeptical destruction by relying on experiential and well known operational mechanisms for acceptance. The three analogies of Chapter 9 have the distinct advantage over purely mythical types in that they are based on the factual existence and properties of the soul.

Chapter 10 responds to the above mentioned conditions for an acceptance status. The method is as Descartes stipulated in Rule III for analyzing any issue: all past preconceived notions must be disregarded.

The basic philosophy of human nature must be redefined with the dual nature as an accepted condition. From that basic beginning, a holistic synthesis of how the human being functions can proceed. A such-as framework is defined in the chapter. It is not an all-knowing, irrevocable approach but a realistic appraisal of the immensity of the undertaking and a realization that there must be a beginning, no matter how sketchy. Humility in accepting new guidelines that replace past conflicting concepts and in admitting the lack of knowledge about spiritual parameters should reduce the friction between all participants in this rethinking.

The final chapter is a report of the complete trial. The importance of the trial, the first of its kind in having verified physical evidence of a non-physical entity, with conflicts of philosophic and conceptual thinking, and supported by deconfirmation of physical causality, cannot be minimized. It is a report that must be reviewed objectively. The entire scientific, philosophic, theological, and secular communities are affected by it. The demand for a complete rethinking of dualism does not negate all the work that has been done by the cognitive scientists. What must be initiated is the integration of the soul with the researched functions of the human brain and the nervous system.

And so ends the challenge. May it have a positive effect on some innovative thinkers to map the course of a new science that embodies the dedicated capabilities of objective researchers to finally producing the system description of the human being.

9

Analogies for Understanding

An aid to understanding the Elephant.

Analogies do not solve problems or prove theories. Through familiar similarities they attempt to bridge the gap between doubt or disbelief and understanding, or at least inject a temporary acceptance of the argument as it progresses. They are a valuable asset in explaining problems or nebulous concepts. But when inappropriate and irrelevant analogies are used, they have a damaging effect by leaving a distorted view of the basic issue. (Remember the mention of this fault in Chapter 2 , the Galileo comparison?)

The goal here is to describe a controversial relationship that has been attempted many times without a credible solution: the relationship between the physical and non-physical entities of the human being. It is the Cartesian philosophy once again, that which scientists and philosophers fail to solidify because they base their assumptions on preconceived but unsupported versions of what reality might be. The principal and overwhelming difference in the base for the analogies in this chapter is that it is supported by fact rather than conjecture. This difference allows the analogies to have more credibility than merely idealistic connotations. Knowing that the soul is a reality places it on the same credibility level as the physiological and neurological data of the physical entity. To say that the properties of the soul are not known is incorrect because much is known about them. Memory, free will, volition, and the mind properties have been studied for decades as physical properties but the requirements for their functionality are equally applicable to the mind-brain problem with a soul. A rethinking of the relationships is required but the goals of defining human behavior are intact. The main differences, aside from the relocation of the properties as real components of an entity which is separate from the body, are the conclusions of their being brain states rather than immaterial properties.

The analogies can proceed with more certainty of the mind-brain relationship because now they are not only supported by a solid factual base but also by the functional locations of the entities that support the analogy's comparisons. An example is the location of the decision function in the soul's non-physical mind, not as a conceptualized creation of a physically caused brain state.

What are the problems of dualism that analogies can shed light on? The main one is the control that the soul exerts on the physical brain to perform physical activity. Another is the encompassing awareness of the related activities of the two entity collaboration. No longer is there a question of how the brain can create non-physical thoughts. The brain is a physical entity that the mind can collaborate with. What an awesome factual change from current uncertainty! After centuries of philosophical and scientific debate, we have arrived at a condition in which understanding is a real possibility. It is this condition in which the analogies can spell out the direction in which the analyses and philosophical rethinking should proceed. The first step is the acknowledgement that there is a new basic philosophy of factual evidence, not conceptual idealism, that must be generated and accepted.

I present three analogies to elaborate on the stated goals above. The first is one that I used in my previous book. It is the analogy of the driver's control and interaction with an auto. Functionally, it not only describes the operating situation but helps to understand the neuroscientific criticisms that have been directed at dualism. The second analogy is the one that is used so often in attempting to explain human behavior: the comparison of the brain control of the body to the software control of computer hardware. It is misleading because it is incomplete. I describe how this analogy, taking into account the new evidence of the soul, is even more effective when applied to a mind-brain control situation. The inclusion of a system operator is the third analogy.

Mind-Body vs. Driver-Car

The analogy of a soul controlling the body as compared to a driver controlling a car has been used before. I used it in my previous book in the search for the meaning of life. However, the analogy has never been used as it is here. For this is the first time that the analogy is being applied to dualism with the knowledge that the soul does exist, therefore, adding more credibility to the comparison than before. I did not realize that my use of the analogy in my first book would become a reality in this book. There is also a by-product of the analogy: a clarifying outlook on the social problem of discrimination.

Everyone understands the general operating parts of a car. Without going into the detailed functions of the various components, a car is a bunch of mechanical and electronic parts that have been designed to perform specific functions so the integrated assembly can provide transportation for a driver. Similarly, the human body is an assembly of physical elements that provide mobility, life sustenance, and association with the environment. Just as the details of the component and integrated functioning of the car's parts are of no consequence to the driver of the car, so also is the irrelevance of how the body parts function to the mind as it decides what it wants to do and where it wants to go. The driver knows which keys to turn, which control knobs to push or turn to transmit the instructions to the car. The car computer, like the physical brain, having received the instructions from the driver, gives the operating commands to the engine according to the operating procedures that are stored in the computer memory. That brings up an interesting conceptual comparison. The car has a memory, but so does the driver. From the evidence that proves the soul's reality, it is a fact that the soul has a memory. But the brain must also have a memory for assuring that the body's life-sustaining functions continue to operate without any conscious commands from the mind. There must be two memories in each human being. This is more logical than depriving the driver of a memory and storing all the information in the car memory. This part of the analogy argues against the neuroscientific approach to memory, an approach that requires transformation of mental (driver) thoughts into physical (car) signals for storage in physical (car computer) synapses.

The car without any driver commands does nothing. The body without the mind's commands also does nothing. As the body is asleep, so is the car. In dreams, the body does not participate in that mind's activity, just as the car does nothing when the driver is asleep in the car.

There are other similarities that are so obvious that they do not merit mention, such as the food and drink for the body represented by gasoline and water for the car, or that the feet can be compared to the car's wheels. There is one major difference that is not consistently applicable. (There may be others because analogies are not perfect explanations.) That difference is in the eyes. The driver uses eyes for driving at all times. But the driver only uses the car's eyes, the lights, at night, although they may be on in the daytime. There may be an important comparison here that the soul has perceptive capability without the body eyes but is able to use them through the body-soul interface. (Remember the car braking problem in Benjamin Libet's delayed awareness analysis in Chapter 7?)

What is the interaction between soul and body that can be shown by analogy? This is difficult because no one knows how the brain and mind do it. There must be some unexplained interface because the human system functions well. The analogy can identify the car-driver interface because it deals with a physical situation. Other than the ignition starting and stopping, the car-driver interface is at the hand and the foot points; the hands steer and the feet do the accelerating and braking. The mind-brain interface is a problem that will tax the combined scientific and philosophic minds.

Body damage or sickness is similar to an operating problem in the car. If it is a physical defect in a body part or the brain, caused by accidental damage or age-related deterioration, the damaged body element is comparable to a faulty part in the car's system. It isn't the mind's fault if it doesn't have a fully functional body to work with, just as it is not the driver's fault that the car will not function.

If a damaged brain can still perform some functions, the mind also has a commensurate ability. In the analogy, if the engine problem is not critical, like occasional stalling, the driver still has transportation.

A person in a coma can be compared to a car with the computer not working or a major part like the alternator inoperable. The parts are all still there but the system must be repaired to function properly.

Considering the analogy philosophically, the soul uses the body for the trip of earthly life. When the trip ends and the body is no longer needed, the soul discards it and continues the journey into eternity. That was the theme of my last book in which I compared the body to a rental car that is rented for a specific trip and is then turned in at the end of that trip. I did not realize that within a few years I could claim that the analogy of a rental car was an appropriate choice and has become a fact of life. With reincarnation a reality, the soul may rent more rental cars in later lives.

There is also an analogy within the car-driver analogy that bears on our dealings with other human beings. As we drive along the crowded freeways of life we are aware of other vehicles moving in the same direction. There are cars, vans, motorcycles, trucks, and SUV's. Although there are many types of vehicles, there is one commonality between all of them and our vehicle: each one has a driver. But if each vehicle represents the driver's body as we considered above, there are implications about that mass of traffic that reflect social attitudes. Each vehicle-driver pair is similar to ours in an operational sense. We respect the rights of other vehicles to be on this freeway, regardless of their shape, color, model, age, or size. As we drive along, these properties have no effect on our driving and we do not discriminate against them for these reasons. The important issue is that we are headed in the same direction together. We may pay more attention to the vehicles that are close to us, noting their color, type of vehicle, and the spacing between our vehicles for driving safety. We move to our destinations without caring about the vehicles, concerned only with the drivers' abilities to drive their vehicles. In other words, we don't discriminate against them until they compete with us for the space we are occupying on the road by tailgating, inappropriate lane changing, and cutting us off abruptly. They endanger our well being and our lives. It is this behavior that merits discrimination, not the appearance of the vehicle. In like manner, so should it be in our daily dealings with other people. *The only discriminating factor should be behavior*, not race, color, religion, gender, or ethnic background.

Science has found that of the three billion genes of the DNA, only thirty thousand affect the life of the human being. Of these, only one tenth of one percent affect the appearance of the human being, that is his/her skin color, facial appearance, color of eyes and hair. The rest are for the common body functions of every human being. Yet we discriminate against people physically on the basis of this incrementally small gene contribution when the discriminating factor should only be behavior.

Brain-Mind vs. Computer Hardware-Software

The computer analogy has been used frequently to explain the mind-brain problem. It is useful only with those cognitive concepts that deny the existence of a mind in favor of the brain's generating the mental states because such a concept is physically oriented. The general concept of physical hardware used with digital informational commands sounds like a credible comparison with the brain's ability to process sensory information. But there are differences that disrupt the comparison and fail the analogy's goal of promoting understanding. Whereas an analogy is intended for further understanding, this analogy fails because it is misleading in its simplistic generality and implied completeness.

A commercial computer uses hardware with specific operational algorithms to run any software program compatible with its capabilities. The software is also limited in what it can direct the hardware to do. This is not the analogy of a dynamic system like the human mind. The detailed failings of the analogy were treated in 5-3-S of Chapter 5.

The main discrepancy in the analogy is the lack of identification of the computer operator. A commercial computer, with operational hardware and software, does nothing until an operator turns it on and then uses the system to obtain information. In operating the system, the operator continues to generate the commands and interprets the resulting information. The computer is always under the control of the operator. This raises the question: Who is the operator in the analogy if only the software represents the mind? If the mind is the operator, the analogy is not only incomplete but misleading.

Now that we know a soul is involved, a different analogy is needed to replace the computer analogy for understanding the relationships between mind and brain because the hidden operator must be exposed. The analogy now becomes:

Body-Soul vs TV-Operator

This new analogy is like the car analogy but now the driver can change the car computer and the interface is between the driver and the car computer rather than the physical car elements, which was unrealistic and incomplete. The interface must change and the continuous awareness of system operation (consciousness) must be represented. An analogy of an operator using a TV set with a video recorder is more revealing of the body-soul relationship. This analogy is different from the others because the mind function is split between the operator and the set unifying the two parts into an integrated unit. In the car analogy, the mind was solely a function of the driver. The split is necessary here to account for consciousness which enfolds all parts of the union.

A TV set represents the body with its computer, the brain. The video recorder represents the undefined consciousness because it remains awake when the TV sleeps and perceives what the TV is doing when it is awake. As that conscious entity it is part of the operating mind since it is the locality for long term memory storage of information generated by the body sensors as they receive the incoming signals. It is also the body clock that is used by the physical memory to regulate the operating physical systems of the TV.

The main difference of this new analogy is that the *implied* operator of the previous computer analogy becomes an external entity replacing the computerized brain control of the body. The mind as part of the soul/operator is the dominant control. The interface is the hand held wireless remote control unit. In the following description I parenthetically inject those non-physical properties and actions that the scientists argue against in their physically caused concepts.

The operator and the TV wake up in the morning and assume their separate roles as controller and responder. Choosing the desired channel is the operator's action (free will). The TV responds with a picture containing much information but the operator selectively concentrates on a main issue of interest (discriminate control). If there is nothing of special interest the viewing is passive (daydreaming). If the interest is sufficiently high, the operator can commit the information to memory by recording it for future recall. The memory is stored as new software on external data storage for recall when desired (free will). If the program being recorded is a political debate, it requires close attention (awareness and rationality) to evaluate the positions of the debaters. This rational activity is an operator activity which is sending some information to internal memory, that portion of the total memory which resides in the operator's functional system for use at any time without need for the TV or recall from the recorder (meditation and creativity).

The operator can modify the TV software format by adding an insert to the main picture (like engaging in two conversations at the same time). Before going to sleep again, the operator decides to record a program while he/she and the TV are asleep. Since the recorder, like the mind, does not sleep, it will wait to memorize the desired program after which it will continue receiving the incoming signals without retaining them in memory (dreaming).

The above analogies demonstrate the understanding that can be added to an apparently doubtful concept or a system that lacks details. They also demonstrate how forceful an analogy can be when the basic supporting facts are known. The mind directs the brain through an unknown interchange although in the TV analogy it is the remote control.

With a hopefully increased understanding of how the human system functions, it is finally time to face the daunting task of answering the skeptic's unavoidable challenge: If you say the current neuroscientific and cognitive concepts are false, what have you to offer as a substitute? My best effort at a response is the next chapter.

10

The Sentence - To Rethink Dualism

How Elephants think.

In previous chapters, I disclosed the pertinent evidence of reincarnation and the soul's existence, the relationship of that evidence to dualism, the background of the dualism debate, and the inconsistencies of the current scientific approaches that attempt to explain the complex nature of the human being. I tried to present the total overview of the situation through analogies in an attempt to show that there is a commonsense approach that cannot be ignored, although commonsense is anathema to a scientist. It is now my turn to take on the responsibility for completing my challenge of rethinking dualism by proposing a systems approach for redirecting the research to include the soul.

This chapter is off-limits to skeptics because this is an unprecedented attempt at opening a new field of inquiry in defining human nature through dualism. Instead of a mad rush to criticize this vast area of unknowns, there is a more productive and meaningful opportunity to appraise the basic message and implement, rather than disrupt, it. The multitude of unknowns is the new area of inquisitive and challenging research. Nor should this venture into the unknown be an opportunity for vindictive retaliation for the derailing of purely materialistic substitutes for non-physical phenomena. The truth in factual evidence has been available for over four decades to be used by anyone as I have; but wasn't. I am merely the messenger who felt everyone should be aware of the message.

Proposed herein are suggested guidelines for a top-down synthesized approach for defining a system level description of the human being. It begins with the question of when a *person* becomes aware that a *personality* is forming. An important aspect of this approach is that some issues and questions, raised by the inclusion of the soul, that might be overlooked or intentionally ignored, are raised to expose their relevance as necessary points of initial agreement. An example is the awareness during pregnancy by the person, the soul, of its becoming a human being . This issue is addressed as a first item in the synthesis, an issue that is controversial in the current pro-choice culture. No longer can these issues be ignored because if they are part of the total life cycle, they must be addressed regardless of their impact on prevailing preferences.

The inclusion of a soul as a distinct part of the human being introduces a need for a new *fact-based philosophy* and a new *discipline of scientific research*. In other words a new and different researching *mindset* is required. It reverses, modifies, or negates the philosophies of David Hume, George Hegel, Immanuel Kant, Bertrand Russell, Sigmund Freud, Karl Marx, etc., with respect to their objections to dualism because they all argued against the reality of the soul. Even metaphysical philosophy, the study of *being*, takes on a new importance because for the first time it can use soul-based facts rather than beliefs or speculation to define that being. For neuroscience and cognitive science it is a major opportunity for the gradual resolution of the many inconsistencies awaiting "eventual scientific verification". Science is capable of astonishing achievements, provided that scientists understand clearly the nature of the system they are studying. A scientist tasked to analyze the properties of living non-physical entities will probably consider it an impossible assignment. Such should not be the case because cognitive scientists have been involved in such research for decades. They have defined the required cognitive properties and the requirements for functional performance. The problem has been that they have tried to explain these properties in physical terms and have found it impossible. As an example, as I've noted before in Chapter 6, they have tried to explain memory as a physical property but in the demonstrated evidence it is part of the soul. A realignment of the thinking is required, utilizing as much of the researched data about memory as applicable. *As a prediction of a major and necessary change in neural thinking, I have proposed the possibility of two memories, one the soul's, the other a physical body memory.* Such an arrangement would allow the continuation of the physical research of memory while initiating a neurorealistic research area of integrating the two. I expand on this unusual but intriguing aspect and the reasons for it later in the chapter.

When, not if, the scientific community accepts the reality of the soul, there will be a need for analytical and philosophical diversity that will create a monumental upheaval of the scientific, philosophical, theological and secular mindsets. This is not a trivial, hasty, presumptuous, irrational, or exaggerated prediction. It is a realistic appraisal of the meaning and effect of the inevitable introduction of the soul into a void that was created by the medieval Enlightenment and has been nurtured by a materialistic science and an apathetic society. I hope that the seriousness of the situation is appreciated by everyone reading this book. My background as an author, engineer, and quasi philosopher is not as relevant to the importance of the book as the potential implication of the subject matter. Neither the scientific nor the secular community may be ready to accept this disturbing information. But the untouched field of research should excite the researcher's mind for it promises a future of innovative opportunities and provocative debate. What I propose in this chapter are not answers to the problems resulting from the inclusion of the soul. Definite answers will not be known for many years or decades. Only a possible direction and content of the new research can be proposed now with obvious expectation of much revision and expansion.

The initial research must be pointed in the appropriate direction. It will not be easy for neuroscientists to redirect their thinking. They have been convinced for decades that they were on the correct path to full knowledge, relying on continued research to verify their materialistic goals. The initial considerations must be: definition of the holistic system requirements (what is the human system?); the requirements for the system and subsystem analyses (what must research produce?); the applicability of the analyses (are they pertinent to the overall goal?); and the limits of the outlined research (do the study boundaries match the timeline requirements?). The research methods should not dictate the subject matter like relying completely on computer models that can verify anything depending on the inputs. The reverse is needed with the subject matter dictating the nature of the studies like coordinating the scientific viewpoints with metaphysical philosophy, an unusual new requirement. Philosophic thinking must not get bogged down with semantic excursions and innovative ventures with idealistic side issues like trying to create life, or a soul, in a test tube. A more pressing problem will be the definition of requirements for experimental tests about the interactions between the soul and the body. It may be difficult to devise experiments for proving the non-physical characteristics of the dualist nature of the human. But this has been done. Dr. Stevenson's research is an historic example of linking the spiritual with the physical. The tests of spirituality through brain scans, as mentioned in an earlier chapter, resulted in some positive data. Future fields of investigation depend on innovation through coordination between scientists, philosophers, and laboratory technicians.

The Cartesian philosophy defines the human being as a composite of body and soul. The evidence of the soul's reality supports this philosophy, making this certainty the base upon which the rethinking of dualism can and must proceed. But this creates a dilemma in that neuroscience, neurophilosophy and cognitive science have moved so far in the opposite direction by assuming there is no soul. Redirection seems like an impossible objective. The situation is one of science's own making, but there is an opening for a reversal which science has also provided. The repeated statements by several scientists and neurophilosophers intimate that if evidence of the soul's reality or of a non-physical mind is ever produced, science will be willing to evaluate the evidence for its applicability to neuroscientific efforts. There are also scientists and psychologists who find it possible for a non-physical entity, a mind or a soul, to function with a physical brain without violating physical laws, such as the conservation of energy and momentum. Others believe that a non-physical entity working with a physical brain is the best explanation for the complex mind-brain problem. Examples of such promises, claims, and convictions follow:

> This is not yet a decisive point against dualism, since neither dualism nor materialism can yet explain all of the phenomena to be explained. But the objection has some force, especially since there is no doubt at all that physical matter exists, while spiritual matter remains a tenacious hypothesis. If this latter hypothesis brought us some definite explanatory advantage obtainable in no other way, then we would happily violate the demand for simplicity, and we would be right to do so. But it does not, claims the materialist. In fact, the advantage is just the other way around, he argues, and this brings us to the second objection to dualism: the relative explanatory impotence of dualism as compared to materialism.[73]

The original publication of the book from which this quote is taken, was in 1984. It was reprinted in 1999. Dr Stevenson's first book with the evidence of the soul was published in 1966, yet there has been no recognition of the evidence in the quote or in the entire book. The reality of the soul is without a doubt a "definite exploratory advantage" for filling the scientific gaps in cognition. The sincerity of the offer to reevaluate the situation is obvious from the choice of words that were used to describe it. To rethink "happily" indicated a lack of any apprehension about doing so. To "violate the demand for simplicity" indicates that the effort will be expended to do whatever is necessary to resolve the issue. With this willingness to reconsider a hypothesis, a much greater resolve should become evident to finally embark on research supported by factual evidence.

> **Hypothesis 1** Mental activity is brain activity. It is susceptible to scientific methods of investigation.
>
> Hypothesis 1 is a front-and-center topic of the entire book. It will be continually dissected, tested, and defended when we address the nature of the self, consciousness, free will, and knowledge. Ultimately, its soundness will be settled by what actually happens when

> the mind/brain sciences continue to make progress. Conceivably, it will turn out that thinking, feeling, and so on are in fact carried out by a nonphysical soul stuff. At this stage of science, however, the Cartesian outcome looks improbable.[74] (2002)

The first quote is by philosopher Paul Churchland. The second is by his wife, neurophilosopher, Patricia Churchland. Evidently husband and wife had discussed the probability of a nonphysical mind and agreed that the possibility could not be discounted.

The standard materialist argument against the credibility of a mind/soul functioning with a material brain because it must produce a detectable increase in energy as a physical indicator is contradicted by the following statement.

> The conservation of energy and momentum is a consequence of the homogeneity of time and space. This is warranted for systems that are causally closed. As to material systems that are open to causal influences from non-material mind, either energy/momentum is/are ill-defined or there is no reason why it/they should be conserved.[75]

Although the reference is to the uncertainty principle of quantum physics, it is an admission that even the physical laws are questioned when an immaterial entity is involved. Since a soul with a mind has no body and occupies no space, it is not subject to physical laws. The interaction with a physical body may be governed by metaphysical laws which do not require physical energy or momentum.

The following statement by the renowned psychologist William James (1842-1910) indicates that after extensive analyses of the mind-brain problems, he had to admit that there was no physically logical explanation of the cognitive process. A century later, the scientific community can still only provide conceptual explanations, but now the soul properties can help in the development of logical explanations.

> I confess, therefore, that to posit a soul influenced in some mysterious way by the brain-states and responding to them by conscious affections of its own, seems to me the line of least logical resistance, so far as we yet have attained.[76]

In spite of knowing that acceptance of what I propose here may be long in arriving, I will supply a such-as framework for the redirection of the search for the true human nature. It is an objective overview of what can emerge and how it could do so. I cannot give answers or produce detailed functional diagrams that only years of research by scientists, philosophers, and experimenters can produce. There are conditions, goals, and facts that must be incorporated in the initial planning, like a completely different philosophy of the dual human nature that must be accepted by all disciplines and communities. Research goals must be identified, like the priorities of interacting studies, which can be highlighted through a holistic functional diagram.

A problem that must have an early solution will be who will assume the leading role in the inter-disciplined studies because there must be a system coordinator and progress monitor, either a knowledgeable professional or an institution. This is a mandatory requirement for a new research area that will need strong leadership.

In the ongoing research, there is no functional diagram of the brain operation. Consequently, the uncoordinated fractioned studies have no cohesive credibility; they are independent concepts developed by specialists for their own areas of interest. The nearest approach to an attempt at coordination is the effort represented by the book *The Volitional Brain* for which articles were prepared in the various areas that might lead to some coordination of research results. The areas covered are: neuroscience, psychology and psychiatry, physics, and philosophy. Such a grouping of disciplines is needed in the new discipline for rethinking dualism. I have left out theology because the need is for a rethinking of the scientific and philosophical viewpoints. The religious and theological problems require a separate and different rethinking; the religious concerning dogmatic effects and the theological explaining the effects on other philosophical viewpoints.

With the soul's reality established, an opportunity presents itself for a holistic approach to define the dualistic system using information generated by a joint team of scientists and philosophers. But where is the starting point in this complex conglomeration of adamantly defended ideas? In keeping with the Cartesian methodology explained earlier, a logical step is to use the rules developed by Descartes for performing analytical studies. His Rule III, defines the mental attitude conducive to exploratory initiation of original concepts and methodology, precisely what is needed here. (This may be difficult for those researchers who abhor the idea of being associated with Cartesian Dualists.) I repeat Rule III here for convenience.

In the subjects we propose to investigate, our inquiries should be directed, not to what others have thought, nor to what we ourselves conjecture, but to what we can clearly and perspicuously behold and with certainty deduce; for knowledge is not won in any other way.

Listing what is known and what is not known is the first step in following Rule III. Compliance demands that the neuroscientific unsupported concepts be ignored. The optimistic assurances that they will eventually be confirmed may be a scientific crutch for continuing the extensive research but can hardly be called credible verifications. They are only philosophic conjecture. In contrast, the evidence of the soul's reality is now scientifically documented, as is the knowledge of the physical brain and body. Both constitute what is known about the human being.

What is known.

- Reincarnation is factual because it has been scientifically verified. The scientific verification substitutes *demonstrated first person performance* for the criticisms of reincarnation as paranormal illusion and mythical fantasy.
- Reincarnation is an accepted occurrence in Eastern cultures.
- The act of reincarnation defines the soul as a separate entity from the body. The presence of the same personality in a different body, with specific detailed references to

the previous body and its past experiences, proves the separation from the earlier body. These references are made by the reincarnated subject prior to the initiation of any investigative effort of their accuracy, eliminating the possibility of investigative prompting to obtain positive or negative information.

- The soul is functionally compatible with the body. The compatibility is so efficient that the soul's presence has not been detected since it cannot be seen or felt. This has led to an assumption that it does not exist, although there has always been a belief and unproven evidence of its existence as in the near death experiences.
- The mind, experiential memory, free will, rationality, and emotion are properties of the soul, not the body, as demonstrated in reincarnation. The investigations of reincarnation have verified the separation of these properties from the expired body.
- Skills and learning are part of the non-physical memory.
- There are different types of memory, like short term and long term, with attendant characteristics.
- The mind controls the brain as demonstrated with the infant inexperienced brain in the reincarnated personality.
- The body, brain, and nervous system are real and have been experimentally proven in great detail. The knowledge accumulated about the functioning of the physical systems is such that it is factual for future research of the integration of body and soul properties.
- Sensory collection of the external environment is clearly defined. The pathways are understood.
- The mental states of thought, volition, emotion, and rationality are not totally dependent on external sensory stimulation, as in meditation, although they utilize these stimuli in the process of normal thought while awake.
- Life support functions like breathing, heart pumping blood, and digestion are not normally consciously controlled; nor

is consciousness in the waking condition. Sleep is also a necessary function, normally a case of volition but imposed without conscious effort in cases of physical exhaustion.

What is not known.

- The non-physical laws and cognitive physical elements that integrate the immaterial with the physical are not defined. These may never be known. The effects are experienced but the operating principles are not known.
- What is perception if it is an immaterial property?
- How does the brain transform sensor information into a usable format for perception by the mind? If such transformed information is not required for perception, how is this information utilized to produce physical activity?
- Is overall perception by the soul independent of brain processing of sensor stimuli (as in the Libet rapid braking case) but dependent on it for reactive motor activity?
- Does the brain alone utilize the sensory information to control physical responses like body balance, conditional reflexes and limiting potentially damaging stimuli or is the mind's perception involved?
- When does the soul enter the body?
- The nature of the soul which would provide insight into the integration with the physical body.
- Are there dual subsystems that coordinate activity and behavior?
- The functional details of memory storage and recall.
- The physical location of memory if there is a physical component of total memory.
- Is experimentation with spiritual involvement possible to understand dualism?
- What is life in a human being? A dead body has lost life and consciousness but the soul retains that same

consciousness and a continuing life. It must be the same life and consciousness for the soul as demonstrated through its continuous memory. Although plants and animals have life, is it the same kind of life as human life and consciousness?

- What is the mechanism, physical or non-physical, that is the interface between the mind and the brain? Obviously, it is effective. Is it some brain component that can be identified as a physical interface, or is it one of the non-physical properties of the soul?
- Is there a separate physical memory for the body's physical needs and if so what is the means for information storage? Are the present concepts verifiable if limited to physical use only?
- Do the limits of physical laws on non-physical performance, and vice versa, determine the degree of control on human existence and behavior? Does the soul leave the body because the body dies or does the body die because the soul separates itself?
- How can the mind control the brain? From the reincarnation evidence, the mind is in control of behavior and activity. It is a controlling action that requires the participation of the brain. Recent experiments have shown that brain commands can be initiated through mental thoughts. Can the mind control the plasticity of the brain?
- How are learned skills stored in the soul's memory transferred into operating motions of physical organs like hands, fingers, arms, vocal cords, etc.?
- Are there distinctions between soul, mind, self, and person for personal, scientific, philosophical, legal, and political reasons? Are they all the same, the soul? This is a philosophical problem.
- What is control and how is it implemented? Control is not a property, a physical element, or spontaneous neural stimulus. It is an action that determines how a human behaves. What is the next step after initiation of thought, free willed decision for or against, and commitment to act?

There appear to be more unknowns than knowns and there will be others as the research continues. It should be obvious from this short list of unknown items that the rethinking of dualism is not strictly a scientifically oriented task. The task must include traditional contributions, not just neurophilosophic conceptual interpretations as in the past. It may even be necessary to rely on paranormal research data, which are currently being derided as fraudulent, mythical or illusory, but may now acquire more credibility and supply clues for research topics. It was such paranormal research that produced the reincarnation reality.

With the above information it is possible to show the difference diagrammatically between monism and dualism in Figure 10.1 by comparing the defining levels. In the monist concept the brain and body are on different levels as two functionally separate parts since the brain is responsible for the non-physical mental states in the absence of a mind. There are various concepts trying to resolve this functional inconsistency.

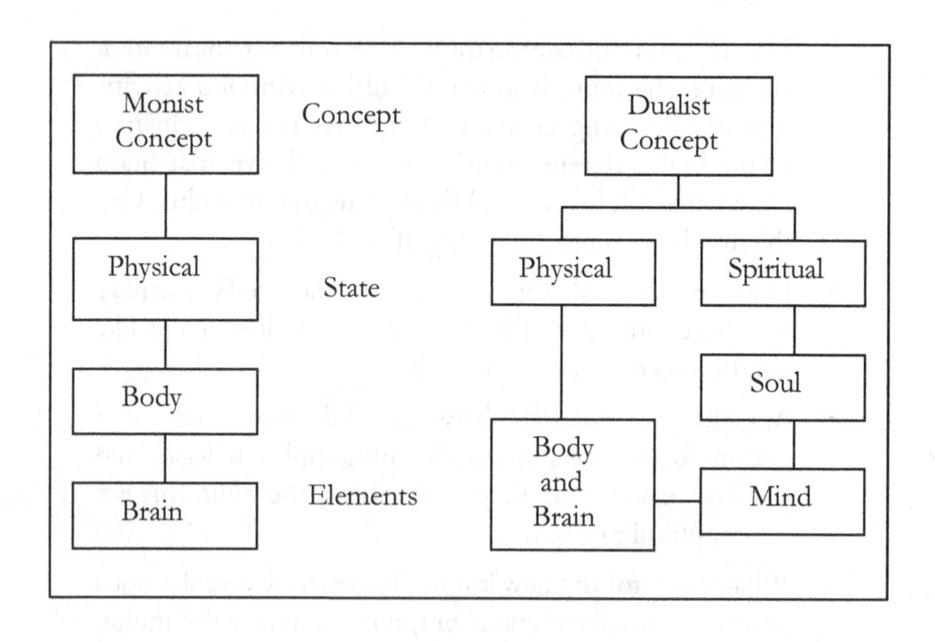

Figure 10.1 Comparison of Monist and Dualist Levels

In the dualist concept, the brain is only a physical element, a part of the body, having no mental functions. It generates the motor signals that control the physical organs through directed commands from the mind or from its physical memory for the automatic life-sustaining operations such as breathing, digestion, etc. The line between the mind and the body/brain denotes the psychophysical interface through which the mind directs the control commands. There is no line between the soul and the body to show that there is a distinct separation between the two entities although they are inseparable functionally. A single box containing these two would not illustrate the separation.

Although the soul's presence eliminates the need for referring to monism, showing the differences between it and dualism is necessary in the development of the new research. The comparisons lead to understanding how the properties of the soul fill some of the gaps in the current approaches.

Having complied with Rule III by showing what is to be retained (known) and what is yet to be made known, we move to Rule V. Again, I repeat the rule here for convenience.

Method consists entirely in the order and disposition of the objects toward which our mental vision must be directed if we would find out any truth. We shall comply with it exactly if we reduce involved and obscure propositions step by step to those that are simpler, and then starting with the intuitive apprehensions of all those that are absolutely simple, attempt to ascend to the knowledge of all others by precisely similar steps.

A key feature of this rule, which is important in the attempt to formulate a new system, is the call for simplicity. At this stage, technical and philosophical esoteric language only leads to possible confusion, especially in referring to non-physical unknowns. Understanding basic ideas is far more important than precise detailing.

In striving for simplicity there will be many unknowns. Since the body-soul union is a fact, there are many new areas of investigation to be pursued. Therefore, I am forced to approach problems without knowing if metaphysical laws affect them and what these effects might be. I can only conceptualize what the mind-brain relationship may be but at least my concepts are based on fundamental facts. It is a thought-provoking beginning, not minimized by convenient exclusions of bothersome issues, but reinforced by the opportunities to raise relevant questions.

As a first step, I do a preliminary evaluation of what must be done to produce a meaningful explanation of who we are. This has not been done before for a body-soul functional system, so I ask your understanding that what is involved here is a first attempt at how to rethink dualism. There is scant evidence on which to base this approach. My framework of what should be done is the initial step for other innovative and knowledgeable scientists and philosophers to grab the ball and run with it. Then, I can only be the cheerleader, hoping for a meaningful victory. Since it will be a long journey into the unknown, it must include a commitment to objective self-examination of who we are.

The remainder of the chapter is full of questions. Not only do they expose the lack of suitable answers but they also emphasize the magnitude of the scientific, philosophic, and theological rationalization they should trigger and demand in many new research areas. As they force you, the reader, to also raise questions, I hope that you will begin to appreciate the changes that will be required in the mindsets of the new organizers and researchers. It will be a new domain of psychological interplays.

10.1 System Considerations

I begin by using diagrams, which set the stage for defining the overall problem and provide the approach for delineating research goals. These diagrams are philosophically based on the soul's reality because they do not attempt to solve problems. Later, functional diagrams will be more detailed about what specific problems must be addressed.

Complying with Rule V for simplicity, the first simple question is how and when are the soul and body joined? The situation is shown in Figure 10.2. (The question is not "simple" in content but in the general approach to defining the dualism concept. Actually the question raises many complex problems.) Since the soul has no physical properties, its connection with the body cannot be investigated or detected before or after birth. When does the soul enter the body; at conception or later? As reproduction is one of the functions of life, the mother's life sustains the fetus.

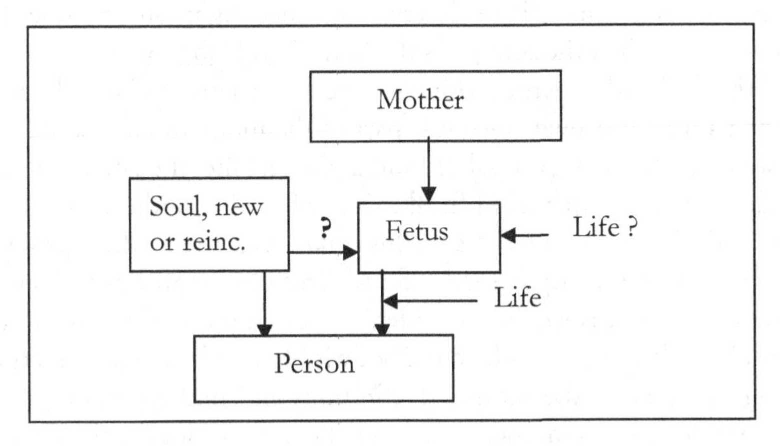

Figure 10.2 Body-Soul Mating

The soul may or may not be involved in the process, which is a philosophical question because there are no physical indications of soul participation. When does life begin in the fetus: at conception, before birth, or at birth? Is the soul involved in this life-inducing process as the creator of a person? In reincarnation, the soul is the person in both the previous and the new bodies. Therefore, if the soul joins the fetus before birth, the fetus is a person. According to the statements of some reincarnated children who had experiences during pregnancy, the soul does join the fetus during pregnancy. These cases are described in Dr. Jim Tucker's book *Life Before Life.*[77] The information is dramatic and disturbing. It is dramatic because of the fetus's "knowledge" of the mother's activities, associations, and even features of her external environment. The mothers acknowledged the accuracies of such knowledge, when claimed by the reincarnated children, although they were surprised by the details. The shocking descriptions of some of the cases should trouble the minds of everyone because they disclose that the fetus can already feel pain and is aware of anything that is causing it. *Especially disturbing was the account of the failure to be born due to an abortion in a previous pregnancy during which the fetus was* <u>*aware*</u> *of being aborted,* <u>*felt*</u> *the pain, and even* <u>*tried to resist*</u>. Although skeptics may refuse to believe such accounts, the scientific investigations that verify them should alert the legal and scientific communities and society in general that the fetus is a person in development requiring moral and legal considerations. These descriptions shed light on the question above about when the soul joins the body. <u>It is before birth!</u>

The soul with a perpetual life is alive in the fetus without the fetus having a life of its own since it is part of the mother's life. Could that mean that there is a spiritual life and a physical life, the latter added at birth, or does the spiritual life take over when the mother's support is removed? There are two sides to this unknown. The medical problem is: when does life begin in the fetus? The philosophical problem is whether the soul is the life provider. Although the soul is alive in the fetus, its ability to instill life in the body is blocked by the life functions being supplied by the mother. Definitions and understanding of life and personhood are needed to answer those questions for theological, philosophical and legal reasons as they are part of the complete definition of human nature and its formation.

Having arrived at the birth point in the timeline of a person's existence, we begin the analysis of the living human being. When the original primordial cell split into two cells, life was transmitted from it to the recipient cell. But life already existed in the first cell to be transmitted through reproduction. So life, whatever it is, existed in the first cell. One of the basic problems to be studied, therefore, is: what is life? A dictionary definition states that it is a property or quality that distinguishes living organisms from dead organisms. This applies only to the physical life and does not address the condition that the soul introduces. There is no universal agreement about what life is but agreement only in that its manifestations are organization, metabolism, growth, irritability, and reproduction. These are the effects of life but they do not define the "what" of life. Is it energy or an undefined spiritual force (energy and force are physical terms used here and elsewhere since there are no comparable immaterial terms) ? Could the soul be life? But plants have life, therefore, such physical life is not the same as that of a soul. The soul, however, does have a life with the principal difference that it does not die. Could it be the injection of this immortal life into the human body that gives that body an added quality, the ability to live? Could there be a duality of lives, a physical life and a spiritual life by which the latter makes the human different from animals and plants? It is the soul that complements the body with rationality, the distinguishing property. If there are two lives, which is the dominant one? If the body life ceases, the soul is forced to leave. But as in NDE, if it can reenter the lifeless body and revitalize it, the indication is that the soul is the one and only life source arguing against a dual type of life in a human being. Plants and animals may have different types of life. This is an issue for philosophical and theological resolution with science as a subordinate participant until facts evolve.

If DNA and RNA structure the life of the physical body through the instructional code, what determines the differences in souls, or are all souls the same in essence? They are different as persons. What choices do souls have as evidenced by predicted reincarnations before the initial death?

Do physical laws like the conservation of energy, entropy, and quantum physics affect the soul's participation in human activity? Although they have been proposed as possible answers to the vexing problems of monistic concepts, the immaterial nature of the soul precludes such applications. There are no experimental methods of detecting the soul's effects on the physical body, except by the definition of the soul's properties. The best information of the soul's effects is through scans of brain activity during specific test programs. These provide information about the locations of brain activity but not its meaning.

What is the composition of the body and soul elements? The breakdown, Table 10.1, is necessary for the rethinking of human nature. Obviously a listing like this has not been made in materialistic research and must be made if Rule V is to be followed for understanding through simplicity. Two memories are included for the separation of functions in the integrated system.

Table 10.1 Body/Soul Breakdown

Body	Soul
Brain	Mind
Sensors	Memory (experiential)
Nervous system	Neurospiritual interface
Memory (physical)	Free will
Life support organs	Consciousness
Torso w/arms and legs	

This simple separation of entity parts immediately reveals the direction that the new analysis must take. There must be an integration of the two sets of components rather than explanations of the soul components as physical properties of the body. By assigning a role to each entity, the study requirements can be formalized. It is clear from this separation that the body properties, with the exception of memory, have been mostly identified and are ready for use in the new study. Note that consciousness is included only with the soul due to its continuous presence even after earthly death. The physical parts of the body are all functional with, but not without, consciousness. A physical interface is not included because there is no experimental evidence of one.

Life support activities are performed without mental effort. Figure 10.3 is a simple diagram that exposes a problem that has not been of any concern to neuroscientists because in their research these activities were all lumped together with the one-body one-memory concept. The location of the mind in the soul deprives the physical body of a mandatory memory function of its own for the life support subsystems. The physical operation is an automatic closed loop operation.

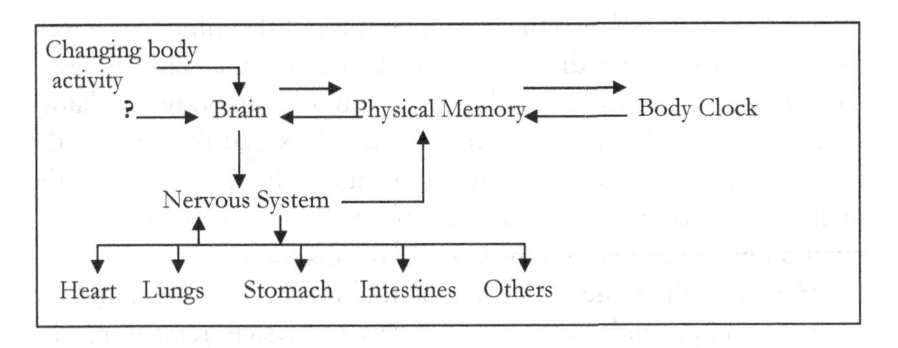

Figure 10.3 Physical Body Functions

A key element in this process must be a memory that stores the sequencing of information for controlling the body rates. As the body activity changes, the brain must adjust the organ rates to provide the necessary support. The memory should have a clock-type base against which to control the system. The brain generates the motor signals and transmits them for the sequential operation of the organs. What happens when the heart stops due to a physical problem like choking or drowning and is started again with a defibrillator or other means? How does it begin to function normally again? It must get an operating reference rate from some stored data bank, which cannot be located in the soul memory which is a separate unit for experiential information. It appears that there is a need for a physical memory for storage of information that is necessary for all body functions not requiring conscious mental inputs. There is the case of "Broken Heart Syndrome" which causes heart stoppage. That situation is better explained later within a system diagram since the mind is involved.

It is possible that the explanations the neuroscientists have been developing for memory as information stored in the weighted synaptic configurations are the description of this physical memory. The physical operation of such a memory for storage of physical information is a logical concept, free of the problem of converting experiential mental information into physically storable and recallable data.

The entire automatic operation is driven by an undefined life force. When this force is present the body functions without external intentionality. But what is this life force? It is not the mind that directs the brain to generate the motor signals because the mind is never consciously concerned with this responsibility for body regulatory functions. In joining the body, the soul provides that life force to the body, in which case one of its functions might be the control of the brain in a special mode for this life support. That is the reason for the question mark leading into the brain in the diagram.

How does the mind act without a need for the body or brain as in abstract thinking and meditation? The physical system remains passive while the non-physical mind is active without spatial or temporal limitations, that is, it can travel mentally into space, into the future or into the past. There is no comparable physical similarity because physical information is sensor oriented, not mentally initiated. So how does a thought or meditation proceed as an activity, rather than as a system oriented information flow, that could be described by a functional diagram? It is an *on demand* system that switches modes as the thought progresses. As an example of such a thinking process, how do I think about explaining the interaction of the soul with the body (which I will have to do later)?

My first step is the desire to do the explaining, realizing its complexity and possibility of failure. With the possibility of failure, I must be motivated to continue. Without motivation, not much productive activity is generated. Motivation is also not a *thing* but a mental commitment. At this point, I am aware of what I am trying to do, so *awareness*, sometimes identified with consciousness, is a condition under which I am making the decision to go ahead. But I am already in a conscious state, so the aforementioned consciousness may be something else. By making the decision to go ahead, I have used *free will.* I could have terminated the pursuit of this motivated activity but I chose to go ahead. Thus free will, another property of the mind is exercised as a non-physical activity. I now begin *thinking* about the problem. What do I know from all the reading and analyzing that I have done? All that I can conjure up without referring to additional material (Rule III) is stored in my memory, or at least I know where I can get confirming information, if needed. The methodology I use, not information but an acquired skill, is from my past engineering experience reinforced with philosophical aspects of the demonstrated evidence of the soul's reality (known fact). From this point, the effort is a mental search for the solution.

I must admit that this is a simple, almost trivial, explanation. So why do I include it here? I do so because it conflicts with and, with the soul's inclusion, contradicts the current neuroscientific substituted concepts for rational thought. As you were reading this simple presentation, you could relate to the continuing process because you have experienced it. To try to understand the scientific presentation you would have to wade through semantic obstacles of brain representations that generate mental states, transformations of representations to be stored in synaptic configurations, information recalled through sensory stimulations, and on and on, with the ultimate caveat that all these semantic conniptions have yet to be proved. Although you were familiar with the explained example through personal experience, according to the neurophilosophers it was all an illusion because it is only an example of having been deluded by folk psychology into believing that there is a soul with an independent mind.

How do I perform the search? The general approach is integrating one entity with another entity as a single unit with functional compatibility. A possible integrating factor is consciousness, as shown in Figure 10.4, a pictorial aid to visualize the separate but related parts under a "controlled" condition of consciousness. The body is involved as the physical "container" of the operation because physical activity, like drawing diagrams, may be needed in the thinking process. The soul is part of the body, not outside the body. But it is consciousness in both the body and the soul-mind that is the overall container involving body presence and mind operation. The brain assists physically but does not contribute to this study effort; it appears as an element of the body but detached from the mind. This is contrary to the ongoing scientific concept that the brain creates the mental states.

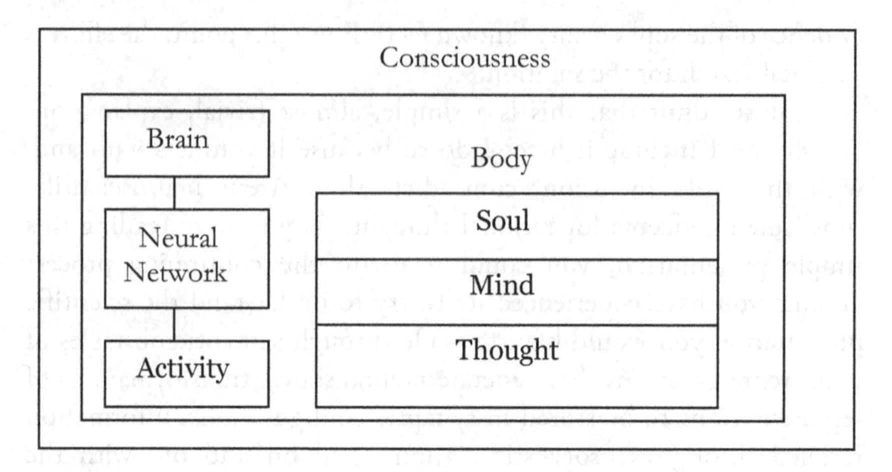

Figure 10.4 The Thinking Envelope

Throughout this thinking process the use of the physical brain for sensory contributions is controlled by the mind. There is no additional *representation* of any external meaning. This example is a simplistic explanation of a complex operation which we take for granted. However, its relevance is to emphasize that it can be visualized as a mental experience that is more in consonance with actuality, with the soul in the picture, than as a purely physical activity.

In the figure, the mind is shown separate from the soul because the soul may have other properties that are not involved in the specific thinking problem such as interruptive blocking of sensory input. For instance, the eyes are closed for concentration by blocking out all visual interference. Keeping the eyes closed is a volitional physical act but it plays no part in the concentrated thinking. Emphasis is on the mind's thinking ability and process. However, there is a possibility that the soul is the mind or vice versa. It is even possible that the soul, mind, self, and the person are one. In recalling the reincarnation evidence, all were included in the transition from the dead body to the new one.

As consciousness is not defined or understood, it is possible that the soul is consciousness since both retain reality and continuity after leaving the dead body, whereas the body loses both. Since the relationship of the mental state to the physical brain is not understood, the soul-mind identity with consciousness could explain the integrating effect. In that case, Figure 10.4 needs to be revised to show the difference this type of integration would involve. This version makes the soul the containing element instead of the body, as shown in Figure 10.5, with some major implications.

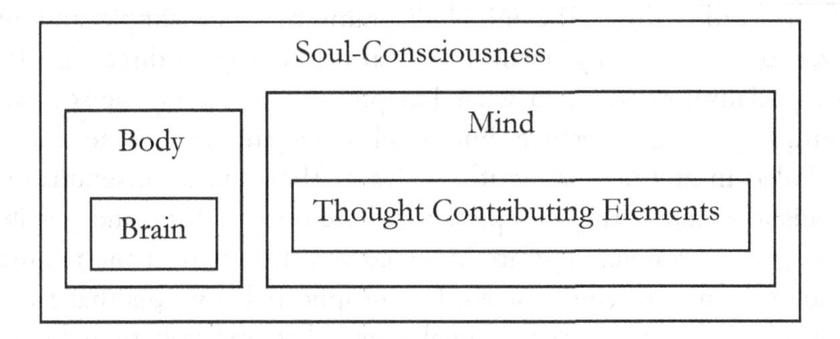

Figure 10.5 Effect of Soul-Consciousness Unity

The primary one is that the soul is the *person*, not the physical body; it is using the body, not the reverse. The soul is the integrating function along with consciousness, whereas consciousness alone performed that function in Figure 10.4. In this case, the soul, the mind, the self and the person can be one, retaining this combined entity after earthly death, while leaving the detached entity without consciousness. This version seems to agree with the psychoneural translation hypothesis (PTH), which I mentioned in Chapter 4, which posits that the brain and the mind, two epistemologically different elements combine to function as one. They are not in the same domain but do interact because they are complementary entities of the same transcendental reality. However, if a thought has to be translated into physical activity and vice versa, which functions carry out that translation? Explaining this is not possible because the definitions of the separate functional responsibilities do not exist. Unless this step is taken there can be no further system progress. So let's try to envision what the next step would be.

10.2 Functional Diagrams

Before a research program is initiated, functional diagrams should be prepared at least at the two top levels to scope the nature of the program. These diagrams define the relationships of the major elements of the system and provide an overview of the completeness of the study effort. The initial diagrams must be simple and top level to give not only scientific but also philosophic direction. This is especially important in what I'm proposing because the soul and mind (as distinct references but possibly one entity) have never been included in any neuroscientific analyses. There are no diagrams of a dualistic system. The ones presented here are simplistic and possibly wrong or incomplete but are intended as a first step in the required system definition. Unlike scientific unsupported concepts that I have been deconfirming, my conceptual approaches are firmly rooted in the soul's demonstrated reality. They are guidelines for anyone interested in being a pioneer in a completely new and different psychoneural scientific investigation. Before even starting, there is an unknown that must be addressed because it exists although the implementation is not known and may be difficult to explain. It is the separation of the physical and non-physical functions as shown in Figure 10.6 with the unknown interface.

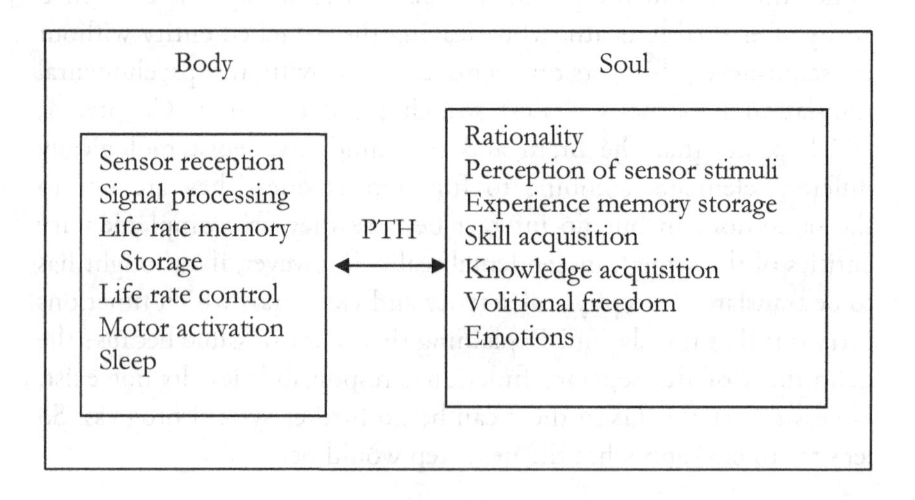

Fugore 10.6 Functional Locations

Perhaps experiments can be designed to ferret out this information, which is what the new scientific effort should be centered around. Is it the mental issue or is it at a higher system level of the relationship of the soul to the body? Perhaps it is a new definition of "self" to obsolete any past definitions of it. For example: There must be agreement on the functionality of each entity. In the simplicity of Figure 10.6, it's still a functional diagram due to the PTH connection which indicates functionality between the two. It does not describe the *how* of the relationships, only the *what. In any functional diagram, the how is the problem for research.* At this top level, the major *how* is the PTH solution, which cannot occur at the system level until the lower level functions are defined. The top level diagram shows the direction the lower level efforts must take. The above system type descriptions point to the need for top level philosophy and direction for the Big Picture. Philosophers and scientists must collaborate in this effort. Unless there is agreement about the top level guidelines for the research effort, the lower level studies will be misdirected.

10.3 Problems for Analysis

The following functional areas are proposed as starting points for expansion into analytical and, if practical, experimental studies. The situations must be addressed in their simplest forms to limit disagreement before zeroing in on the required research. These are not solutions of functionality; they are merely conceivable solutions subject to continued updating. The diagrams explain the functionality better than verbal descriptions alone. Combined, they are also an invaluable aid as working group papers for coordinating the organization of related studies and in locating specific questions or problems.

Situation A

How does the brain relate to the mind during meditation, or vice versa? The principal action is one of decoupling the inflow of some sensor information (sight, hearing and motor activity) from the brain. The commands from the mind to the brain are: close the eyes, assume a meditative position, ignore noises and sounds, and let me meditate and think without disturbance. Figure 10.7 shows how the communication channels allow uninterrupted meditation.

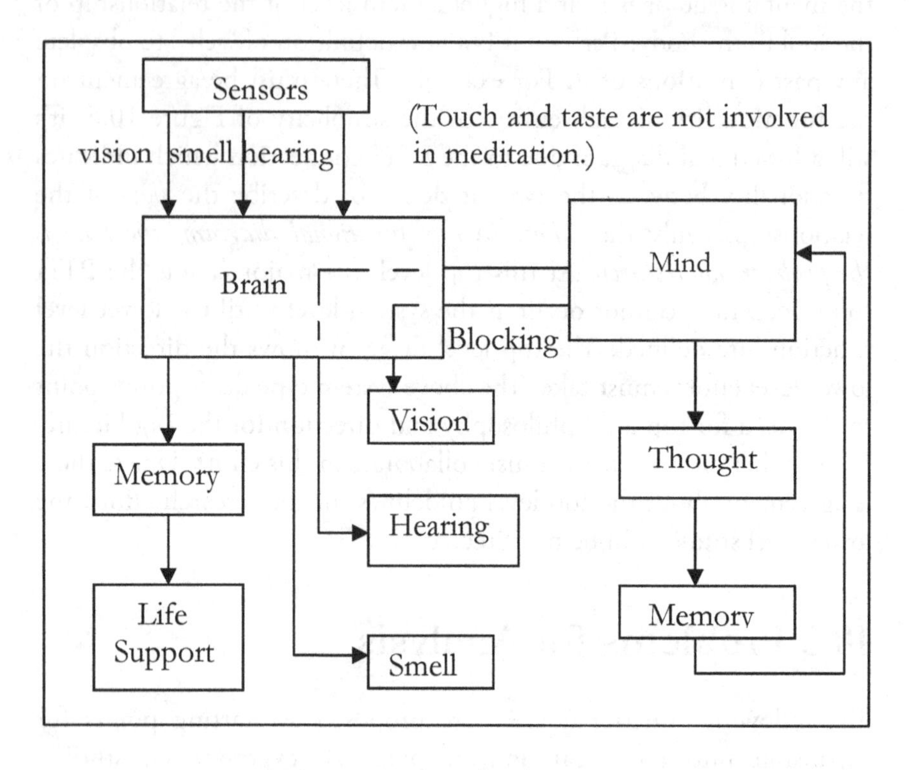

Figure 10.7 Meditation

How is the sensor blocking achieved? Thought control from mind to brain cannot be in physical atoms, molecules or quarks. It must be in electrical energy that the nervous system understands. But how are these signals generated and transmitted as calls for usable motor commands? The visual channel is closed by closing the eyes but there is no stopping sound without mechanical devices. How well can the mind block the effects of sound in intense meditation? Is there perception of the sound by the mind because that is where perception resides? If the perception is through the brain, how does it transmit the notion of sound to the mind? Perhaps the soul, rather than the mind specifically, is always aware of the sensor inputs.

Are any parts of the brain affected in meditation as distinguished from normal thinking? There is a similarity between meditation and dreaming. During meditation, memory is being used for storage and recall of the meditative thoughts. This memory is the experiential memory of the mind, not the body control memory of the brain. The brain memory may be affected in some way by the possible emotional experience during meditation which might call for an increase or decrease in the heart rate or blood pressure. Taste and touch are not involved through inactivity and smell cannot be intercepted due to required breathing, therefore, these three sensory stimuli do not contribute to meditation in this diagram. They do in situations that disrupt meditation.

There are candidates for research even in such a relatively uncomplicated activity. Experimental studies with brain scanning (described in the book *The Spiritual Brain* by Beauregard and O'Leary) have been conducted to identify unusual brain activity during meditation with surprising results.

Situation B

How does the mind relate to the brain in the undisturbed condition of thought, that is, with continuing sensor inputs but no discriminatory demands for imminent decisions, and no need for specific responses to sensory triggers? Thinking is basically the same as meditation except that the intercept commands become discrimination options. It is more likely that there will be a need for isolating specific items from the general array of incoming sensory information, like walking toward an acquaintance in a group of people, or concentrating on the sound of a certain instrument in an orchestral performance. The visual input is always a complete picture and the aural one is always the composite of all the sounds. Without any reason for choosing a specific item or area of the picture, like in day dreaming, the mind relinquishes the visual need to the other sensory priorities like listening to music, or to other activities like planning or analyzing past and future events.

Thinking is a first person activity. However, it can become a confused issue through explanations of it as third person behavior. This is an important point in the attempts at explaining human nature.

My confusion in the neuroscientific explanations of human behavior is that in trying to explain third person cognition devoid of a distinct functional mind, as in lacking free will, the first person is *making choices* in how to explain the third person's deficiency. If thinking is the result of sensory representations to the brain, then how can a book be written if the writing depends on sensory stimuli rather than creative exploration of original thoughts? And why does an author, or musician, or poet demand recognition for the content in a book or poem, or a musical performance, if they were not responsible for the creative products? The answer is evident but apparently not to the scientists or philosophers so engrossed in trying to explain away the very faculty they use to do the explaining. It is rationality being used as a first-person process to deny that same rationality in a third-person subject. How can free will be denied by an analyst to a subject, while the decision to do so is exercised by the analyst?

Situation C

How do the mind and brain function when they react to sensor inputs such as voices, noises, sights or pain? This is a crucial question in dualism because there must be an interface between the two entities (as hypothesized by PTH). This allows the shuttling back and forth of commands and signals for interpretation or translation, then processing, with feedback for confirmation. In the two previous situations, the brain continued its functions but the mind was acting within its own capabilities. (This may not be true, but as an initial approach, it sets the stage for modification, which is the purpose of the functional analysis.)

Is it possible to experimentally locate the area or the physical element(s) that are this interface? I have named it the psychoneural interface for compatibility with PTH. Perhaps there is no need for a physical interface if there is an undefined metaphysical influence that transmutes the information between the domains.

What and where is perception in this scenario? Does the brain do the perceiving, or is it the mind using its spiritual sensing ability? (Do you remember in the evidence of a near death experience that the out-of-body soul saw and heard the operating room activities although a spirit does not have physical eyes or ears?) The mind does the selective determination of how much of the sensor stimuli it wants to use. In vision, the entire picture is always available for use, but the mind picks the specific area of the picture that is of interest, like its own car in a parking lot. It can also select what one person is saying in a mix of several conversations as well as distinguish the sound of the double bass out of a symphonic composite of musical sounds.

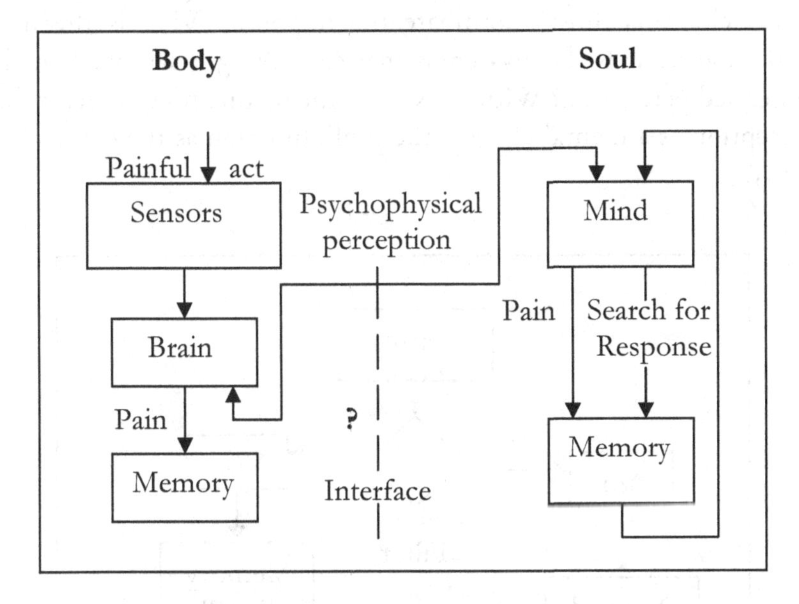

Figure 10.8 Perception of Pain

Figure 10.8 illustrates the pain problem as a schematic because it does not clarify how pain is perceived. Does the brain perceive it for future correlation with similar pain for reflex activity or is it perceived by the mind since emotional response is involved? The diagram shows the dual nature of the problem which appears to be complex but the two memory concept seems to apply adequately. If perception is a brain-mind issue, where is it located? Which path does the pain impulse take? Obviously, pain is a physical occurrence, which calls for awareness by the brain. But the mind is also aware of the pain, and the memory of the pain is stored as an experience for possible future recall. What happens when the mind can ignore the perception of pain under stress or danger? Is perception turned off, and where, in the brain or the mind?

Situation D
If the eyes, ears, or nose sound an alarm of possible danger or harm, what is the coordinated reaction by both brain and mind? A basic functional flow diagram of what must happen, Figure 10.9, shows which elements must coordinate the response. What is the filter in the figure? It is the awareness that something is wrong. Is it the undefined perception? With the soul in the picture now, it seems that perception as a mental effort is the soul's function as the controlling entity.

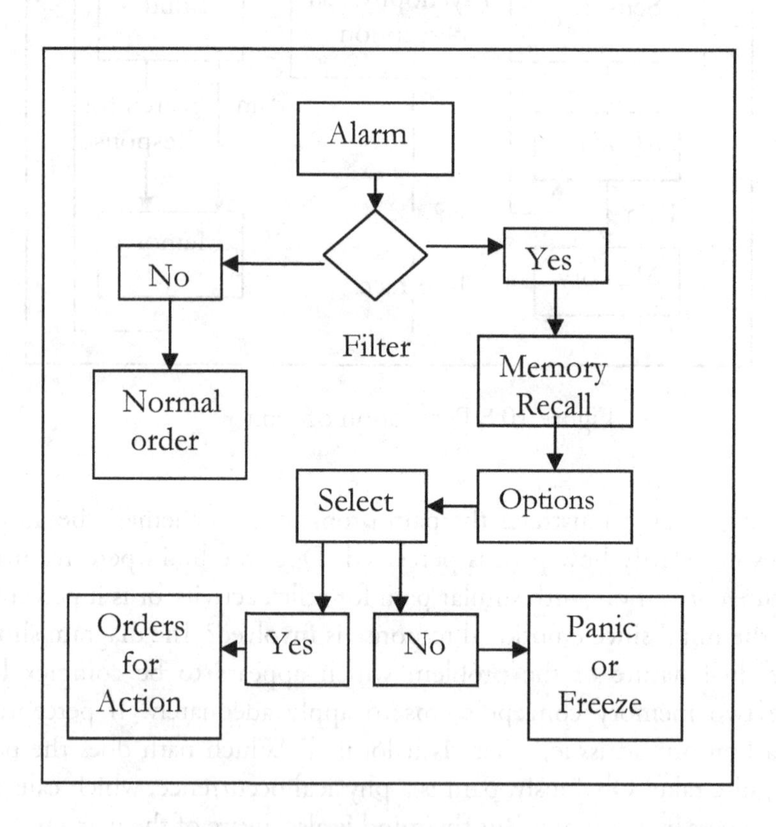

Figure 10.9 Danger Functional Flow

Which functions of the brain can send alarm signals based strictly on sensors' inputs? The physical stimuli have no danger threshold limits to determine which signal content or level would exceed such a threshold. (The current scientific approach is that the brain generates representations [ideas, concepts, etc] of what the sensors are detecting to the "mental state"). The sensor information that the mind (soul?) needs to evaluate as possible danger must come through the brain or through the soul's as yet unknown ability to see, hear, and detect dangerous situations. (Two examples are near death experiences and Dr. Libet's description in Chapter 3 of the unexplained reaction to a child running in front of a car before the brain's awareness of the danger.) The nature of the danger is discerned by the soul from discrete parts of the total sensor input.

In vision, it is the unusual activity in the viewed scene, not the visual signal content that produces the alarm; in hearing, it is not the sound level but the specific content of the sound that generates the alarm. A dangerous odor's composition, not the intensity, determines the alarm although the more intense the level, the more detectable the danger signal becomes. So there is a functional need for perception to be inherent in the soul not in the body, where the brain processes all the sensor data. Functionally, therefore, it becomes clear that the soul, without eyes, ears or a nose (presumably) must be able to take in the nature of the immediate environment through an unknown ability that I am forced to call perception, in spite of all the explanations and concepts that are currently proposed. For perception is an evaluation by the mind, not a signal processing of sensor stimuli by the brain. This is an opportunity to grasp the great difference between the current scientific approach and the problems that must be rethought by the introduction of the soul.

In a scientific concept, the explanation of perception is that it results from sensation. The brain-mind (scientific version of mind, not soul mind) must analyze the sensor stimuli. The object is recognized by extracting the features of the stimuli. These features are compared with stored records of objects in memory. The analysis is performed in the cerebral cortex which is divided according to the senses. These sectors are connected physically to each other and to associated areas where the comparisons are made. (How?) The object recognitions are performed by neurons. (How?) If more recognition information is needed, it is obtained through attention, a more careful processing in the thalamus, where the sensor stimuli are collected and from where they are distributed. This description is an explanation by physicalism. The initial analysis is performed by the brain, a physical element that extracts the features of the object. (How?) It must happen because that is what is needed for comparison with objects stored in the memory. But here is a major flaw. The memory involved here is a physical memory according to neurological concepts. However, the experiential memory that has "remembered" the characteristics of the objects needed for comparison resides in the soul, as demonstrated by the soul evidence, not in a physical memory. This has never been a conceptual problem because the soul with the experiential memory does not exist for the scientist. The physical neurons have been doing all the perceiving. When more recognition information is needed, the information must be obtained from the thalamus by the physical neurons that are the *intelligent* entities that do the recognition. But the key question that is never asked or answered in the complexity of the analysis is: For whom is the object being recognized and who then is the perceiver? The scientific version of perception, therefore, becomes a concept based on expected future verification whereas the soul version identifies the perceiver as the mind and is based on demonstrated evidence. Both have unanswered questions, but only one is based on reality.

Research Problem: How are physical sensor stimuli converted into perceptive information for the soul's mind if the suggested unique ability of the soul to perceive independently does not exist?

The reason that perception, rather than awareness, sensation, or attention, is the relevant activity is that perception is the integration of cause and effect. The possible harm in the problem of Figure 10.9 is composed of both: the cause is the alarm, while the effect is the knowledge of the possible consequences. The knowledge may be a recall of past experiences or past observations of similar situations.

Once the alarm is sounded, the action is all mental as to how to react, even in directing the brain to refrain from a reaction in the absence of potential danger. If action is necessary, the brain requires the information to generate motor signals for the correct physical response.

This functional diagram, as simple as it is, still raises basic questions about how the system must perform in the given situation. The diagram may be wrong, but that is precisely why it is effective. It allows discussion of its parts and its flow to arrive at an agreed approach to understanding the action and what must be analyzed to arrive at correct feedback to the top level functional diagram. From this point, we proceed to a lower level diagram, Figure 10.10, to further define the responsibilities of the subsystems. This figure expands the four principal functions of the previous figure: sensing the alarm, filtering, selecting action, and taking action.

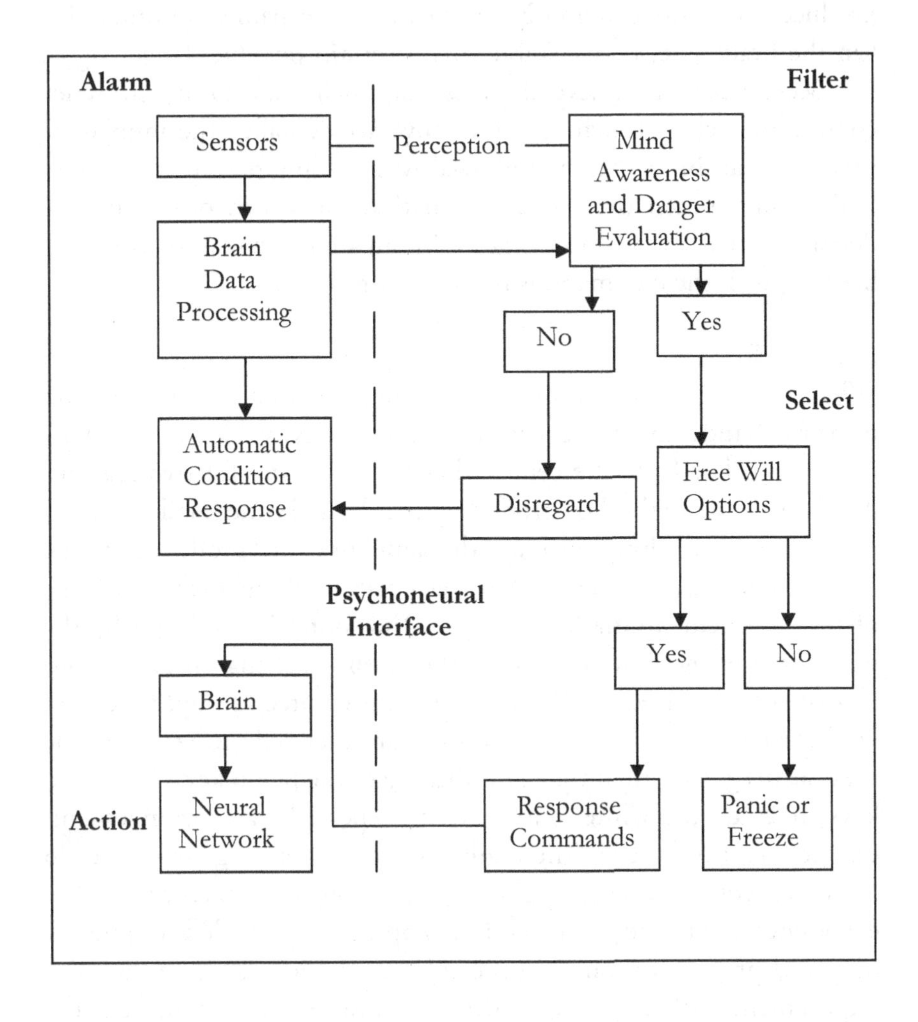

Figure 10.10 Expanded Danger Functional Flow

The figure shows the two principal problems that must be investigated for not only this condition but for the general problem of dualism. The first is the perception issue. It is a troublesome issue because it cannot be proven experimentally, being a first person activity in non-repeatable, ever-changing situations. Its reality is obvious from personal experience. The other obvious problem is the psychoneural interface, as mentioned previously.

The selection process indicates that in the absence of good decision information, the mind simply does not know what to do, which produces no positive action but results in either panic or immobility. On the brain side, it is possible that a certain possible danger signal has been experienced several times producing an automatic reflex condition, such as touching a hot stove accidentally. The immediate reaction is an immediate withdrawal without any perceptive train of evaluation. When there is an evaluation process involved and the decision is that there is no danger, without a conditional reflex having been applied, the command is to stop any reaction.

Situation E

What is the coordination of mind, brain and physical manipulation of arms, hands and fingers in a complex activity like a musician's playing a cello? I use the cello rather than a piano for two reasons. The first is the method of producing the desired tone. Striking a key on the piano will always produce the same tone. But producing a tone on the cello requires the exact finger position along a length of one of four adjacent, unmarked strings. The sound is produced by the movement of the bow across only the intended string unless a chord of two tones is needed. The coordination of precise finger location with the controlled bowing to produce an intended pure tone on the cello seems greater than the mere striking of the piano key. The piano keys are black and white to identify the repeated scalar arrangement; easy to identify by eye. There are no similar markings on the cello strings except that each string is tuned precisely to a specific tone. My experience is with the piano and I can appreciate the added complexity of sound production on the cello. However, there is also a complex requirement with the piano. Although the keys are identified, the pianist must correctly strike from two to ten keys simultaneously in playing chords. This requires a different coordination of mind, brain and fingers than with the cello.

The second reason for my choice of the cello is that while I watched the renowned cellist Yo Yo Ma perform the Dvorak Cello Concerto with the New York Philharmonic Orchestra, it occurred to me that I was watching a perfect performance of a mind-brain cooperative effort. Seeing the performance on television with close-up coverage gave me a much greater appreciation of the required coordination of technical proficiency and emotional concentration than if I had viewed it from a more distant seat in the concert hall. Seeing the total effect of the artist's performance convinced me that it exhibited the complex integration of physical and non-physical activities. How do the mind and body coordinate the use of skills requiring physical performance of a memorized concerto with symphonic accompaniment? It was an analytical exercise to do a functional analysis of the complete process from the first look at the musical score to the final performance with symphonic accompaniment.

The process beginning with the first look at the musical score consists of three phases: (1) learning the composer's intent and committing it to memory, (2) acquiring the interpretive skill, and (3) performing the concerto with symphonic accompaniment.

The first phase of learning requires translating the score from musical notation on paper to memory. It is translation in addition to processing the information for memory storage, unless the mind does not rely on brain processing for its memory. This translation is different from the normal processing of other sensory inputs. Seeing an apple identifies the apple, eating a peach identifies the taste of a peach, and hearing a voice identifies the speaker's unique quality. In contrast, the musical instructions of the score must be converted into sound—a complex visual to interpretive conversion through the mind's application of skilled experience. This cannot be explained as the physical brain's using the musical notation as representations of musical sound for the brain's generation of physical sound producing activity, as the current neuroscientific concepts might explain it. The reason this can't be so is because skills are stored in the soul's memory, not in the physical brain.

The preliminary evaluation of the composition's phrasing and complexity is accomplished through exploratory playing (sight reading) for combining visual interpretation with finger and hand movements. This skill-demanding procedure is the initial commitment to memory for later ease in maintaining performance continuity, especially through the technically difficult passages. Figure 10.11 is my understanding of how a current conceptual materialistic description of this learning process might look. I use the materialistic approach in this figure for later comparison with a version based on a soul in the loop. It may not be accurate but then it is based on concepts that are only materialistic assumptions which are unsupported, so my guessing is adequate. These concepts were argued in earlier chapters so their explanations here would be superfluous.

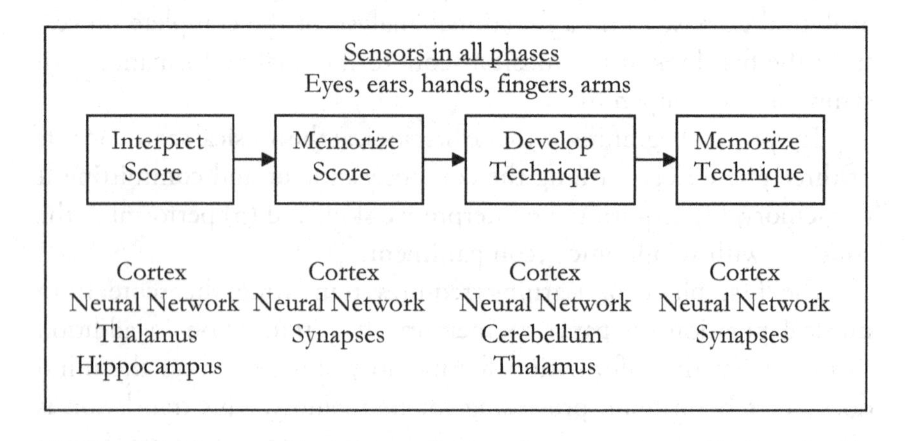

Figure 10.11 Learning a Concerto

Within this simple diagram lie many unknowns. The brain elements shown are based on interpretations of laboratory tests and scans. There is no certainty about how learning is acquired or how it is committed to memory. The synapses are included under the second and fourth boxes because that is where memory is supposedly stored. The cortex is under each box because it includes all sensory reception areas.

Other problems that add to the uncertainty are: what is the exact function of the hippocampus; how is the consolidation of mental and physical tasks committed to memory; what are the mechanisms for translating the coordinated learning into motor signals for the physical activity; and, finally, how is the correlation of emotional response to the composer's intent achieved? In spite of all these unknowns, the learning is accomplished. The complexity is conveniently trivialized by the simple progression in Figure 10.11. The following Figure 10.12 is also trivialized because there have been no analyses of the same learning process with a soul in the loop. This is my opportunity to conceptualize the same learning process but with one assuring advantage. I begin with a certainty that the physical and mental aspects are separate. Unquestionably, the main unknown complexity is the interface between the two entities.

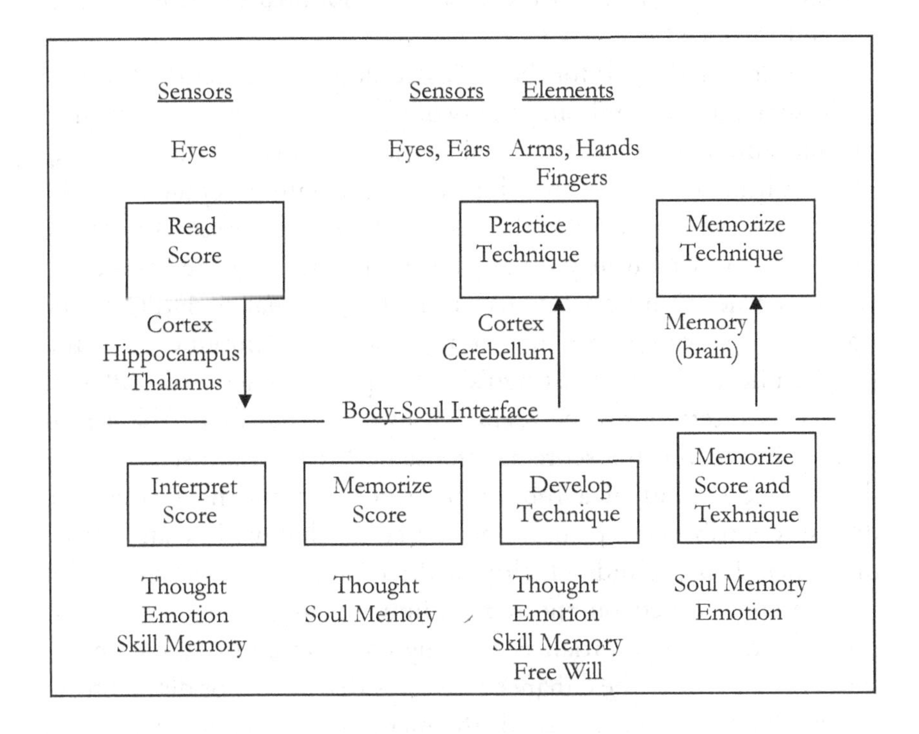

Figure 10.12 Learning a Concerto with a Soul

In the Figure 10.12 scenario, the body is merely the physical means for initiating the learning: reading the score and the physical activity in sight reading. The soul is the center for developing the fundamentals of the artistic interpretation. The development includes emotion, evaluation, and decisions on how to interpret the composer's intent and the addition of personally preferred embellishments. The functional diagram shows why I believe this is one of the two reasons for a dual memory: the "program" of the physical movements must be stored somewhere for rapid or automated recall, like a conditional reflex, while the performer is concentrating, in this phase, on the accuracy of the sound with respect to the technical placement of the fingers and the bowing. The physical technique must be stored somewhere, especially for the ultimate performance with the orchestra. (The other reason for the dual memory is its monitoring of the unconsciously generated motor commands for the bodily functions of breathing, digesting, heart beat, etc. caused by the emotional excitation during the performance.)

The first phase calls for the basic decisions that affect the learning of the score. It is not only an activity of translating symbolic notes into mental representations of musical sounds that inspires the planning of the ultimate performance but also deliberate acceptance of the composer's will in creating the intended musical effect. An artistic interpretation which may differ (not by much) from the composer's intent is a decision by a free will to add a personal signature to the upcoming performance. That and the artist's technique of sound production are what distinguish the performances of different musicians interpreting the identical notes of the score. The sensor stimuli in reading the score are the same for the monist as for the dualist but instead of a robotic interpretation by the former, the dualist contributes an emotional variance that makes an artistic difference. It is the individuality of the mind, not a consequence of brain activity based on the same stimuli of an original score that produces the artistic variance. Lacking emotion and free will, brain states cannot create the enhanced nuances that add to or detract from the quality of a performance. If the method of how a brain creates a mind state cannot be explained or validly analyzed, the presence of a soul with the attributes that explain these nuances is more credible than conceptually nebulous explanations.

Research Problem: How does the mental intent to create a sound instantly trigger the memorized finger and bowing positions without a conscious effort (as in sight reading)?

The second phase, consisting of complete commitment to memory, with developing artistic interpretations of the composer's intent into a meaningful musical message, brings into action the developing emotion that the artist intends to inject into the performance. This is a deliberate act, one not capable of initiation by a robotic brain state. The awareness that there will eventually be orchestral accompaniment with which the solo performer must comply, or the need to request some personal preferences, further complicates the learning process. Perhaps it involves a study and memorizing of sections of the orchestral score. There is a possibility that the musician can find a recording of the orchestral accompaniment without the solo part with which to practice. The functional activity in phase two is similar to phase one with some modification. The artist begins to rely less and less on the eyes as the score is committed to memory and the fingering and bowing become synchronized with the mental concept of the musical effect. Whether the musician needs to look at the instrument at all or not depends on experience and skill.

The analysis of the third phase is of the soloist's internal coordination with the external accompaniment for a balanced performance. Figure 10.13 shows the integration and the active sensors in the final preparations and in the concert. The external inputs consist of the conductor's comments about his interpretation of the score and the actual sound of the orchestra. These inputs may be compatible with the soloist's expectations or may cause concern about the balance of sound between the solo instrument and the accompaniment. The voice, as the communication link for coordinating the desired effect becomes an important physical necessity in the rehearsal. The eyes and ears are still necessary sensors in the dialogue. In the final integrated performance, the importance of the eyes diminishes for the soloist while the ears, arms, hands and fingers become major sensors for feedback to the coordinating mind. It is in the final performance that the total integration of physical and non-physical coordination is achieved.

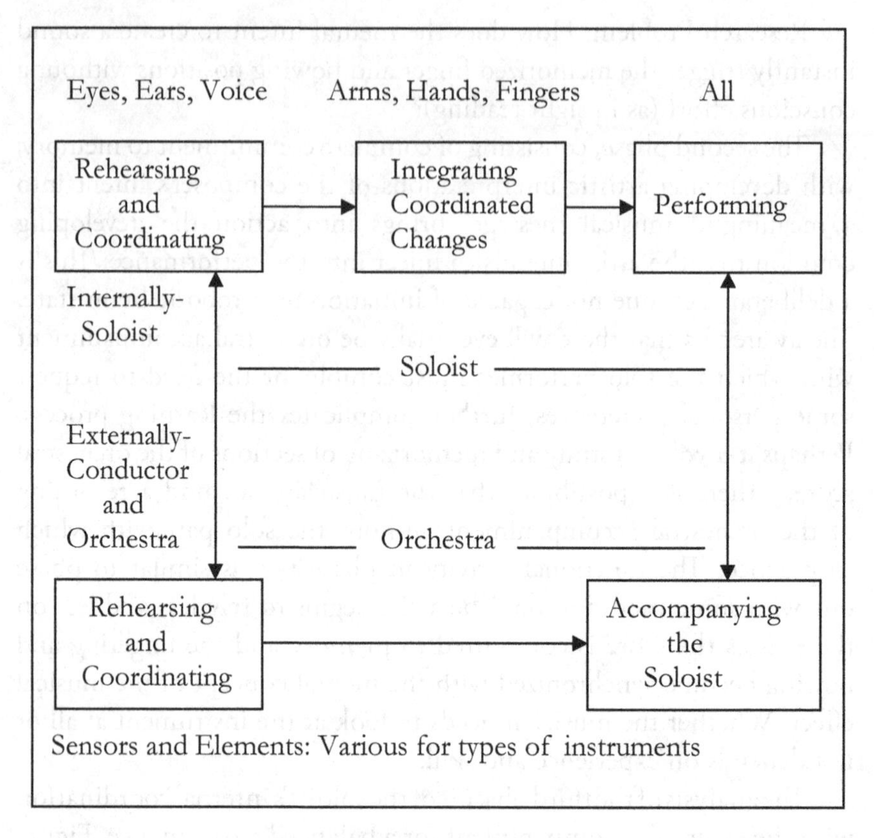

Figure 10.13 Performance

As I watched Yo Yo Ma perform, I was struck by his complete emotional concentration. Although he used his eyes to make occasional contact with the conductor, they remained closed during the major playing of the concerto. He never looked down at the cello to confirm the positions of his fingers on the finger board, or at the position of the bow on the strings. He was totally committed to the emotional interpretation of the music which was now augmented by the full orchestral sound. The performance was not a robotic performance of a set of musical instructions by a brain-created mental state. Instead, it was a dynamic interpretation of those musical instructions by a mind that coordinated its efforts with a conductor's control of up to a hundred minds to entice an audience's participation in a musical and emotional experience. The mind directed but the brain performed!

Research Problems: Activities such as this three phased operation pose many questions about the integration of the physical and non-physical activities. The mind, either as a soul element or the soul itself, is without a doubt in control. But how does the mind coordinate the instantaneous complex multiple activities of the physical parts of the body as it is emotionally involved in generating the musical flow? The possibility of two memories again becomes apparent.

The above are examples of the types of analytical diagrams that can be generated about any behavioral or performing activity. Developing such preliminary diagrams should become standard procedure when initiating a new analytic or synthetic study. Such "talking paper" is valuable in confining and assigning problems for the study participants and in monitoring a unified analytical approach.

10.4 Timelines

Activity is time-related, as is the functional performance during that activity. Functional performance analysis to determine whether a response to a situation, as in the case of a danger alarm is adequate, as in Figure 10.10, requires a definition of the sequential interaction of the involved functions. In that problem, a time analysis could show that if the reaction time of each functional activity was excessive, the resulting response could not prevent the impending damage. For instance, the memory search for an effective response could be too long (What should I do?) to react effectively. It is always a race between time allowed and time spent in the reactive performance. The sequencing of the functional activity plotted against time in a linear plotting is a timeline analysis. It will show the conflicts between the expected and the actual potential of a solution.

An example of such a timeline analysis was discussed in Chapter 3 where a problem was revealed in Benjamin Libet's analysis of brain awareness with respect to a necessary reaction. The revealed problem was that the brain became aware of the need for a reaction a time interval after the brain had already reacted. Such inconsistency must then be explained if there is to be any assurance that a functional response is adequate to solve the problem. If you remember, the Libet explanation was not scientifically credible due to its dependence on subjective referral of a sensory experience *backward in time,* although the presence of a soul adequately explained the inconsistency.

Timelines can, and should, be plotted as analytical checks of system feasibility. As an example: if feedback information is required to continue an activity, which functions must await that information before proceeding to the next step in system operation? Could alternate paths or combining complementary sensor options (sight and sound) be designed into the system to facilitate the operation? Such pinpointing of conflicts helps to define the requirements for system procedures and components early in the system design.

10.5 System Diagrams

Unlike the earlier diagrams, which described how a specific situation is resolved, system diagrams attempt to give the whole story of what the system is composed of, how it operates, and the relationships of the subsystems to each other. A system diagram should precede any analysis of the system's subsystems to reveal the integration requirements which determine how a subsystem must operate to achieve the intended goals. The neuroscientific and cognitive analyses are deficient in this respect. They attempt to explain the subsystems without regard to the demands of the integrated relations. There is no system diagram of monism. There are only questions of why the innumerable subsystem concepts do not have a cohesive explanation. If the holistic approach has not been used with monism, it is obvious that there never has been a system diagram with the soul as a contributor to the functioning of a total system.

For the rethinking of dualism due to the soul's reality, a system diagram is needed to understand the differences between monism and dualism. But since there never was a monist system diagram, I present one for comparison in Figure 10.14. My version with a soul, that follows the monist presentation, bases its credibility on the soul's demonstrated properties and it illustrates problems that the new scientific effort will encounter by the inclusion of a non-physical mind, as D. L. Wilson claims in his statement that I used as the opening line of the Introduction to the book.

Monist System

The system diagram in Figure 10.14 separates three issues that are compared with the dualist version that follows. First, it separates the physical aspect from the mental brain states. The physical part of the system is well defined from the extensive research that has been done by biologists, cognitive scientists, neuroscientists, and physiologists. It must be the same in the later dualist diagram because the neurophysiological facts are known and cannot be changed. The second issue, the concept of the memory, that I believe and have stated before, is a major difference between the two systems. In the monist system it is not a property but a physical element (synaptic storage), which causes problems that have no credible answers to date. The third issue for comparison is the conceptual generation of a mental state by the brain that is the subject of many attempted explanations. This part also constitutes a difference between the two systems.

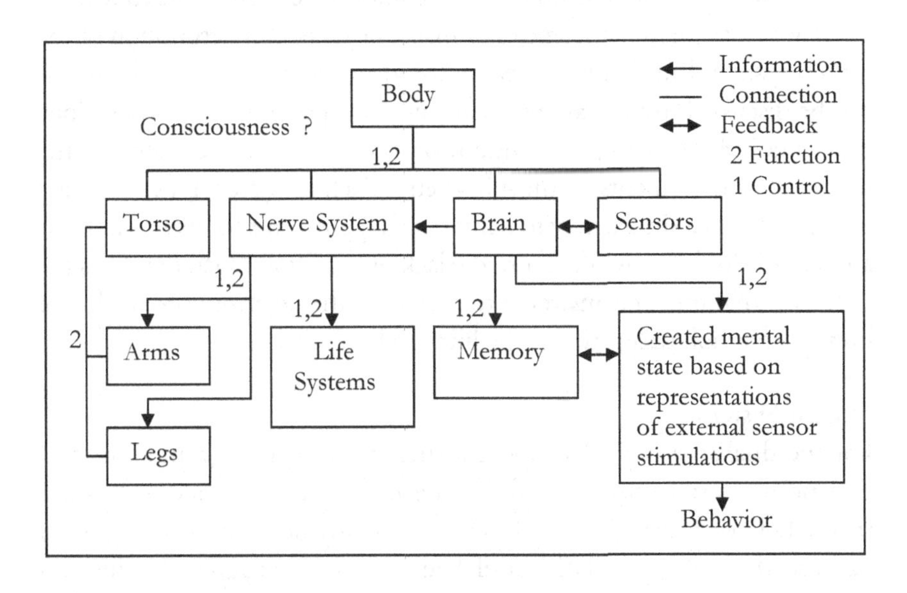

Figure 10.14 System Concept of the Materialistic Monism

In addition to identifying the top level elements of the system, the diagram also illustrates the functional integration of these elements. Designation is possible by coding for the operational integrating lines;

a simple line shows connectivity; a unidirectional arrow is information flow; numerals 1 and 2 identify controlling or merely functioning tasks. Where feedback is necessary for continued operation, the feedback arrows indicate a more complex operation than a single informational relation. One property that is undefined, and for which I have no idea where to locate it or how to designate its effect on the system, is consciousness. There are many definitions of what it might be but to me it is a condition and it applies to all the elements of the system without any specific integrating assignment. It is like heat in a room, or smog enveloping a city. It is there affecting every element without doing anything specific.

Note that behavior is driven by the "created" mental state. Representations of the external environment with extracted characteristics are supplied to the brain. From this source, the mental states derive their abilities to make "decisions" about behavior. The only conclusion that can be drawn from such a concept is that the result is deterministic, if not robotic, behavior that has no override ability since there is no volition.

There may be inconsistencies or inaccuracies in this description of the monist system. But that is the reason for identifying a specific system that can be discussed, modified and finally agreed upon or found impossible. The tentatively agreed upon solution, always amenable to revision, defines the general requirements for the subsystems. Each subsystem block can then be expanded in a similar way to define the requirements for its solution. As an example in the above diagram, the lack of a volition block identifies the questions that must be answered to overcome the deterministic quality of human behavior, unless that is the intended causality.

Dualist System

For the dualist system, I propose a such-as system diagram similar to the monist system with the following caveat: It is a system diagram that I believe has not been developed by any scientist (or engineer) because the evidence of the soul had not been recognized. This is a first of its kind. Consequently, its controversial content may be subject to criticism. But criticism should not be the main response to this proposed system because it is based on factual evidence of the soul's reality. What is needed is a systematic evaluation of the need for a logical development of the system diagram to pave the way for a meaningful acceptance of this unprecedented, but factual, definition of human behavior. Undoubtedly, there are many unknowns about the mind-brain and body-soul relationships. But the eventual resolution is inescapable because of the evidence. As the neurophilosopher said in her book about accepting the brain as it is because of biological evolution (quoted in Chapter 4), so should the soul be accepted as part of the human being because of spiritual evolution.

In the eventual newly defined scientific discipline, the first step should be the evaluation of all past research efforts that are applicable to the new direction. With that understanding of the applicability of what has been done, the rethinking can begin and the following system diagram, Figure 10.15, can be "toyed with" to establish a firm foundation for the new research effort.

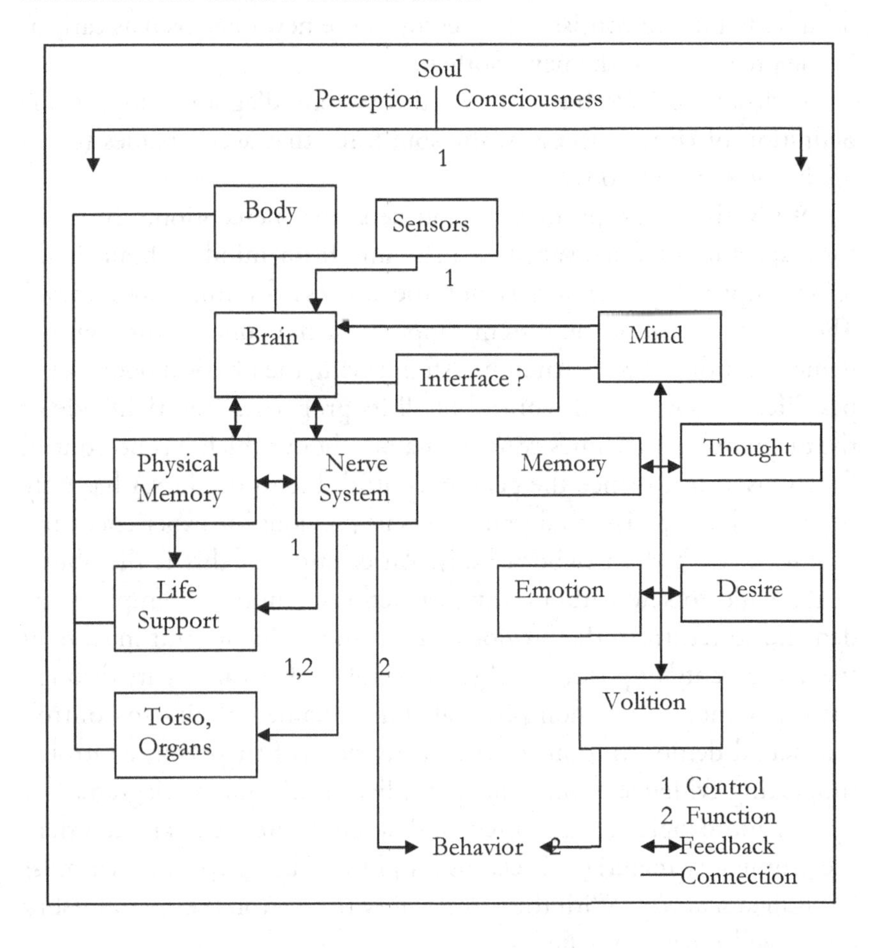

Figure 10.15 System Diagram of Dualism

The major difference between this diagram and the previous monist version is the dominance of the entire box (system). In Figure 10.14, the body controls the box. In Figure 10.15 the soul is in charge of the entire operation. Whereas, consciousness could not be attributed to any specific function in the body system, it has a commanding effect on the soul system, with perception as an accomplice in interpreting the external stimuli. Could these two attributes be a non-physical interface rather than a physical element of the brain, like the pineal gland that was proposed by René Descartes in his *Meditations*? How this interface is accomplished is the unknown factor, but from the evidence of reincarnation, consciousness and perception are part of the traveling soul. It is obvious that consciousness and perception leave the dead body but are consistently present in the new one, even as early as in the pregnancy of the new mother.

Philosophical Problem: Does the system diagram support the definition of consciousness as the soul-force that accompanies (or is) life in the physical body?

With this concept of consciousness and perception, the need for a specific interface vanishes. (The line from mind to brain is for sensor input if perception is not the process for mind awareness.) The soul's control of the system supports such a concept through the demonstrations of reincarnation. After leaving the physical body of the past life, the soul retains control of all its properties (the right side of the diagram) and assumes control of a new body which has no control functions at birth since the genetic control due to the DNA has only begun to develop. The soul remembers its previous life experiences and begins to teach the newborn body. An example of this is the ability of the child to sew without any instructions. Another example is the demand to return to the previous family and to insist on it in spite of the new parents' opposition. This is control of physical activity through the experience of the non-physical soul's retained ability to control. The actual demonstrations of such activities as first person control is supporting evidence of the general validity of this system diagram. It is by far a more logically supported, and demonstrated, explanation than the promissory materialistic claims for physical brain control that must be verified *someday*. With the soul's ability to control, these promissory claims will *never* be verified.

The explanation is not one of how this control is exercised. That too may never be explained, but it must be accepted because in the evolutionary development of Homo sapiens it did happen and we *must accept it as it is*. Perhaps the research in the rethinking of dualism will eventually unravel this intriguing relationship that we all possess. Research in neuroplasticity is also revealing that the mind can and does exert a control over the actions of the brain. Ailing conditions have been reversed through programs of commanding the brain functions to alleviate pain and discomfort as well as reverse the progress of a developing defect.

The physical part of the system is similar to the monist version. Since the memory of past experiences and learning of skills and languages resides in the soul and not the physical body (as described in Chapter 4), there must be a physical memory to control the physical body functions of life support. Such control is not a conscious effort, therefore, it must be regulatory information that is stored somewhere for automatic reference and, if necessary, recall as in cardiac arrest or cardiac fibrillation. In such cases, the heart has either stopped pumping blood or is twitching uncontrollably. It must be shocked into normal activity by a defibrillator. For the heart to return to its normal pace it needs instructions for the pumping rate. From where does it get the proper rate signal if it is not a conscious command? It seems to me that this is the type information that should be stored in the weighted synaptic configurations, not *all* information as the neuroscientists claim. This regulating information contains physical signals that control the automatic physical activities and cannot change. Because it must not change, it can be stored in a synaptic memory without the complication and competition with ever changing experiential storage and recall. It is predetermined by the individual's genes in the DNA instructions for varying body conditions, such as accelerated activity in physical exercise, or in stressful conditions, or in digestive problems. It is a given set of instructions for the various body activities. This is a minimum requirement for such a memory; there may be others as in the case of Broken Heart Syndrome in which there is cardiac arrest through emotional shock of losing a loved one. The cause here is not a physical one but an emotional one and there must be a message from one entity to the other. The line between the mind and the brain in the diagram is the pathway for the transmission of the cause from the emotional side to the physical side. But scientific research confirms that life support functions are a nonconscious physical operation. Therefore, the restart of the heart function is a physical responsibility. Is it possible under these conditions for the mind to be responsible for a failure of the heart to restart due to an emotional desire to "end it all"?

In the diagram, the properties of the mind are lumped together in a feedback network because it is not possible to define their fluctuating contributions in a continuing thought progression. Constant reference to memory might include emotional content and lead to a decision for action. The net effect of this mind complex defines the direction of behavior, which has no feedback because the final decision has been made. Any subsequent approval or regret of that behavior is perceived by the overall cognizant awareness. There are, however, physical contributions to behavior, as shown in the diagram, for without them behavioral activity would not be possible.

Identifying control and functional responsibilities in the diagram is a first intuitive evaluation of how the system can function. Future modification is a foregone conclusion. For instance, continuing analysis might change the control, function, and feedback designations, or additional blocks may be required.

How does the diagram explain what happens when there is a brain defect or damage? In the monist version it is difficult to explain because the damaged brain initiates the mental conditions that control behavior. The dualist diagram presents a different explanation. The brain damage occurs on the left side of the diagram. This does not affect the right non-physical side. However, at the behavior block, there is an input from both sides. If the input from the left side is defective, the behavior suffers. The right side being non-physical is not affected by the physical damage. In the behavior action, the mind's unaffected contribution cannot perform normally due to the incoherent signals from the physical part. In the case of amnesia, the effect may be total or partial. The level of incoherency of the signal arriving at the behavior block determines how much blockage there is to the undamaged memory bank. In the monist version, the memory bank itself is damaged because of where it is presumed to be located.

In meditation, the activity is all in the right side without any demands on the left side. This is a simple explanation of pure thought or meditation. In an emergency, the diagram supports the reaction to the external sensory stimuli by doing the evaluation on the right side with the left side performing its required changes to the heart rate, the breathing rate, and the contraction of the stomach muscles (butterflies). The action signal from the right side is not a behavior signal but an alarm signal from the mind to the brain for immediate expedited body movement. Behavior is a steady state operation, while an emergency input is an accelerated response state.

This system diagram of a dualist nature is a such-as attempt at understanding how the dualist complex acts. It is intended as a working diagram, a first pass at setting up a system of analytical studies, both scientific and philosophic. They should search for the answers that will be more difficult to reach than the purely physical ones that are testable by physical experimentation. But the scientific verification that established the reality of the soul is an example of how the non-physical domain can also be investigated if the specific items to be investigated can be defined. An example of such definition is Dr. Stevenson's goal of verifying reincarnation. In finding that birth marks and birth defects provided medical identification in addition to supporting his main goal, he began to correlate the changes in these defects with the decrease in reincarnation memory as the subjects grew older. Thus he expanded his investigations because he found a new source of medical information related to reincarnation. He continued this correlation, in spite of the skeptical views about paranormal effects. They may be paranormal according to terminology, but they are real according to investigation. Therefore, analytical objectives about the soul's interaction with the body should be defined in spite of the possible skepticism about their probable success. For the soul is a reality and human behavior results from that reality.

10.6 Soul, Self, Person, Mind, and Consciousness

The above diagrammatic analyses show that there are many questions awaiting resolution about who we are and how we function. What is it that motivates us and how is this motivation carried out to achieve goals? According to the current scientific explanations, it doesn't matter whether it is a brain or a mind that is responsible for our actions because the two are the same or some conceptual variation of the one entity. But the above analyses show that the mind and the brain are separate with associated functional responsibilities. Does that separation affect our understanding of who or what we are? It should.

The terms associated with identity that have various meanings are: self, mind, brain, soul, person, personality, and consciousness. Creating one's own version of identity depends on the beliefs associated with these terms. Thus, an uncertainty about the mind-brain relationship confuses the issue of which of the two controls one's behavior. If materialism has the upper hand, identity is a physical conclusion that goes along with physical appearance, body language (he/she always makes that gesture), color of eyes, size of ears, etc. If the dualistic mind is selected, these characteristics do not apply. Thinking ability, rationality, social relationships, volition, forgiving nature, are more reflective of individuality. If the soul matters, is it as a dominant factor like the reason for immortality or as an acquiescing spiritual attitude toward life's demands? Consciousness, undefined as either physical or non-physical, needs rethinking due to the soul's reality. It is part of me although I cannot exercise it having no control over it. It like the life that sustains me allows me to exercise the other properties and is therefore also part of me.

Self is a subjective awareness of *own* existence. It can only be a first person experience of here and now and in that sense it is that person's being. It is not the same as *ego* which has a relativistic meaning which implies that there is a psychic need in everyone for a balance between impulse and conscience. If self is to have personal importance in identity, it must have a functional contribution to existence or be some part of an individual that is identifiable with other humans. It does neither, so it is relegated to convenient first person recognition of total being, physical and spiritual. Self, therefore, is a personal look at oneself.

As pointed out in an earlier chapter there are many definitions of mind. In the context of a physical entity, there are many concepts of its existence as a brain generated mental state. With the evidence of the mind as part of the soul's reality, it takes on a commanding responsibility for human activity. Therefore, I ignore all the materialistic concepts of the mind as unproven explanations, unscientific, and irrelevant in my certainty of who I am. I cannot envision my mind being driven by any external control without my permission. If that should happen through some unforeseen power, I would no longer be me; I would be a different identity. I am in control of my actions and behavior and I take responsibility for my actions. With this attitude, my problem of identity is whether the mind is sufficient or are there other elements that define me. Since I know that I also have a soul, I'm faced with the indecision about the relationship between the two, since I "feel" no difference in considering either as the total identity. Which of the two is responsible for causal priming, the initiation of an action which has not been necessitated by antecedent causality? This indecision leads me to the conclusion that the two are one, in the basic understanding of who I am. That is identity, different from the definition of self which includes the awareness of existence in the here and now.

I do not exclude my body from the personality, but I give preference to the mind-soul-self as identity rather than the physical body. How does the body add to the totality of identity? Others identify me by my physical characteristics, therefore, the body is part of identity. But that is how others identify me, which is not my understanding of me. Why is this not so? The clearest explanation is that when I refer to it I say *my* body, not I body. The possessive reference is to my ownership of the body; I am the owner. The distinction reflects my subconscious relationship to my body, which is confirmed when I die and leave that body. I will continue the immortal life as the person that I am. I also acknowledge that it is possible I was another personality in a past life with another body.

By shedding or having to leave the body, the soul is the surviving identity. The soul in its totality, that is, with mind, free will, emotions, and whatever else it possesses, or is, like a life source, leaves the body as is demonstrated in reincarnation. If that is I as a surviving identity, then all is one, and that is the person that I am for all eternity and will be in any other personality I may assume in later lives. Since reincarnation has been proven, I may or may not be allowed to return to drive another "rental car". But if I do, I will have retained all those properties of the soul that I now possess. And with that assumption, I must acknowledge that soul, mind, and person are one, with personality an addition for an earthly life. I leave self out because I don't know how it applies after earthly death. The difference lies in how self is defined in earthly existence. If the body is included, then self ceases at death: if it is not included then self, as person, continues into eternity. This *here and now* aspect of the definition holds true.

Is there a difference between a person and a personality, whereby they should not be used interchangeably? Reincarnation defines that difference. Personality is the characterization of a person in a specific existence on earth. The reincarnated personality remembers the previous personality and may want to return to the environment of that existence, but it must accept its new individuality. However, the entity with the same memory, consciousness, and whatever immortal properties are involved is the same person in both lives. If more reincarnations are to be experienced, the same person will be involved in all succeeding personalities. However, shedding a personality does not erase its lasting value: it remains after death, remembered as the description of a doer of things after that presence is gone. A deceased being remains in the memories of surviving relatives and friends as the *personality* they knew but he/she lives on as a *person* in the afterlife until a new personality is born. So, personhood is a surviving attribute of identity, not like personality that changes with each life. Identity, as a person to that person, therefore, is the one "I" forever.

What is not clear to me at this point is the role that consciousness plays in dualism. Its general nature is illustrated in the system diagram, Figure 10.15. From the reincarnation evidence, it accompanies, or is part of, the soul, since the previous body lost consciousness when the soul departed. It seems to be a condition, a state of "livability", like life energy, rather than a unique property. Or is consciousness this life energy that the soul is endowed with? If consciousness is a unique feature of the soul and applicable to union only with the human body, then if animals have consciousness, is that of a different nature? This may be why the definition of consciousness is difficult to pin down. In relation to identity, however, the definition can be narrowed to a more specific meaning, which is, providing a being's existence. Without it, a human is a robot depending on other controlling factors to guide its behavior. Consciousness argues against that explanation.

The inescapable conclusion, after evaluating the various meanings and implications of the above six terms, is that although I strive for an Identity, I cannot identify knowingly with any specific one. If I have a mind, I don't know how to verify that except by its use. I am a person but I can't see or feel personhood. I believe I have a soul as has been proven but I don't feel it or have any conflicts with it. I have consciousness when awake or asleep, but because it is always there I don't question its various definitions. Self is a semantic reference to a present existence that is unquestionable since I am writing this explanation of my identity. This inability to resolve the differences and the similarities of these non-physical variously defined entities forces me to conclude that the soul, mind, person and self are one: the identity I. Why quibble philosophically about the meanings when it all funnels down to the one realization that I am a person with all these attributes that wants to be recognized and later remembered for what I have accomplished with these attributes. When I depart this earthly life, what can I take with me? I will be a soul with a mind and consciousness and a memory of this life. I may retain my ability to create system diagrams, or to critique conceptual explanations in the new neurorealistic science, but I will go as a personality, that is with the characterization of who I had been until it evolves into a different personality. With this admission, I have no problem in meditating without brain assistance on the meaning of life.

10.7 Summary of the New Approach

The evidence of the soul's reality is sufficient to invalidate the concept of materialistic physical causality. The demonstrated properties of the soul in reincarnation like memory, free will, desire, rationality, and control of the human brain exhibit sufficient credible detail to justify the development of a plan for a *holistic system* approach to dualism. Such an unparalleled plan for the initiation of new research of the effects of an immaterial substance on a physical brain defines the necessary procedure for redirecting the ongoing research. It may be necessary to organize a uniquely different research effort due to the contrasting objectives between physical causality and immaterial mind control of the human brain.

The methodology for synthesizing a holistic system demands a base of philosophic and scientific cooperation. A philosophic determination of when life begins in the human embryo, which may be tantamount to the soul's taking control of the body may be necessary to understand the subsequent development of the brain and its dependency on this immaterial control. Such problems will be the norm in the new integrated research of the interfacing between physical and non-physical relationships. To plan for the coordinated efforts of scientific and philosophic researchers in molding this new program, there must be agreement on the direction the research effort will take. The awareness of such basic understanding is the reason for the initial phase of the planning in this chapter. Pregnancy is the staring point of the understanding of human nature. The reincarnation information about cognition in the womb may be difficult for a scientist to accept, but since the repeated evidence by reincarnated subjects about these pre-birth experiences has been corroborated, the awareness and acceptance by researchers of such evidence should lead to major social consequences.

Based on these considerations, this chapter is a progressive plan of mindset revision, analytic methodology, and steps to a systems approach for defining dualism as a functioning human being. Simplistic situations are used consistently to comply with the methodology of the Descartes Rule V. The extensive experience of learning and performing a cello concerto as a subjective experience hopefully exposed the complexity of the dualistic interfaces.

The system diagram with the separated but integrated entities shows how the problem of analyzing a seemingly impossible reality can be defined to allow discussion of the compatible relationships. A byproduct is the ultimate definition of who we are. Since this has never been done, this chapter is the first venture into what is required in the rethinking of dualism when it is finally accepted as a scientific fact. Consequently, being the first attempt at a completely new philosophically viable systems approach to the description of the human being, this chapter is full of questions that cannot be answered in this first attempt at a system description. It may be decades before some of the answers emerge. As a first effort based on information that the scientific community is not aware of, or has intentionally disregarded, the approaches presented here are original, logical, and based on the demonstrated first-person characteristics of the reincarnated persons in the Stevenson research.

This chapter described a sequential operation which will need elaboration with time. It is a suggested procedure for those who, being unfamiliar with a methodology to perform a system analysis, need a workable beginning. A number of brief reminders of the contents of the chapter should suffice in assembling a mental image of a systematic search for how man thinks and acts.

Tabulate the components, both physical and non-physical, of the human system. Most of the components are known, even the non-physical ones because they have been the subjects of conceptual substitutions.

Prepare a top level system diagram, like Figure 10.15 to define major interfaces. Subsequently, detail the organization within the component boxes. An example is the synapses of the nervous system box or the memory part of the mind box.

Define the functional requirements for all components of the system. These requirements have also been defined in the conceptual studies but they have only been applied to physical components.

Perform functional analyses of system responses to typical and specific conditions or situations to detect interaction flaws. Include timeline analyses for confirmation of operational integrity. Integrate the results of situation analyses with the total system analysis.

Prepare verbal descriptions of the diagrams for dissemination to researchers who will be assigned research tasks in the component analyses.

Initiate a documentation policy which will allow tracking the development of the system. Make all the reports available to all researchers.

Find a systems director who understands the required coordination between scientists and philosophers and who can monitor the progress according to the goals. This could be an individual, a think tank, a university, or a scientific company.

A brief discussion of the meanings and effects of the soul, its mind, personhood, and self resulted in my conclusion that they are all one entity: an individual who requires a physical body for earthly life but lives eternally when deprived of such a body. This is the person that survives earthly death and may reappear in another reincarnated personality. Consciousness still defies definition. With its inclusion as a component of the soul, it takes on a new meaning, which may help in understanding what it is.

This chapter has been full of questions about a newly defined entity, its properties and its participation in the human operating system. Some of the research problems have been specifically called out. Actually, almost every question can lead to analytical or experimental research for plausible answers. There will be many more questions even before any meaningful actions are taken to rethink dualism. Philosophers with their propensity for idealistic and metaphysical excursions should find this new field of research a bonanza of challenges.

As a closing thought, doesn't the above reevaluation of the nature of the human being, especially the identity of an individual, bring to mind the René Descartes statement, "***I think, therefore I am***."?

CONCLUSIONS

A new sanctuary for Elephants to thrive in.

From the first sentence of the Introduction to this sentence, the message has been consistent and conclusive: In spite of the neuroscientific denial, the soul has been shown to be real through scientific physical evidence of reincarnation. By the inclusion of the soul with its independent mind, the nature of the human being must be rethought and redefined as an immaterial *person* occupying a physical body. This combined dual nature establishes the personality for an earthly lifetime. Between earthly lifetimes, the person exists as a spirit in an eternal spiritual domain. The scientific and secular mindsets must accept this dual human nature, in spite of the deeply embedded resistance to it. Every human being is destined to an eternal existence. Cognitive research must be redirected to include the soul's immaterial properties of mind, memory, free will, and emotion.

"I told you so" is what René Descartes would say if he were alive today. But he is alive today in spirit if not in body, just as he tried to convince everyone about the immortality of the soul. He has probably added some "word-thoughts" when meeting each newly arrived philosopher who had denied the existence of the soul while on earth. The additional words might be, "What you are experiencing is what I tried to explain to you in my philosophy". And with humility, if there be such a part of the mind in its transition beyond the Pearly Gates, the new arrival has to agree that *Descartes was right.*

Getting back to earthly matters, this is the time to review and conclude what this challenge to rethink dualism has attempted to do.

The challenge is directed not only at the scientific community that has, in general, refused to include the concept of a soul in its research, but also at the philosophers who have not been able to apply their idealistic and rhetorical talents to broaden the materialistic boundaries to include dualism. For the secular world, the challenge is to rethink the meaning of life as a preparation for immortality rather than merely a physical burial. Each person, as a soul, will live on in a hereafter. Admitting this fact should affect the meaning of life for everyone.

The discovery of the evidence of the soul is an historical event comparable in its effects on humanity to the Newtonian explanation of gravity and the Copernican definition of the sun/earth relation. How can such an apparently disproportionate statement be made? It is true for two main reasons: (1) it dispels the uncertainty, like in the other two, about human nature that has been questioned since Grecian philosophic inquiry, and (2) it affects every human being's mindset of the inevitable afterlife. Newton and Copernicus dealt with the physical part of the universe. The effects of the soul as the interface with the spiritual universe are as binding on the individual as gravity and spatial orientation. It is a spiritual integration with intellectually binding physical consequences. This historic event of the disclosure of soul's reality has been cloaked in specialized medical research and ignored by everyone else. It is time for the Newtonian "apple" to fall!

The advocates of dualism have been waiting for this book for over 350 years. Their needs consisted not only of the verification of the soul's existence but also of the non-physical rational and volitional properties associated with that soul. Since the evidence could not be produced, the explanations eventually narrowed to a purely physical base. But the evidence has been found, which sets the stage for a revolutionary change in the explanation of how the human being functions. What are the explanations which will begin the development of this worldly stage play that will involve all humanity in its cast? The following four areas define a rationale that explains the conclusions drawn in this book: (1) The nature of the evidence, (2) the application of the evidence to the mind-brain problem, (3) the effects of the soul's presence, and (4) the repercussions of introducing the soul into the world's mindset.

Nature of the evidence

The basic evidence is that reincarnation is an incontestable reality. Not only is it scientifically verified but verified as physical evidence of a non-physical entity. All previous allusions to it as fraud, a paranormal myth, or illusory fantasy must be relegated to the trash bin. What reincarnation does is demonstrate the passage of an entity from a dead body to a new body with retention of all its experiential properties. Such an entity, the spiritual tenant in a physical body, is understood by the term soul. As reincarnation exposes and defines the soul's existence, the soul in turn is the reason reincarnation can occur.

The validity of the reincarnation evidence is supported by the scientific documentation of the investigations of about 2500 cases, all positively verified case by case. Cases with inconclusive information were discontinued. The goal of the research was identification of birth marks and birth defects attributed to reincarnation. The precise shapes and locations of wounds due to injuries or accidents in the previous lives were found on the reincarnated bodies. The reporting of the evidence was confined to the research goals and did not elaborate on the definition of a soul. However, the evidence demonstrates properties that are not part of the physical body because these properties of mind, memory, desire, volition and ability to control physical activity accompanied the soul to the reincarnated body. They continued the experiential relationships with the past but without the previous body. Had they been physical properties, they would have terminated in and with the dead body.

What is the basis for claiming that the evidence of reincarnation is the physical evidence of a non-physical soul? What would be the most extreme requirement for proving the existence of a soul? It would be for the soul to appear physically and be recognized as a physical entity. As impossible as it seems, this is precisely what happens in reincarnation. The soul *spontaneously as a first person entity without any prompting* demands recognition as a personality from a previous life. Its recollections of activities, persons, locations, and specific items from the past life are verified by thorough investigations of personnel, on the spot comparisons, and official and media records. This scientifically researched evidence satisfies the extreme requirements of an immaterial entity appearing physically. The soul *spontaneously verifies its own existence*, unlike any other test item that must be planned, set up, tested, observed and interpreted by a third-person observer.

Science demands that proof of a theory must be repeatable as in a laboratory test. Application of such a requirement is only relevant in a planned test but irrelevant in a spontaneous act. Neither is it relevant in tests that involve living entities as test items because the *precise* conditions, activities, and reactions cannot be duplicated as by brain scans. In reincarnation, repeatability is demonstrated in the continuing insistence by the reincarnated subject of its past personality and in some cases the desire to return to that environment. This is continuing spontaneous repeatability unlike the controlled laboratory tests. Further repeatability is demonstrated by the thousands of individual confirmed claims of reincarnation.

The reappearance of a personality in a new body after leaving a deceased body demonstrates that the soul does not die although the body does. There is continuity of the soul's life in a non-material domain, a spiritual universe, until the soul enters another body. Eternal life is every human's destiny after earthly death.

The bottom line of the reincarnation evidence is that reincarnation has been scientifically verified, the soul does exist, there is a life after death, the soul may have a second earthly life (or multiple lives), and that it retains its functional properties in the transmigration.

Application of the evidence to the mind-brain problem

The reincarnation event proves that an non-physical entity separates from a physical body and migrates to another physical body. The existence of this spirit, the soul, and its separation from the body negate the neuroscientific and cognitive assumptions that there is no soul and that the physical body is the only natural entity of the human being. **This exposure of the false basic assumptions negates all the scientific conceptual explanations of the mind-brain problem that rely on these fundamental assumptions.** No longer is there a mental state created by a physical brain, for it is an immaterial and independent mind that is the mental state. The concepts of sensor driven behavior, synaptic storage of experiential memory and skills, the computer brain, etc. are meaningless; merely examples of innovative thinking to explain physical substitutes for non-physical properties. How could such a farce continue for over four decades and be accepted by the scientific and secular communities simply because of a bias against the possibility of a spiritual universe? The inability to understand or explain so many of the functional problems of the human system and to be forced to resort to convenient and assumed naïve concepts should have been sufficient cause for the scientists to search for the flaw in the basic rationale.

Although the above reason is sufficient to negate the scientific approach, there are more definitive reasons to support the charge that the scientific analyses are ill-conceived. The principal issue is that of memory. The neuroscientific explanation of *total* memory being stored in the physical synapses is not only false but incredibly naïve since unintelligent physical neurons cannot translate mental phenomena into physical storage. And the memory's functioning and location have, by scientific admission, never been understood. The reason for the lack of understanding is *that memory is not a physical element but a property of the non-physical soul.* As the reincarnation evidence has demonstrated, if memory were a physical element, it would cease to exist as the previous body dies. But that does not happen. The memory remains alive and is transmigrated to the new body where it is repeatedly exercised, and verified, in the demonstrated call by the soul for recognition. The insistence by the reincarnated child to return to a previous distant location and family (Chapter 3, Table 3.1) in spite of parental disapproval indicated that the free will to insist existed with the soul. The new brain could not generate desires about distant locations and people without having any connections with those surroundings. This verification of memory and free will, in addition to the baseless denial of the soul, are three of the most obvious reasons for the refutation of the neuroscientific claims of the past four decades. The research effort continues *although the evidence of the scientific viability of reincarnation was published in 1966!*

Since the above statements explain that experiential memory is not part of the brain, the need for physical memory functions remains. I have, therefore proposed the unique idea that there are two memories with the brain's memory supplying the storage for the physical data requirements of body basic functions of breathing, digestion, repair, etc.

Although there are many definitions of the mind, the above statements define the mind as the soul's mental equivalent to the body's brain. The mind controls the brain, not the opposite as science would have it. Control is an activity, not an element or a function, and can only be exercised by one causal principal. That must be the mind because that is where desires, plans, and decisions are generated. As the evidence indicated (Chapter 3, Table 3.2), learning and skills are part of this mind, not the brain's representation of the external world for computational reduction of significant information. Furthermore, there is no scientific explanation of how the newborn brain learns to generate mental states, whereas the mind entering the body with mature thinking experience from the previous life is a logical explanation of evolving brain control leading to brain maturity.

The effects of the soul's presence

The primary effect of the soul's presence is that monism is dead. It reverses the dominance of monism in the dualism debate. An official statement to this effect might be viewed as an attempt to impose a religious belief on the secular community. But as scientific reality, it must be accepted like any verified evidence of a natural law. A major self-incriminating effect of such acceptance, however, will raise a formidable obstacle to its issuance because it would: a) convincingly exemplify science's vulnerability to fallacy, b) expose the lack of scientific compliance with its own guidelines for establishing an inclusive base for an extensive research effort, and c) call attention to the wasteful expenditure of resources and talent in pursuing a misleading approach that admittedly lacked scientific supporting data. Such an unsupported base normally demands cancellation of projects that can only produce negative results.

Another major effect, after the acceptance of the soul's presence by the scientific and philosophic communities, will be the need for a new research direction, which will utilize existing physiological knowledge for integration with newly developed philosophies and scientific findings. This completely redirected effort is inevitable.

The acceptance by the secular society will take much adjustment, not only on an individual basis but in communal redirections. For the individual, the effect will depend on his/her current belief in a soul. In addition to the impact the reality of the soul will have on the behavior of the individual, it is the inevitability of an afterlife that should concern everyone. Individual awareness should change from association of self with a physical body to one with a spiritual person destined for eternity. The mindset of man's ultimate destiny is due for a change. There may be a resurgence of religious activity as a real soul adds meaning to life and religion.

The communal effect will be felt noticeably in the educational system. The current refusal to allow any religious teaching, due to the separation of church and state mentality, will be an obstacle even if the soul's presence is scientifically accepted. Revising the text books will be a complicated issue considering the animosity that the evolution/Intelligent Design controversy has caused. The teachers' abilities to relate their own acceptance of duality to the mixed beliefs of the students will be a problem at all levels of education, especially at the university level where individualized philosophies are major and controversial subjects.

The acceptance of the soul will probably be very difficult for the atheist who denies the probability of an afterlife. His religious belief, for atheism is a religion, is also on shaky ground because a soul and an afterlife are associated with a high probability of a divine presence.

The judicial mindset should change with the knowledge of the soul's existence. The definition of a person can now be based on the eternal quality of each human being, not merely on an unclear definition of personhood. The inclusion of a soul in the birth process reverses the assumption that the fetus is only a mass of tissue. The right to die takes on a different meaning; it is no longer a termination but a change, as is capital punishment.

The soul's effect at the governmental level is unpredictable. It is possible that there could be far reaching effects as in striving for peaceful international relations, avoiding wars, working for the common good, etc., if heads of states realize that there is a spiritual universe that they will eventually visit. The soul introduces a moral aspect into the legislative process which may not exist in a purely physical outlook on life.

The soul's reality forces awareness of the credibility of paranormal phenomena. The near death experiences could become events for scientific interest in search of possible information about the properties of the soul and the afterlife.

Repercussions of the introduction of the soul to the world's mindset

The scientific and secular communities are not ready to integrate a soul with a physical human being. The dedication to materialistic tradition cannot be overcome easily. Some, not all, religions have included the soul in their teachings. Philosophic reaction may be mixed. Although the soul may be acceptable to theologians, reincarnation may not. The initial acceptance of dualism will be hesitant if not negative. But since the evidence proves its reality, eventual acceptance must follow.

The immediate repercussions will be the reactions to this book. The response will probably be immediate due to the length of time that the scientists have been promoting the opposite viewpoints. The credibility of science's analytical ability in this neurophysical area is cast into serious doubt due to the positive position science has taken on the materialistic approach. Furthermore, its inability to assess the probability of an immaterial entity in spite of all the philosophical assurances of such an existence may continue to hamper future soul-body interfacing research. This may create a major problem for the cognitive research effort.

The disruption of the research efforts of the unproven neurophysical concepts demands new explanations of the body-soul and brain-mind relationships. The demonstrated properties of the soul should be an illuminating advantage since those properties of memory, free will, desire, and emotion answer the questions hampering the solution of the mental aspects of the brain-mind problem. This can only be accomplished in a joint effort of scientists and philosophers. When such explanations begin to mature, consequent repercussions will be of major proportions. All education in the affected fields from grade to university levels will have to be reoriented with revised texts. All scientific books that exclude the mind and soul in a materialistic explanation will be subjected to critical review.

The integration of the soul will require the definition of its properties. Some of these are known through the corrective effects they can have on the unresolved problems in the materialistic research. A major unknown requiring definition is the interface between the mind and the brain, and in an overall sense the consciousness relationship between body and soul.

These major problems will probably lie dormant in a confused and disorganized interdisciplinary conflict for a long period unless a research leader is found who can organize the new effort and provide direction to the program.

The repercussions in the religious domain are few. The principal one is in the resolution of the reincarnation denial. There are scriptural references that can assist the theologians in the resolution of this issue. In the secular community the soul's implication that there is a Divine Being in the spiritual universe may cause increased conflict in the ongoing debate on the existence of God.

There will be many other problems raised by the revelation of the soul's existence. The above sections are only the "tip of the iceberg" which prompted me to say that this irrevocable evidence of reincarnation, defining a soul's cry for recognition, is of the magnitude of a Newtonian disclosure. Its universal intervention in current and future thinking and activities by formalizing realism where doubt was in control is beyond comprehension. I'm reminded of the statement that the German philosopher Arthur Schopenhauer made about new realistic conclusions, "All truth passes through three stages: first it is ridiculed, then violently opposed, and finally accepted as self-evident".

My Connecting Contributions

I did not produce the evidence of reincarnation; I only found it and recognized its application to the dualism debate. The two scientific disciplines, medical and cognitive, were either not aware of the influence of the evidence on the broader fields of body-soul and mind-brain interaction, or ignored such influence in favor of more immediate interests. The dots were all in place but no one was connecting them. What I did to connect them is the subject of this section. I alone did this analysis and take responsibility, and credit, for the net result.

Although much of the descriptive material in this book is the work of others from René Descartes to the modern philosophers and neuroscientists, my principal contribution has been the linking of the various elements of this material to create the path from belief in, or denial of, a soul to its scientifically researched reality. In the process, the concept of monism was reduced to a mythical state of folk psychology. This disruptive damage demanded a replacement by a logical substitute of a dualistic system. The following steps are the stepping stones I used to construct this pathway to the truth.

1. Through interest in the connection between reincarnation and the meaning of life, I found Dr. Stevenson's book, published in 1966, which describes his initial twenty investigations of reincarnation, and includes the scientific reports that validated his investigative methods. The reports convinced me of the reality of reincarnation.

2. I concluded that for reincarnation to be real by demonstrated connectivity to a previous life, the migrating entity must be the immaterial (spiritual) portion of a previous union of body and soul. Since that body was dead and remained in the past, this migrating entity must be the soul with a continuing life and consciousness.

3. Since reincarnation is not immediate between earthly death and a new physical existence, there must be a domain in which the continuing existence of the soul is possible. This afterlife is the domain that all souls must enter after earthly death. The soul's life, therefore, is eternal.

4. From the information that the reincarnated soul was revealing, it was clear that it had to possess a memory, a desire to be recognized as a previous personality, a desire (in some cases) to return to the previous surroundings, and free will to persist in its demands in spite of parental objections. I recognized this as a soul's cry for recognition, especially since it was a spontaneous, unprompted effort. The soul possessed a mind, a memory, free will, and desires.

5. The soul's possession of a memory of its past-life experiences completely negated the neuroscientific concept of a physical memory that could store non-physical information in the neural synapses. I concluded such a solution was impossible because the memory that was alive and functioning in the reincarnated soul would not exist if it had been a physical part of the body and died with that body. The use of free will by the reincarnated soul negated the scientific denial

of volition. Due to these very obvious conflicts, the entire conceptual approach based on physical causality lacked credibility.

6. Since the physical body must rely on some memory for the automatic bodily functions that are not consciously controlled, like breathing and digestion, I propose that there are two memories; one in the brain and one as part of the mind. To my knowledge this has never been proposed due to the general belief in only one memory. This dual memory approach, however, utilizes the solution of a physical memory with synaptic storage for that physical information used in the body's physical functions. Such a physical memory can also be the storage location for physical reflex data and memorized/practiced procedures.

7. Conflicts between other scientific concepts had to be resolved by showing that the physical approach was not credible. An impersonal method for objective credibility in support of the criticisms was to use rebuttals by scientific and academic peers. By adding my comments to the conflicting professional arguments, I could inject the effects of a soul's contribution to solving the problems.

8. The negation of the monistic concept demanded a substitute approach to fill the gap. The absence of a holistic plan in the ongoing research resulted in a disjointed and fragmented effort. A redirected effort should be planned to assure that the ultimate goal is known and approved before any activities are initiated. An acceptance (not understanding) of the philosophical and scientific interactions in the plan will be mandatory. To initiate the understanding of this complex issue with the myriad of unknowns involved, I prepared a rationale for methodically delving into the interactions of the dual human nature beginning with pregnancy and developing into mature behavior.

9. The synthesis of the dual system begins with the top level system diagram for understanding the interaction of the two entities. I developed the first-ever (to my knowledge) top system diagram of dualism. Though it may be flawed, it is the first because the properties of a soul have never been included in a system diagram. Furthermore, to my knowledge there never has been a comparable diagram for the current research.

10. There is a similarity between the evidence for reincarnation and the evidence for the existence of God, which I mention but leave as a subject for a separate book.

This rationale with original contributions proposes a logical transition from a misleading scientific approach to a new fact based search for the true nature of Homo sapiens. This is the basis for my challenge to the scientific community to rethink dualism. The disruption of the ongoing research can be supplanted by the challenging prospects of a new and exciting search for knowledge about the uniting factors of the physical with the immaterial domains.

This concludes the reasons why dualism replaces monism as the new philosophy of human nature; new because after 350 years of question-begging the soul's reality, the evidence Descartes sought is finally available. This brings the newness to the research effort that can now begin, based on fact rather than speculation and optimistic conceptualization. There may be flaws, omissions, and semantic misunderstandings in this book, but considering the mass of meaningless information that has been accepted as representative of the best neuroscientific thinking, I consider my lone venture into this jungle of words and concepts as a journey with meaningful and realistic results. At my age of 87, I will not be around for the many arguments and criticisms I will have caused. Hopefully, there will be others who think as I have, who will respond to the skeptical tirades. With tongue in cheek and shedding all humility, I say I don't think there should be any major criticisms to this matter-of-fact presentation. There is one more thought with which I must conclude.

AFTERTHOUGHT

The search for the evidence of the soul has been going on for over 350 years since Descartes philosophized its existence in his *Meditations*. This fruitless search has numbed humanity to an acceptance of the impossibility of obtaining physical evidence of a non-physical entity. In this atmosphere of denial and apathy, why should I be inclined to disregard the accepted worldview and be able to ferret out precisely such impossible physical evidence of the soul's reality? Why should I be the one to recognize the applicability of medically produced evidence to the refutation of monism? Why should I find it logically possible to show how the reincarnation evidence confirms the Cartesian concept of dualism? Was my engineering background in logical systems analysis a predestined necessity for detecting this applicability?

My introduction to philosophy and Thomas Aquinas was a coincidence since my major at the Catholic University was engineering. Or was it? What prompted me to become interested in the neuroscientific research that has a mandated objective of replacing the soul with meaningless substitutes of promissory materialistic concepts? Furthermore, what guided me in writing this book, which not only reveals the evidence and applies it to dualism, but also opens the door to a revolution in philosophic and scientific unified thinking about the human being as a final confirmation of the Cartesian philosophy?

In finding the confirmation of reincarnation as a real phenomenon instead of a misconstrued paranormal oddity, I have become aware of the possibility of my own return to become involved in the debate. It seems appropriate at this point in time that I should return to find that the evidence is finally produced in an undeniably scientific manner by someone who was not interested in the philosophy of dualism. Although I cannot recollect any medieval environments or peculiar grammatical usage, I have felt the need to fulfill this mission of completing the unfinished argument. (In reading the *Meditations*, I recognized one passage that was identical to the one I had already written for the book.) Perhaps without being aware of it now but knowing that reincarnation is attainable, I have come back to present this final conclusive argument. The only remaining argument is the physical scientific proof of my philosophy about the existence of God, to which the soul's reality is a major contribution. Perhaps I have a few years left to find that evidence—or it may take another lifetime.

R. D.

(Endnotes)

1 David L. Wilson, Mind-Brain Interaction and Violation of Physical Laws, article in The Volitional Brain, Edited by Benjamin Libet, Anthony Freeman, and Keith Sutherland,1999, Imprint Academic, Exeter, UK, 196.

2 Patricia Smith Churchland, Brain-Wise, Studies in Neurophilosophy, The MIT Press, Cambridge, Massachusetts, 1.

3 Paul M. Churchland, Matter and Consciousness, 1984, revised 1988, The MIT Press, Cambridge, Massachusetts, 18.

4 Jeffrey M. Schwartz and Sharon Begley, The Mind and the Brain, Neuroplasticity and the Power of Mental Force, Regan Books, New York, NY, 17-18.

5 René Descartes, Meditations on First Philosophy, Meditation II, Great Books of the Western World, Vol. 31, 70.

6 William Hasker, The Emergent Self, Cornell University Press, Ithaca, New York, 1999, 190.

7 Nick Bamforth, Duality Into Unity, Amethyst Books, 1992, 223.

8 Walter Gratzer, The Undergrowth of Science, 2000. Oxford University Press, New York, 1, 111.

9 Antony Flew, A Dictionary of Philosophy, Second Edition, 1999, Gramercy Books, New York, 283

10 Ibid, 321

11 Ibid, 41

12 Benjamin Libet, Mind Time, 2005, First Harvard University Press, 134.

13 Ibid, 11.

14 Gary E. Schwartz with William L. Simon, The Afterlife Experiments, 2002, Pocket Books, New York.

15 Mario Beauregard, Ph.D. & Denyse O'Leary, The Spiritual Brain, A Neuroscientist's Case for the Existence of the Soul, 2007, Harper Collins Publishers, New York.

16 Ian Stevenson, M.D., Twenty Cases Suggestive of Reincarnation, first published as Volume 16 (1966) of the Proceedings of the American Society for Psychical Research, Second Edition, revised and enlarged, 1974, University Press of Virginia.

17 Tom Shroder, Old Souls, 1999, published by Fireside, New York.

18 Patricia Churchland, Brain-Wise, 2002, The MIT Press, Cambridge, Massachusetts, 394.

19 Jacques Maritain, The Degree of Knowledge, 1959, Revised edition 1995, Notre Dame Press, Notre Dame, Indiana.

20 Paul K. Moser, Philosophy After Objectivity, Making Sense in Perspective, 1993, Oxford University Press, New York, New York, 3.

21 Michael J. Behe, Darwin's Black Box, The Biochemical Challenge to Evolution, 1996, The Free Press, New York, New York, 145.

22 Patricia Churchland, Brain-Wise, Studies in Neurophilosophy, 2002, Massachusetts Institute of Technology, 40.

23 Ibid, 30.

24 Great Books of the Western World, Vol. 49, Darwin, Encyclopedia Britannica, Inc., 243.

25 Michael J. Behe, Darwin's Black Box, The Biochemical Challenge to Evolution, 1996, The Free Press, New York, New York, 232.

26 Ibid, 233

27 Paul M. Churchland, The Engine of Reason, the Seat of the Soul, A Physical Journey into the Brain, 1996, Massachusetts Institute of Technology, 93.

28 Paul M. Churchland and Patricia S. Churchland, On the Contrary, Critical Essays, 1987-1997, 1998,The MIT Press, 13-14.

29 Paul M. Churchland, The Engine of Reason, The Seat of the Soul, 1996, The MIT Press, 123.

30 Ibid., 42.

31 Ibid., 6.

32 Ibid., 95.

33 Joseph Levine, The Purple Haze, The Puzzle of Consciousness, 2001, The Oxford Press, New York, 4.

34 Ibid., 7.

35 Paul M. Churchland, The Engine of Reason, The Seat of the Soul, 1996, The MIT Press, 93.

36 Ibid., 21.

37 Ibid., 92.

38 Ibid., 26.

39 Ibid., 99.

40 Ibid., 101.

41 Ian Glynn, An Anatomy of Thought, The Origin and Machinery of the Mind, 1999, The Oxford University Press, Oxford, New York, 107.

42 Ibid., 119.

43 Paul M. Churchland, The Engine of Reason, The Seat of the Soul,1996, The MIT Press, 6.

44 Ibid., 8.

45 Ibid., 11.

46 Ibid., 13.

47 Ibid., 42.

48 Ibid., 42.

49 John R. Searle, The Rediscovery of the Mind, 1992, The MIT Press, 1.

50 Ibid., 205.

51 Ibid., 200.

52 Ibid., 200.

53 Ibid., 79.

54 Great Books of the Western World, Vol.53, Encyclopedia Britannica, 116.

55 Ibid., 119.

56 Ian Glynn, An Anatomy of Thought, The Origin and Machinery of the Mind. 1999, Oxford University Press, Oxford, New York, 329.

57 Ibid., 327.

58 Daniel L. Schacter and Endel Tulving, editors, Memory Systems 1994, 1994, Massachusetts Institute of Technology, Chapter 11, Working Memory: The Interface between Memory and Cognition, 354.

59 Ibid., 352.

60 Ibid., 351.

61 Benjamin Libet, Anthony Freeman, and Keith Sutherland, Editors. The Volitional Brain, Towards a Neuroscience of Free Will, 1999, Imprint Academic, Exeter, UK, 56.

62 Ibid., 73.

63 Ibid., 63.

64 Benjamin Libet, Mind Time, The Temporal Factor in Consciousness, 2005, Harvard University Press. Cambridge, Massachusetts, 93.

65 Ibid., 90.

66 Patricia Smith Churchland, Brain-Wise, Studies in Neurophilosophy, 2002, The MIT Press, Cambridge, Massachusetts, 2.

67 John G. Taylor, The Race for Consciousness, 1999, The MIT Press, Cambridge, Massachusetts, 121.

68 Ibid., 125.

69 J. Allan Hobson, Consciousness, 1999, Scientific American Library, New York, 71.

70 Ibid., 229-30.

71 John R. Searle, The Rediscovery of the Mind, 1992, The MIT Press, Cambridge, Massachusetts, 14/

72 Patricia Smith Churchland, Brain-Wise, Studies in Neurophilosophy, 2002, The MIT Press. Cambridge, Massachusetts, 135.

73 Paul M. Churchland, Matter and Consciousness, Revised Edition, 1988, The MIT Press, Cambridge, Massachusetts, 18.

74 Patricia Smith Churchland, Brain-Wise, Studies in Neurophilosophy, 2002, The MIT Press, Cambridge, Massachusetts, 30.

75 Ulrich Mohrhoff, The Physics of Interactionism, article in The Volitional Brain, edited by Benjamin Libet, Anthony Freeman, & Keith Sutherland, Imprint Academic, Exeter, UK, 181.

76 William James, The Principles of Psychology, The Great Books of the Western World, Vol. 53, Encyclopedia Britannica, Chicago, Illinois, 119.

77 Jim Tucker, Life Before Life, A Scientific Investigation of Children's Memories of Previous Lives, 2005, St. Martin's Press, New York, 114, 167.

Bibliography

Books

Bamforth, Nick. *Duality into Unity*. New York: Amethyst Books, 1992.

Beauregard, Mario and Denyse O'Leary. *The Spiritual Brain, A Neuroscientists's Case for the Existence of the Soul*. New York: Harper Collins, 2007.

Behe, Michael J. *Darwin's Black Box, The Biochemical Challenge to Evolution*. New York: The Free Press, 1996

Churchland, Patricia Smith. *Brain-Wise, Studies in Neurophilosophy*. Cambridge: MIT Press, 2002.

Churchland, Paul M. *Matter and Consciousness*. Cambridge: MIT Press, 1988, ninth printing 1999.

Churchland, Paul M. *The Engine of Reason, the Seat of the Soul, A Philosophical Journey into the Brain*. Cambridge: MIT Press, 1996.

Flew, Antony. *A Dictionary of Philosophy*. New York: Gramercy Books, second edition, 1979

Glynn, Ian. *An Anatomy of Thought, The Origin and Machinery of the Mind*. Oxford: University Press, 1999.

Gratzer, Walter. *The Undergrowth of Science, Delusion, Self Deception, and Human Frailty*. Oxford: University Press, 2000.

Hobson, J Allan. *Consciousness*. New York: Scientific American Library, Number 68 of series, 1998.

Levine, Joseph. *The Purple Haze, The Puzzle of Consciousness*. Oxford: University Press, 2001.

Libet, Benjamin. *Mind Time, The Temporal Factor in Consciousness*. Cambridge: Harvard University Press, 2004.

Maritain, Jacques. *The Degrees of Knowledge*. Translated from the fourth edition. Notre Dame: Notre Dame Press, 2002.

Moser, Paul K. *Philosophy After Objectivity, Making Sense in Perspective*. Oxford: Oxford University Press, 1993.

Schwartz, Gary E. *The Afterlife Experiments, Breakthrough Scientific Evidence of Life After Death*. New York: Pocket Books, 2002.

Schwartz, Jeffrey M. and Sharon Begley. *The Mind and the Brain, Neuroplasticity and the Power of Mental Force*. New York: Regan Books, 2002.

Searle, John R. *The Rediscovery of the Mind*. Cambridge: MIT Press, 1992.

Shroder Tom. *Old Souls, Compelling Evidence from Children Who Remember Past Lives*. New York: Simon & Schuster, 1999.

Stevenson, Ian. *Twenty Cases Suggestive of Reincarnation*. First edition in 1966 by the American Society for Psychical Research. Second edition revised and enlarged in Charlottesville: University Press of Virginia, 1974.

Taylor, John. *The Race for Consciousness*. Cambridge: MIT Press, 1999.

Articles

From the Great Books of the Western World, Encyclopedia Britannica, Robert Maynard Hutchins, editor in Chief, Chicago: William Benton Publisher, 1952.

Darwin, Charles. *The Origin of Species by Means of Natural Selection.* Vol. 49.

Descartes, René. *Rules for the Direction of the Mind.* and *Meditations on First Philosophy. Vol. 39.*

James, William. The Principles of Psychology. Vol. 53.

From Volitional Brain, Towards a Neuroscience of Free Will, edited by Benjamin Libet, Anthony Freeman & Keith Sutherland. Exeter: Imprint Academic, 1999, first imprint 2004.

Gomes, Gilberto. *Volition and the Readiness Potential.*

Libet, Benjamin. *Do We Have Free Will?*

Mohrhoff, Ulrich. *The Physics of Interactionism.*

Suggested Reading

Biblical History

Wilson, Ian, *The Bible Is History*. Washington, D. C.: Regnery Publishing, 1999.

A study of documents and archaeological findings to determine whether the disputed authenticity of the Bible as a historical document is warranted. Narrative supported by stunning color photographs, maps, and reconstructions of cities and buildings. Encompasses history, philosophy, and religion.

Wilson, Ian, *Jesus: The Evidence*. Washington, D. C.: Regnery Publishing, 2000.

A study of the period and life of Jesus Christ to establish the authenticity of the Gospel accounts. Recent findings tend to confirm the accounts. Done with beautiful photographs of evidence and locale.

Dualism

Cooper, John. *Body, Soul, and Life Everlasting. Biblical Anthropology and the Monism-Dualism Debate.* Grand Rapids: Eerdmans Publishing, 2000.

A review of the monist-dualist debate in the context of biblical references.

Corcoran, Kevin, editor. *Soul, Body, and Survival, Essays on the Metaphysics of Human Persons*. Ithaca: Cornell University Press, 2001.

Essays by various professors of philosophy on the different aspects of dualism.

Evolution

Behe, Michael. *The Edge of Evolution, The Search for the Limits of Darwinism*. New York: The Free Press, 2007.

A radical redefinition of the debate about Darwinism from its accounting for all aspects of life to a question of *how* it happened. It was nonrandom, not random mutation.

Dyson, Freeman. *Origins of Life*. Cambridge, U. K.: Cambridge University Press, 1985. second revised edition 1999/

Discussion of the main theories of how life began, either through replication or metabolism.

Johnson, Wallace, *The Death of Evolution*. Rockford: Tan Books and Publishers, 1982, third printing 1987.

Scientific, religious, and commonsense reasons why evolution doesn't make sense, is fraudulent, and is intellectually bankrupt.

Wells, Jonathan, *Icons of Evolution, Science or Myth?* Washington, D. C.: Regnery Press, 2000.

What some biologists know and are not telling you about evolution. Some of its best known icons are false or misleading.

Wiker, Benjamin, *Moral Darwinism, How We Became Hedonists*. Downers Grove, Illinois: Inter Varsity Press, 2002.

Intelligent Design

Dembski, William. *Intelligent Design, The Bridge Between Science and Theology*. Downers Grove, Illinois: Inter Varsity Press, 1999.

The author addresses the concerns of science that Intelligent Design is creationism in disguise, and the concerns of theology that it misunderstands divine activity. He challenges the influence of naturalism in science and reinstates design in it.

Witham, Larry. *By Design, Science and the Search for God*. San Francisco: Encounter Books, 2003.

How science, an unlikely agent, has reopened the case of the existence of God through the cutting edge of research in physics, biochemistry, genetics, information theory, and neuroscience.

Logic and Argument

Engel, Pascal. *The Norm of Truth, An Introduction to the Philosophy of Logic.* Toronto: University of Toronto Press, 1991.

As an introductory book, it requires only a small amount of knowledge of logic. Technicalities are kept to a minimum and although logic is a technical subject, philosophically its problems are not always clear and transparent.

Koslowski, Barbara. *Theory and Evidence, The Development of Scientific Reasoning.* Cambridge: MIT Press, 1996.

The author addresses research about the beliefs people hold about what evidence is meaningful in scientific reasoning and how people deal with disconfirming evidence.

Lepore, Ernest. *Meaning and Argument, An Introduction to Logic through Language.* Oxford: Blackwell Publications, 2000.

The author's goal is to teach how to use clear language in formalized arguments before drawing conclusions. A primary task for logic skill development is to provide tools for expressing arguments in natural language.

Mind-Brain

Baudry, Michael, Davis, Joel L. and Thompson, Richard E., editors. *Advances in Synaptic Plasticity.* Cambridge: MIT Press, 2000.

Many neurons exhibit plasticity changing structurally or functionally. The recent trends are described of the analyses at the molecular to cellular and network levels. The book is technically oriented.

Corbi, Josep E. and Prades, Josep L. *Minds, Causes, and Mechanisms, A Case Against Physicalism.* Oxford: Blackwell Publishers, 2000.

The brain is a living organism that can actually change its structure and function, even into old age, through neuroplasticity. A collection of cases is presented.

Doidge, Norman. *The Brain That Changes Itself.* New York: Penguin Group, 2007.

The author questions the internal consistency of causal physicalism and vindicates a novel approach to mental causation. Stories are presented of personal triumph from the frontiers of brain science.

Fauconnier, Gilles and Turner, Mark. *The Way We Think, Conceptual Blending and the Mind's Hidden Complexities.* New York: Basic Books, 2002.

An analysis of the imaginative nature of the human mind. Conceptual blending, already widely known, is at the root of the cognitively modern human mind.

Hasker, William. *The Emergent Self.* Ithaca: Cornell University Press, 1999.

The author challenges physicalist views of human neural functioning and advances the concept of the mind as an emergent individual. He shows that contemporary forms of materialism are seriously deficient in confronting crucial aspects of experience.

Maass, Wolfgang and Bishop, Christopher M., editors. *Pulsed Neural Networks.* Cambridge: MIT Press, 1999.

In recent years data from neurobiological experiments have made it increasingly clear that biological neural networks, which communicate through pulses, use the timing of the pulses to transmit information and perform computations. The book presents the complete spectrum of current research in pulsed neural networks.

<u>Philosophy, Psychology</u>

Feyerabend, Paul. *Farewell to Reason.* New York: New Left Books, 1987, reprinted 2002.

This is a vigorous challenge to the scientific rationalism that underlies Western ideas of progress and development. The author insists that an appeal to reason is empty and must be replaced by a notion of science that subordinates it to the needs of citizens and communities.

Jardine, Nicholas. *The Scenes of Inquiry, On the Reality of Questions in the Sciences.* Oxford: Clarendon Press, 1991, 2000.

It is time that scientists break with science, freeing themselves of the mythology of science and becoming more perceptive of their varied practices and of the workings of their own social and political institutions.

Kilpatrick, William. *Psychological Seduction, The Failure of Modern Psychology.* Ridgefield, CT: Roger A. Mc Caffrey Publishing, 1983.

The author identifies some of the important ways in which psychology has become the secular religious system of our age.

Silverman, Hugh, editor. *Questioning Foundations, Truth/Subjectivity/ Culture.* New York: Routledge, 1993.

The series of essays addresses the problem of a contemporary world devoid of foundations, exploring a context of subverted truths, shattered identities, and fragmented cultures.

Stich, Stephen. *The Fragmentation of Reason.* Cambridge: MIT Press, 1990.

The author explores the nature of rationality and irrationality; what distinguishes good reasoning from bad. Rejecting the widely accepted approaches, he proposes an alternative that leads to a radical epistemic relativism,

<u>Reincarnation</u>

Bowman, Carol. *Return from Heaven, Beloved Relatives Reincarnated Within Your Family.* New York: Harper Collins Publishers, 2001.

The empirical book is based on direct observations of very young children whose statements and behaviors are reminiscent of the past lives of deceased relatives.

Prophet, Elizabeth Claire. *Reincarnation, The Missing Link in Christianity.* Corwin Springs, MT: Summit University Press, 1997.

This is a history of reincarnation in Christianity from Jesus and early Christians through Church councils and persecutions of so-called heretics. The case is made that Jesus taught reincarnation.

Stevenson, Ian. *Where Reincarnation and Biology Intersect.* Westport, CT: Praeger Publishers, 1997.

Dr. Stevenson describes cases in which birthmarks are related to experiences of remembered past lives, particularly violent deaths. Collected memories of former identities have been matched with data of the former families, residences, and manners of death. Illustrations of birthmarks are included.

Ten Dam, Hans. *Exploring Reincarnation, The Classic Guide to the Evidence for Past-Life Experiences.* London: Rider Books, revised and updated in 2003.

A guide to past life recall prepared by one of the world's leading experts includes case histories and theories about the relationship between body and soul. Extensive lists are given for English as well as other language reincarnation texts.

Tucker, Jim. *Life Before Life, A Scientific Investigation of Children's Memories of Previous Lives.* New York: St. Martin's Press, 2005.

Additional insight is given of the investigative work of Dr. Stevenson with author participation that qualifies as valid scientific research. The methodology is described and criticisms of the results are responded to with logical explanations.

Index